Applied Mathematical Sciences
Volume 135

Editors
J.E. Marsden L. Sirovich

Advisors
S. Antman J.K. Hale P. Holmes
T. Kambe J. Keller K. Kirchgässner
B.J. Matkowsky C.S. Peskin

Springer
New York
Berlin
Heidelberg
Barcelona
Hong Kong
London
Milan
Paris
Singapore
Tokyo

Applied Mathematical Sciences

1. *John:* Partial Differential Equations, 4th ed.
2. *Sirovich:* Techniques of Asymptotic Analysis.
3. *Hale:* Theory of Functional Differential Equations, 2nd ed.
4. *Percus:* Combinatorial Methods.
5. *von Mises/Friedrichs:* Fluid Dynamics.
6. *Freiberger/Grenander:* A Short Course in Computational Probability and Statistics.
7. *Pipkin:* Lectures on Viscoelasticity Theory.
8. *Giacoglia:* Perturbation Methods in Non-linear Systems.
9. *Friedrichs:* Spectral Theory of Operators in Hilbert Space.
10. *Stroud:* Numerical Quadrature and Solution of Ordinary Differential Equations.
11. *Wolovich:* Linear Multivariable Systems.
12. *Berkovitz:* Optimal Control Theory.
13. *Bluman/Cole:* Similarity Methods for Differential Equations.
14. *Yoshizawa:* Stability Theory and the Existence of Periodic Solution and Almost Periodic Solutions.
15. *Braun:* Differential Equations and Their Applications, 3rd ed.
16. *Lefschetz:* Applications of Algebraic Topology.
17. *Collatz/Wetterling:* Optimization Problems.
18. *Grenander:* Pattern Synthesis: Lectures in Pattern Theory, Vol. I.
19. *Marsden/McCracken:* Hopf Bifurcation and Its Applications.
20. *Driver:* Ordinary and Delay Differential Equations.
21. *Courant/Friedrichs:* Supersonic Flow and Shock Waves.
22. *Rouche/Habets/Laloy:* Stability Theory by Liapunov's Direct Method.
23. *Lamperti:* Stochastic Processes: A Survey of the Mathematical Theory.
24. *Grenander:* Pattern Analysis: Lectures in Pattern Theory, Vol. II.
25. *Davies:* Integral Transforms and Their Applications, 2nd ed.
26. *Kushner/Clark:* Stochastic Approximation Methods for Constrained and Unconstrained Systems.
27. *de Boor:* A Practical Guide to Splines.
28. *Keilson:* Markov Chain Models—Rarity and Exponentiality.
29. *de Veubeke:* A Course in Elasticity.
30. *Shiatycki:* Geometric Quantization and Quantum Mechanics.
31. *Reid:* Sturmian Theory for Ordinary Differential Equations.
32. *Meis/Markowitz:* Numerical Solution of Partial Differential Equations.
33. *Grenander:* Regular Structures: Lectures in Pattern Theory, Vol. III.
34. *Kevorkian/Cole:* Perturbation Methods in Applied Mathematics.
35. *Carr:* Applications of Centre Manifold Theory.
36. *Bengtsson/Ghil/Källén:* Dynamic Meteorology: Data Assimilation Methods.
37. *Saperstone:* Semidynamical Systems in Infinite Dimensional Spaces.
38. *Lichtenberg/Lieberman:* Regular and Chaotic Dynamics, 2nd ed.
39. *Piccini/Stampacchia/Vidossich:* Ordinary Differential Equations in $\mathbf{R}^n$.
40. *Naylor/Sell:* Linear Operator Theory in Engineering and Science.
41. *Sparrow:* The Lorenz Equations: Bifurcations, Chaos, and Strange Attractors.
42. *Guckenheimer/Holmes:* Nonlinear Oscillations, Dynamical Systems, and Bifurcations of Vector Fields.
43. *Ockendon/Taylor:* Inviscid Fluid Flows.
44. *Pazy:* Semigroups of Linear Operators and Applications to Partial Differential Equations.
45. *Glashoff/Gustafson:* Linear Operations and Approximation: An Introduction to the Theoretical Analysis and Numerical Treatment of Semi-Infinite Programs.
46. *Wilcox:* Scattering Theory for Diffraction Gratings.
47. *Hale et al:* An Introduction to Infinite Dimensional Dynamical Systems—Geometric Theory.
48. *Murray:* Asymptotic Analysis.
49. *Ladyzhenskaya:* The Boundary-Value Problems of Mathematical Physics.
50. *Wilcox:* Sound Propagation in Stratified Fluids.
51. *Golubitsky/Schaeffer:* Bifurcation and Groups in Bifurcation Theory, Vol. I.
52. *Chipot:* Variational Inequalities and Flow in Porous Media.
53. *Majda:* Compressible Fluid Flow and System of Conservation Laws in Several Space Variables.
54. *Wasow:* Linear Turning Point Theory.
55. *Yosida:* Operational Calculus: A Theory of Hyperfunctions.
56. *Chang/Howes:* Nonlinear Singular Perturbation Phenomena: Theory and Applications.
57. *Reinhardt:* Analysis of Approximation Methods for Differential and Integral Equations.
58. *Dwoyer/Hussaini/Voigt (eds):* Theoretical Approaches to Turbulence.
59. *Sanders/Verhulst:* Averaging Methods in Nonlinear Dynamical Systems.
60. *Ghil/Childress:* Topics in Geophysical Dynamics: Atmospheric Dynamics, Dynamo Theory and Climate Dynamics.

(continued following index)

Donald A. Drew   Stephen L. Passman

Theory of Multicomponent Fluids

With 40 Illustrations

 Springer

Donald A. Drew
Ricketts Professor of Applied Mathematics
Department of Mathematical Science
Rensselaer Polytechnic Institute
Troy, NY 12180-3590
USA

Stephen L. Passman
Sandia National Laboratories
P.O. Box 5800
Albuquerque, NM 87185
USA

Editors

J.E. Marsden
Control and Dynamical Systems, 107-81
California Institute of Technology
Pasadena, CA 91125
USA

L. Sirovich
Division of Applied Mathematics
Brown University
Providence, RI 02912
USA

Mathematics Subject Classification (1991): 76T05, 35B20

Library of Congress Cataloging-in-Publication Data
Drew, Donald A. (Donald Allen), 1945–
  Theory of multicomponent fluids / Donald A. Drew, Stephen L.
Passman
    p. cm. — (Applied mathematical sciences ; 135)
  Includes bibliographical references and index.
  ISBN 0-387-98380-5 (hardcover : alk. paper)
  1. Multiphase flow. 2. Continuum mechanics. I. Passman, Stephen
L. II. Title. III. Series: Applied mathematical sciences (Springer-
Verlag New York, Inc.) ; v. 135.
TA357.5.M84D74 1998
620.1'06—dc21   98-18392

Printed on acid-free paper.

© 1999 Springer-Verlag New York, Inc.
All rights reserved. This work may not be translated or copied in whole or in part without the written permission of the publisher (Springer-Verlag New York, Inc., 175 Fifth Avenue, New York, NY 10010, USA), except for brief excerpts in connection with reviews or scholarly analysis. Use in connection with any form of information storage and retrieval, electronic adaptation, computer software, or by similar or dissimilar methodology now known or hereafter developed is forbidden.
The use of general descriptive names, trade names, trademarks, etc., in this publication, even if the former are not especially identified, is not to be taken as a sign that such names, as understood by the Trade Marks and Merchandise Marks Act, may accordingly be used freely by anyone.

Production coordinated by Robert Wexler and managed by Francine McNeill; manufacturing supervised by Joe Quatela.
Photocomposed copy prepared by The Bartlett Press, Inc., Marietta, GA, using the authors' LaTeX files.
Printed and bound by Maple-Vail Book Manufacturing Group, York, PA.
Printed in the United States of America.

9 8 7 6 5 4 3 2 1

ISBN 0-387-98380-5 Springer-Verlag New York Berlin Heidelberg SPIN 10657516

Contents

List of Figures — ix

Introduction — 1

I Preliminaries — 11

1 Physical Reality, Corpuscular Models, Continuum Models — 13

2 Classical Continuum Theory — 20
- 2.1. Kinematics — 20
- 2.2. Balance Equations — 22
- 2.3. Jump Conditions — 31
- 2.4. Frames, Frame Indifference, Objectivity — 31
- 2.5. Constitutive Equations, Well-Posedness of Boundary-Value Problems — 35
- 2.6. Representation Theorems — 35
- 2.7. Thermodynamic Processes — 36

3 Viscous and Inviscid Fluids and Elastic Solids — 41
- 3.1. Fluids — 41
- 3.2. Solids — 44

4 Kinetic Theory — 48
- 4.1. Collision Operator — 49

	4.2. The H-Theorem	51
	4.3. Maxwellian Distribution	53
	4.4. Slight Disequilibrium	53
	4.5. Dense Gases	56
5	**Classical Theory of Solutions**	**59**

II Continuum Theory   63

6	**Continuum Balance Equations for Multicomponent Fluids**	**65**
	6.1. Multicomponent Mixtures	65
	6.2. Kinematics	67
	6.3. Balance Equations	68
	6.4. Multicomponent Entropy Inequalities	74
7	**Mixture Equations**	**81**

III Averaging Theory   85

8	**Introduction**	**87**
	8.1. Local Conservation Equations	88
	8.2. Jump Conditions	89
	8.3. Summary of the Exact Equations	91
9	**Ensemble Averaging**	**92**
	9.1. Results for Ensemble Averaging	99
10	**Other Averages**	**105**
	10.1. Multiparticle Distribution Functions	106
	10.2. Time Averaging	114
	10.3. Volume Averaging	116
	10.4. Desiderata	120
11	**Averaged Equations**	**121**
	11.1. Averaging Balance Equations	121
	11.2. Definition of Average Variables	122
	11.3. Averaged Balance Equations	126
12	**Postulational and Averaging Approaches**	**131**

IV Modeling Multicomponent Flows   135

13	**Introduction**	**137**

14 Closure Framework — 140
- 14.1. Completeness of the Formulation — 140
- 14.2. Constitutive Equations — 141
- 14.3. Forms for Constitutive Equations — 144
- 14.4. Entropy Restrictions — 147

15 Relation of Microstructure to Constitutive Equations — 153
- 15.1. Averaging Techniques — 154
- 15.2. Cell Model — 159
- 15.3. Inviscid Fluid Flowing Around a Sphere — 162
- 15.4. Viscous Flow Around a Sphere — 177

16 Maxwell–Boltzmann Dynamics — 193
- 16.1. Collision Effects — 194
- 16.2. Fluid Velocity Effects — 194

17 Interfacial Area — 199
- 17.1. Geometry Models — 200
- 17.2. Evolution of Geometric Statistics — 205
- 17.3. Average Geometrical Properties — 210
- 17.4. A Coalescence and Breakup Model and Geometry — 217
- 17.5. Conclusion — 220

18 Equations of Motion for Dilute Flow — 221
- 18.1. Constitutive Equations — 221
- 18.2. Dispersed Flow Equations of Motion — 231

V Consequences — 235

19 Nature of the Equations — 237
- 19.1. Special Cases of the Equations — 238

20 Well-Posedness — 243
- 20.1. Formulation — 243
- 20.2. Characteristic Values — 244
- 20.3. The Simplest Model — 248
- 20.4. Effect of Viscosity — 249
- 20.5. Inertial Effects — 251
- 20.6. Summary Observations — 252

21 Solutions for Shearing Flows — 254
- 21.1. Field Equations — 255
- 21.2. Kinematics and Dynamics of Shearing Flow — 258

22 Wave Dynamics — **273**

- 22.1. Introduction — 273
- 22.2. Acoustic Propagation — 274
- 22.3. Volumetric Waves — 282
- 22.4. Characteristics and Linear Stability — 284
- 22.5. Inlet Step Response — 288
- 22.6. Nonlinear Waves — 293

References — **297**

Index — **303**

List of Figures

1.1	Mass-to-volume ratios for various sized balls	15
2.1	A "flake".	25
2.2	A tetrahedron.	25
2.3	A typical configuration with a cut.	26
2.4	A typical configuration.	27
2.5	Another configuration with a cut.	27
2.6	Cylindrical volume illustrating balances across interfaces.	31
4.1	Collision geometry.	50
4.2	Cylindrical volume for potential collisions.	51
4.3	Excluded Volume	57
4.4	Shielding effect	58
6.1	Configurations and mappings.	68
9.1	Realizations.	93
10.1	Excluded volume	112
10.2	Time history	115
15.1	Nearest-neighbor and "top hat" distributions	160
15.2	A "cell".	160
15.3	Cell averaging of the interfacial force.	166
15.4	Volume average.	176
15.5	Cell approximation to the volume average.	176

15.6 Effective viscosity 190

17.1 Steady-state values of s versus α_d. 205
17.2 Principal coordinates and radii of curvature. 207
17.3 Surface coalescence. 216

19.1 Momentum balance for mixture 239

20.1 Normal and tangential coordinates. 244
20.2 Characteristic values 252

21.1 Simple shearing flow. 262
21.2 Volume fraction profile. 271
21.3 Shear stress 272

22.1 Speed of sound versus volume fraction. 279
22.2 Speed of sound versus frequency. 282
22.3 Attenuation versus frequency. 283
22.4 Spreading characteristics. 285
22.5 Converging characteristics, leading to a shock. 285
22.6 "Density" of the conserved momentum. 290
22.7 Evolution of an initial discontinuity 291
22.8 Shock and rarefaction loci. 292
22.9 Traveling wave profile. 295

Introduction

In this book, we give a rational treatment of multicomponent materials as interacting continua. We offer two derivations of the equations of motion for the interacting continua; one which uses the concepts of continua for the components, and one which applies an averaging operation to the continuum equations for each component. Arguments are given for constitutive equations appropriate for dispersed multicomponent flow. The forms of the constitutive equations are derived from the principles of continuum mechanics applied to the components and their interactions. The solutions of problems of hydromechanics of ordinary continua are used as motivation for the forms of certain constitutive equations in multicomponent materials. The balance of the book is devoted to the study of problems of hydrodynamics of multicomponent flows.

Many materials are *homogeneous* in the sense that each part of the material has the same response to a given set of stimuli as all of the other parts. An example of such a material is pure water. Formulation of equations describing the behavior of homogeneous materials is well understood, and is described in numerous standard textbooks.

Many other materials, both manufactured and occurring in nature, are not homogeneous. Such materials are often given names such as *mixtures* or *composites*. Here there is a dichotomy of behavior. Some inhomogeneous materials are solutions, that is, they consist of several substances but the substances are mixed so well that they cannot be distinguished readily. Others are composite, that is, the mixing is at a much larger scale. This dichotomy is not entirely unique. For example, human blood is, to first approximation, a suspension of red cells in a fluid plasma. When this material flows in large vessels it may be treated as a homogeneous fluid. In smaller vessels the size of the red cells is appreciable with respect to the size of the vessels and the treatment of the blood as homogeneous may lead to gross error. In even smaller vessels the size of the red cells is much the same as the size of the vessels; then yet another theoretical treatment is needed.

There is a rational theoretical treatment of solutions, and it is confirmed well by experiment. In contrast, for composite materials both theory and experiment are still developing. Solid composites, such as carbon fibers embedded in a solid polymer, have been the subject of extensive study. The materials and the theoretical treatment are very complex. Still, the complexity is somewhat mitigated by the fact that the components are stationary with respect to one another, that is, one component does not diffuse into another. We do not consider solid composites in this book. Fluid composites are composites which can be made to flow. Generally, for such materials, diffusion is of physical significance. Such materials have often been called "multiphase flows" in the older literature. Here we concentrate our study on fluid composites, and call them *multicomponent fluids*.

Multicomponent fluids occur very commonly both in nature and technology. For example, clouds are droplets of liquid moving in a gas. Oil, gas, and water coexist in rock. Near the surface of the Earth, particles are moved by interacting with air or water; this results in the shaping of geological features. In the realm of human endeavors, boiling heat transfer is the workhorse of the energy industry, involving gas bubbles nucleating, growing, and coalescing. Chemical processing involves mixing, emulsifying, and catalysis in a myriad of scales of flow. Multicomponent wastes include aerosols and soot. We drink carbonated beverages from soda pop to champagne, and eat emulsions and suspensions such as mayonnaise and peanut butter. The widespread presence of multicomponent fluids suggests the utility of a general technique of description to understand their behavior well.

The governing equations for multicomponent fluids should provide a tool for predicting mean features of the motions of natural multicomponent materials, as well as the behavior of practical devices. These features may be gross, such as the pressure drop across some flow component, or may be local, such as the peak temperature on a bounding surface. To predict a particular phenomenon, the equations should be adequate to resolve on a scale smaller than the scale of that phenomenon. This philosophy can be overdone; it is possible to provide so much detail that the useful part of the information is lost. Thus there must be a balance: We must provide a theory sufficiently detailed so that significant phenomena are exhibited fully, but not so detailed that significant phenomena are hard to comprehend.

Science is based on the premise that features of the behavior of a system, device, or material can be predicted, in the sense that if the conditions that lead to the behavior are reconstructed, the behavior will be the same. Furthermore, the foundation of science implies that the prediction and physical behavior should be independent of who observes the behavior, no matter where, and no matter how many times the realization is observed.

All physical systems have some uncertainty in their specification. Take as an example a simple pendulum, consisting of a point mass on a rigid massless rod attached to a frictionless support. The length and mass of the pendulum can be determined only to a certain accuracy. Furthermore, if we specify a realization of the motion of the pendulum, the initial displacement and velocity can be specified only to a certain degree of accuracy, as well. Even the concepts of point mass, rigid massless rod, and frictionless pivot are approximations to the description of the behavior seen by any observer. Nonetheless, the mathematical theory for the behavior of an ideal pendulum yields important insights into mechanics, and becomes the paradigm from which "improved" theories, such as the effect of a flexible rod and friction in the pivot, can be studied.

A theory for the behavior of a system, device, or material is a description of the behavior—in words, concepts, equations, inequalities, diagrams. Mathematical theories have taken on a greater significance since the invention of the calculus and the computer. All mathematical theories are based on models of reality, which have, in addition to the complications of specification alluded to in the pendulum paradigm, assumptions about the physical properties of the materials being described. Nonetheless, a theory provides a description of the behavior of each part of a system, device, or material that can be compared to observations of a physical realization of that system, device, or material, and thereby assessed for validity. The theory can then be used to predict the behavior of the system, device or material when it is subjected to other conditions. It is this ability to predict behavior, however imperfect, that provides the utility and excitement of science.

Scientists, as a rule, do not accept observations or predictions that are not *repeatable*. For a concept or model or prediction to become an accepted scientific theory, it must be possible to reproduce the behavior that it asserts. But, as we have seen, no system, device, or material can be reproduced in exact detail. Moreover, the predictions of a model concern an idealized situation that cannot be realized in a physical system, device, or material. Thus, for any system, device, or material, any theory applies to an ideal situation, any realization will differ from that idea, just as it will differ from any other realization. This idea of imperfect correspondence between the theory and any physical realization of the system, device, or material, pervades science.

There are two different ways to view this conundrum. First, if we conceive of the physical system of an approximation to the ideal one, then we can think of the prediction about the behavior as an approximation to the behavior. The closer the physical realization is to the ideal physical system, the better the agreement between observations and prediction should be. Then the question of repeatability is answered by noting that if two processes are approximations to the ideal, then they should approximate each other. Then all repetitions of the physical system should be approximations to all others.

As an example, consider the stress field in an elastic sphere, subject to a specified external loading. "Exact" solutions to such problems exist [59]. It is impossible to construct a physical body that is exactly spherical. Rather, a body that is within some tolerance of spherical can be constructed. Indeed, given a tolerance, an infinite number of bodies can be constructed that are within that tolerance of a sphere. If it actually were possible to measure the stress field in the real body, we would expect to find it "close" to the theoretical value. We also would expect that the smaller we make the tolerance, the closer the measured stress would be to the predicted stress, and the closer the stresses measured for two different physical objects would be.

The first way to deal with the variability of physical systems, then, is to try to control it.

The second view of the interpretation of the meaning of predictions when the physical system has a large amount of variability is in terms of *expected values* and *variances*. The language even suggests the concept that there are many different events, or realizations, and that the expected value is an average over all of them. If a physical system has some uncertainty as to what will be observed at a particular point at some time after the flow has started, it may be appropriate to deal with an average of the motion. Thus, we can conceive of any physical realization as one member of a set of realizations, all having validity as realizations of the specified event. In this interpretation of the role of variability in physical systems, the ideal system is almost never achieved.

The second way to deal with the variability of physical systems, then, is to average over it.

Each observer must recognize the variability in the system, and judge whether the expected values and variances provide suitable predictions, or whether an attempt must be made to control the variability in the system, device, or material.

We take the approach here that we wish to predict averages over ensembles of flows. If, indeed, an average is justified, it does not matter that we allow the particles or bubbles or droplets to be rearranged in space within reason. Implicit in this discussion is a concept of "more likely" and "less likely" realizations in the ensemble. This suggests that each ensemble of realizations, corresponding to a given physical situation, has a probability measure on subsets of realizations.

The average we choose to use in this book is the ensemble average. The ensemble average is the generalization of adding the values of the variable for each realization, and dividing by the number of observations. We shall refer to a "process" as the set of possible flows that could occur, given that the initial and boundary conditions are those appropriate to the physical situation that we wish to describe. We refer to a "realization" of the flow as a motion, one that could have happened. This is what is usually meant as a process, but when we wish to use this terminology, we shall say "microscopic process." Generally, we expect an infinite number of realizations of the flow, consisting of variations of position, attitude, and velocities of the discrete units and the fluid between them.

The ensemble average is an average that allows the interpretation of phenomena in terms of the repeatability of multicomponent flows. Any particular exact experiment or realization will not be repeatable; however, any repetition of the experiment will lead to another member of the ensemble.

Usually in multicomponent materials, we think of the ensemble corresponding to a given motion as a very large set of realizations, with variations that we can observe. Indeed, the set can (theoretically) contain rearrangements of the atoms within the components; however, these different realizations are not different within the context of continuum mechanics, and we do not consider them further here. Thus, we shall start from the equations of continuum mechanics. The first task in studying the behavior of multicomponent materials is to find appropriate principles of balance. These consist of, at least, balances of mass, momentum, and energy. In addition, statements about entropy often have some utility. Two techniques for finding balance equations are called *"averaging,"* and *"postulation."* *Averaging* is a process whereby a quantity which can take on a set of values (which we shall call outcomes) is converted to a quantity representing a mean value. For a field, which is a function on a set of points and a time interval, the set of outcomes consists of fields, and the average is also a field. For multicomponent fluids, combinations of time and space integration often have been used as averages. Thus, for example, the average body force per unit volume at a point and at a particular time may be defined to be a normalized integral over a time interval of the body force per unit volume at that point. In this work, we use a general type of averaging called *ensemble averaging*.

For the technique of *postulation*, the equations are taken as primitives. However, there is more content to the theory than this. A comprehensive theoretical basis is needed to motivate and structure the postulation of useful and acceptable equations. A principal result derived in this book is that the averaging and postulation yield the same results.

For ensemble averaging, the equivalence class of motions that form the ensemble depends on the application. The situation is not fundamentally different from the approach of the classical kinetic theory of monatomic gases. There, it is assumed that gases are collections of individual units called *atoms*. Some understanding of the dynamics of these individual units is assumed; Newtonian dynamics of "hard spheres" subject to elastic collisions is a possible model of the behavior of the units, and one that is commonly used. In some situations it is necessary to include the deformation of the particles in the model. In the former case, the ensemble consists of a large number of possible trajectories of each molecule. In the latter case, in addition to the trajectories, the deformations of the individual particles must be described to complete the definition of the ensemble.

Study of the microscopic processes does not supersede the continuum approach. Indeed, there are specific theorems that link kinetic theory and continuum concepts. These require commitments on the interpretation of the continuum equations as equations for particular averages of certain quantities.

Multicomponent materials differ from gases in that their basic "molecular" units have vastly different scales. In particular, as opposed to the case of kinetic theories of gases, the microstructure in most multicomponent materials is easily accessible to direct observation.

One way to study the behavior of multicomponent materials is to follow the details of the evolution of all of the constituents and all of their interfaces. Except for simple systems, this is not practical. Computers allow the study of rather larger systems than otherwise could be studied. One viewpoint is that such "exact" results obtained from computation are the standard by which other results are judged; any model that smooths out and consequently loses some of the details of the flows is considered to be an inferior model. That viewpoint is not usually the most useful one. In fact, seldom are exact details of initial and boundary conditions known or even interesting. Likewise, exact details of pressure and velocity fields are more information than is needed for most cases. In fact, such excessive information is difficult to interpret and may mask, rather than elucidate, important physics. However, information about the microscopic fields is important to ensure that the macroscopic model sufficiently reflects the essential physics. Thus, both the microscopic viewpoint and the macroscopic, or averaged, viewpoint are complementary. We focus on the macroscopic viewpoint in this book. It is a useful fact that, once averaged motions and parameters are obtained, it is often possible to compute microscopic effects by using the averaged fields as asymptotic limits of the microscopic motion. For example, given a solution for the motion of a particle-laden gas through a turbine, we may use the average velocity field as a boundary condition to solve for the motion of a single particle striking the turbine blades, use the concentration to estimate the probability that a given particle will assume the trajectory computed, and thereby estimate the rate of wear of the blades.

It is possible to bypass the averaging process and to postulate balance, or conservation, equations directly. Rarely do these equations differ from the equations resulting from averaging in anything but minor detail. If they did, each approach would share a burden of proof: Those who postulate equations must argue why they have ignored a feature present in the averaged equations; those who average their equations must explain the physical significance of, and provide models for, the terms so produced. As a specific example, we point to the entropy inequality. Whereas the postulated formulations commonly accepted for mixtures have only one entropy inequality, averaging indicates that for multicomponent materials there is an entropy inequality for each constituent. The individual entropy inequalities involve interfacial entropy fluxes; these are difficult to quantify. The fluxes contribute to a total entropy increase, as seen in the mixture entropy inequality, which is the sum of the inequalities for the individual phases.

Whether the balance equations are derived from an averaging process or are postulated, they always have unspecified terms. One way both to specify some of them and to understand their physical meaning is to construct simple motivational models for microscopic phenomena. It is at this point where the difference between postulated equations and equations derived by averaging is clearest. If we assume a continuum and deny any underlying structure, it is consistent to assume forms of constitutive functions and to measure coefficient constants or functions. In the averaging approach, the undetermined parameters often have forms relating them to the exact or plausible approximate solutions of the microscopic equations of motion. This allows computation of values for the undetermined parameters using the approximate solutions.

Indeed, it is actually at the point of the prescription of constitutive equations where the microscopic detail is lost. To see this, consider the following argument. It is plausible that all averaging procedures lead to the same averaged equations. Then averaged equations are correct even if the average is trivial, that is, if the average is done over but one element. Thus, for example, the constitutive equations for the stresses are the exact constitutive equations for the microscopic materials, and the interfacial interaction terms become generalized Dirac delta functions, acting at the exact locations of interfaces. Then the "averaged" motion predicted by the averaged equation is the exact motion. The key is the correct prescription of constitutive equations. If constitutive equations valid for some other situation are used, since the averaged equations are the same, the motion predicted will be that appropriate for the other constitutive equations. Of course, the initial and boundary conditions must be defined to a commensurate level of resolution. Defining constitutive equations and initial conditions appropriate to a certain scale determines the resolution of the predictions of the equations. Hence, the constitutive equations are the key to proper predictive behavior.

There are several complementary methods which can be used to determine the constitutive equations for a multicomponent fluid. First, the microstructure satisfies the equations of continuum mechanics, with appropriate jump conditions across the interfaces. Then, any instantaneous pointwise solutions could be averaged to obtain constitutive equations. The problem with this approach is that if general exact solutions were known, they could be used directly and there would be no need for averaging. Usually general exact solutions are not within reach. Then the compromise is to employ solutions to simpler situations to obtain motivation as to what the constitutive equations should be. For example, Stokes flow around a rigid sphere in an infinite fluid provides an expression for the interfacial force which is valid for dilute suspensions of small bubbles for sufficiently small relative velocity. The basis for using the concept of ensemble averaging in this situation is as follows: The ensemble contains an extremely large set of possible motions. For dilute flows, a sphere which is far from other spheres is a situation that occurs fairly often. The flow of the fluid around such a sphere is only approximated by the flow in an infinite fluid; however, the information obtained by averaging over the sphere having many different positions in the flow samples a series of flow events that have sufficient resemblance to actual events that the average over the much smaller set approximates the average over the whole ensemble. Only limited information can be obtained from these and similar calculations about higher concentrations and bubble deformation at higher Reynolds numbers. However, empirical observations can be used to extend this relation to higher relative velocities, even though it cannot be derived from averaging any known microscopic flow.

The second method of obtaining constitutive equations is empirical. A material is assumed to have a particular constitutive equation. An experiment is devised from which the values of constituted variables can be directly or indirectly deduced. A drawback to this method is that it alone is not predictive. Specifically, if a new experiment is contemplated, the constitutive equation must be deduced anew. This implies that if the experiment exactly fits the results of the theory, then the material is of the type assumed *in that motion*. Otherwise, there is no conclusion. One problem with this approach is that predictions can never be made in new situations. There is no need for predictive averaged equations if the purpose of the experiments is to duplicate the application.

The approach that we use in this book is an extrapolation of the first two methods to obtain a putative predictive model. The variables that need specification through constitutive equations are assumed to be dependent on variables that specify the "state" of the multicomponent mixture. Representations of the constituted variables in terms of the state variables give general dependencies within the class of multicomponent mixtures described by the state variables. The assumed form for constitutive equations includes constants or functions to be determined. Statements about entropy provide some restrictions on the constants or functions; however, their determination for a given multicomponent mixture must be determined by appropriate calculation from the microstructure, or from appropriate experiments over a certain range of parameters. Again consider the example of drag on a sphere in a slow flow. Plausibly it should depend on relative velocity between the sphere and the fluid, as well as concentration. In fact, the dependence of the drag on relative velocity can be inferred from Stokes flow calculations. However, for the dependence of drag on concentration, observation of experiment is the only choice. Rules or principles of constitutive equations may be used to restrict dependencies or reduce the number of variables involved.

Once properly established and coupled with initial and boundary conditions, the field equations should yield predictions for the dependent functions. As a practical matter, that is not always the case. The predictions may be inadequate in several ways. The range of validity of the constitutive equations may be violated. An important predictive variable may be missing from the constitutive equations or the solution method may be inadequate. It is possible that the inadequacy of the prediction may not be discovered soon or correctly attributed to a flaw in the model. It is improper at this point to adjust parameters to force the predictions to correlate quantitatively with observed quantities. Rather it is proper to identify the cause of the inadequacy of the model, obtain a better model, and reassess the constitutive equations and predictions.

In normal situations, the motivation for using a model that describes an averaged motion is that it should be simple to apply, yet should provide sufficient information for many engineering purposes.

We expect the reader of this book to have sufficient background to be able to follow our use of the major results in continuum mechanics. The short introduction to the subject we give here is intended only to fix our notation and our basic ideas. Moreover, many of the calculations we do for multicomponent materials are straightforward generalizations of those done in the mechanics of an ordinary single continuum. Thus the material on a single continuum is a reference source for the rest of the book. We shall have occasion to refer to specific results from this material in several places where explanation of the mathematics describing a continuum would cause loss of continuity in our arguments concerning multicomponent materials.

Our intent is not to condense all of the knowledge about continuum mechanics into a few pages. Rather, we present the material we will use in later chapters. For those who wish to pursue the matter further, there is a wealth of reference material. For the classical material there are the excellent treatises by Truesdell and Toupin [87] and Truesdell and Noll [88]. A concise exposition of many of the important points is given by Gurtin [39]. The book by Truesdell [90] is a more comprehensive elementary exposition. The book by Chadwick [19] is also highly useful. Excellent explanations of the modern tensor analysis relevant to this subject are given by Bowen [9], Noll [65], and Bowen and Wang [10].

With respect to notation, we use different notations for different purposes. For the most part, we use a slight modification of the direct vector notation of Gibbs [11],[35]. We use boldface type for vectors and tensors. The particular usage should be clear from the context. Occasionally we use other notation. When we do, we either explain it when we use it, or give appropriate citation to the literature. For example, writing out the components of a vector with respect to a fixed Cartesian coordinate system, often called "Cartesian index notation" is simple, and has the advantages that invariant gradient, divergence, and curl are defined easily and generalized to tensors of arbitrary order in the obvious way. The use of such notation may give the impression that the ideas or proofs have something to do with coordinate systems. That never is the case. Moreover, such notation is intrinsically confusing to the eye, and becomes almost overwhelmingly so for multicomponent materials, where it is normal to denote the different components by indices. Nonetheless, sometimes we find it convenient to use such notation.

We use subscripts to name the components. When we deal with general components, we use a subscript k to denote a component. Most of the time, k can range over an arbitrary number of components. When we discuss specific multicomponent mixtures, we usually consider materials which are comprised of two components. When we discuss specific materials for the components, we use the subscript g for a gaseous component, l for a liquid component, and s for a solid component. In some circumstances, we use the subscript p for particles and f for fluid. We also use the subscript d for a dispersed component, and c for a continuous component.

Acknowledgments

We owe gratitude to many for their support and encouragement. Our respective institutions, Rensselaer Polytechnic Institute and Sandia National Laboratories, have provided moral and material support. It is indeed difficult to name the many people at these institutions that influenced this work. We also spent time in the Department of Theoretical and Applied Mechanics at Cornell University and the Institute for Mathematics and its Applications at the University of Minnesota.

Part I
Preliminaries

1
Physical Reality, Corpuscular Models, Continuum Models

One view of physical reality is that the matter commonly perceived as filling space in fact consists principally of empty space, with an occasional bit of matter. Such a bit may be called an *elementary particle*. The distance between the bits of matter and the structure in which they are arranged, as well as their particular arrangement, dictates whether the gross material is perceived as a solid, as a liquid, or as a gas. A bit of matter may itself have a very rich structure, in which case the bit is no longer an elementary particle, but rather is in itself composed of elementary particles. This kind of structure is worthy of study. The scale of such structures is such that they are not accessible for purposes of describing the behavior of materials normally encountered in everyday experience. Neither is such structure interesting for most such purposes. The exceptions are, for example, that a solid is perceived as amorphous or as having a particular crystalline structure based on the symmetry of the arrangement of the matter in it. This affects its gross symmetry, that is, the mechanics of the macroscopic body reflects the arrangement of the bits of matter. The difficulty of scaling such theories up to the size of a body that interests us here makes it desirable to search for an alternative to description at this level of detail. Moreover, theories of such small structures are often revised, because they are not yet completely understood, and both deeper theoretical understanding and more sophisticated experimental work continue to appear. On the other hand, our experience of the gross behavior of the materials of everyday life, such as steel or water, is quite constant. Thus it is extremely rare for us to revise its mathematical description. The result is, for the purposes of modeling phenomena on the scales ordinarily perceived, it is appropriate to devise models that are independent of modern theories of atomic and subatomic physics. Two types of such models are *corpuscular models* and *continuum models*.

It is possible and useful to devise formal structures describing such models. For each, two types of measure are needed. These are *volume V* and *mass M*. Also for each, there is an extensive formal mathematical structure [90]. Rather than repeating the formal detail that appears in many excellent books, here we take the ideas as intuitive. Thus, assume we have a region of physical space. The concept of a "body" is rather an abstract idea, and is thus appropriate to an abstract mathematical structure. However, the image of a body in physical space is intuitive. Consider such a body $\mathcal{B}$, at time t. Let $\mathbf{x}$ be points in physical space. Our interest at present is points "inside" the body. Map that body into a region of space with a mapping χ, and let $\mathbf{x}$ be positions in space within the range of that mapping. To define mass density at a point $\mathbf{x}$ formally, we let $b(r, \mathbf{x})$ be a sphere of radius r, with center at $\mathbf{x}$. Let V represent volume and let M represent mass, so that, e.g., $V(b(r, \mathbf{x}))$ is the volume of the sphere of radius r, with center at $\mathbf{x}$. Let us examine the behavior of the function $R(r, \mathbf{x})$, defined by

$$R(r, \mathbf{x}) = \frac{M(b(r, \mathbf{x}))}{V(b(r, \mathbf{x}))}. \tag{1}$$

Up to this point, the theory does not distinguish between a corpuscular description and a continuum description. A continuum is a body that is infinitely divisible into smaller parts which, in some sense, are identical to the body itself. The mass density ρ at the point $\mathbf{x}$ in $\mathcal{B}$ is given by

$$\rho(\mathbf{x}) = \lim_{r \to 0^+} \frac{M(b(r, \mathbf{x}))}{V(b(r, \mathbf{x}))}. \tag{2}$$

By the hypothesis that mass is an absolutely continuous function of volume for a continuum, the function $\rho(\mathbf{x})$ exists almost everywhere in $\mathcal{B}$.

The density $\rho(\mathbf{x})$ has the property that

$$m(\mathcal{B}) = \int_{\chi(\mathcal{B})} \rho(\mathbf{x}) \, dv \tag{3}$$

is the mass of the body $\mathcal{B}$, where $\chi(\mathcal{B})$ is the configuration of the body $\mathcal{B}$ at time t.

One description of a "corpuscular" model of a body is as follows. Assume that most of the interior of a body $\mathcal{B}$ is empty. Since we are discussing such models so generically, it is difficult to make unequivocal statements about the assumption of the distribution of mass. Indeed, it is difficult to define an appropriate surface to be the boundary of a configuration of $\mathcal{B}$. Assume that somehow this problem is solved. Then a fairly typical assumption would be that there exists a reasonably evenly distributed finite set of subbodies, of small but finite volume measure, in which all of the mass is concentrated. An even more severe assumption is that the masses occupy no volume, and are "mass points." Then, there are values of r for which R changes discontinuously.

Agreement with modern physics leads to the assertion that $R(r, \mathbf{x})$ has the following character as r decreases. For large values of r, R has an unpredictable character due to the inclusion into the sampling volume of macroscopic variations in the mass. There exists a range of r, say $\underline{r} < r < \bar{r}$, in which R is an approxi-

1. Physical Reality, Corpuscular Models, Continuum Models   15

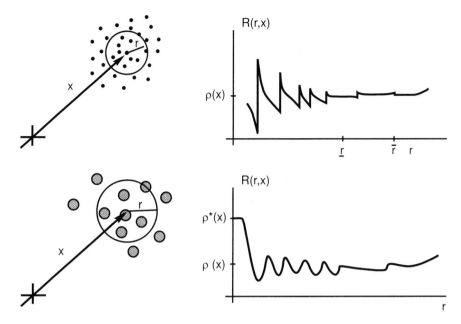

FIGURE 1.1. Mass-to-volume ratios, for point masses and for finite-sized uniform spheres.

mately constant function. This value of $R(r, \mathbf{x})$ is identified with the mass density $\rho(\mathbf{x})$ in the continuum theory. For yet smaller values of r, $r < \underline{r}$, R oscillates in the following manner. As r shrinks in such a fashion that the surface of the ball b does not cross any of the mass points, the "density" R increases, since the volume in the denominator decreases, and the numerator does not change. At a value of r where a point mass is lost as the radius of the ball shrinks, the mass changes discontinuously, while the volume changes gradually. At such values of r, the density decreases discontinuously.[1] Ultimately, r attains a sufficiently small value so that for all r less than this value and almost everywhere in $\mathcal{B}$, b contains no mass points. Then for smaller values of r, R vanishes. However, if the point $\mathbf{x}$ is the location of a point mass, then when the ball is sufficiently small, it will contain exactly one point mass, and for all smaller values of r, the mass will remain constant, while the volume will shrink to zero. For these points, R is infinite. These ideas are illustrated in Figure 1.1(a).

The case that the body $\mathcal{B}$ consists of subbodies of finite volume and of uniform density $\overline{\rho}$, as defined above, is of interest in the study of heterogeneous media. The useful concept of density is defined as in the previous argument. In this case, the result is that the limit of mass per unit volume exists for almost all $\mathbf{x}$, and is given

---

[1] Note that this process of jump decreases in the mass, and gradual decrease in the volume occurs even for large values of r. However, it is only when there are a few mass points left in the ball that the effect becomes appreciable.

16   1. Physical Reality, Corpuscular Models, Continuum Models

by

$$\rho^*(\mathbf{x}) = \lim_{r \to 0} R$$
$$= \begin{cases} 0 & \text{if } \mathbf{x} \text{ lies outside all of the subbodies,} \\ \overline{\rho} & \text{if } \mathbf{x} \text{ lies inside any of the subbodies.} \end{cases} \quad (4)$$

However, the limit is not the mass density ρ of the body that would be measured in a laboratory if the body were homogeneous. Moreover, $\rho^*(\mathbf{x})$ is not a smooth function. In fact, according to this type of corpuscular theory, the value of r for which $\rho^* = R(r, \mathbf{x})$ is so small compared to the typical dimensions of a boundary-value problem of physical interest, that for the practical purpose of using the field equations $\rho^*(\mathbf{x})$ does not have a derivative almost everywhere in $\mathcal{B}$. Moreover, this property is inherited by many of the other fields of interest. Thus, for example, the divergence operator in Cauchy's First Law (3.35) would not be defined.

In all of this book, and indeed in most of the work in the theoretical foundations of mechanics, our principal goal is to obtain local equations of motion and boundary conditions, so that boundary-value problems can be formulated and solved. Intrinsic to this process is the existence of numerous limits of the type (2). Moreover, not only must such limits exist, but the resulting functions must be sufficiently smooth so that the normal analytic operations used in formulating and solving boundary-value problems can be performed. One of these operations is involved in deriving local equations from integral balance statements, formulated for arbitrary subbodies $\overline{\mathcal{B}}$ of $\mathcal{B}$. A sufficient condition for the existence of local equations is contained in the *Dubois–Reymond lemma* [80]:

Lemma (Dubois–Reymond). *If $f(\mathbf{x})$ is continuous in some domain D and if, for every smooth subbody $\overline{D} \subset D$, $\int_{\overline{D}} f(\mathbf{x}) \, dv = 0$, then $f(\mathbf{x}) = 0$.*

Thus, for mathematical rigor using this argument, we must define quantities that allow the limit as $r \to 0$.

Accordance with the opinion that some particular corpuscular theory is a "true" model of reality, while continuum theories are "idealizations," causes difficulty. Occasionally assertions are made that the difficulties can be overcome by changing the limiting operations intrinsic to the differential calculus. For example, we could assume that there exists a range of relatively small[2] $\overline{r}$ such that for $r < \overline{r}$, $R(r, \mathbf{x})$ is constant. Then $\rho(\mathbf{x})$ would be defined by

$$\rho(\mathbf{x}) = \lim_{r \to \overline{r}} R(r, \mathbf{x}). \quad (5)$$

We find two obvious objections to this definition. First, there is no assurance that such a limit exists everywhere (or even anywhere) in $\chi(\mathcal{B})$, or that if it does exist, that $\overline{r}$ is uniform in $\chi(\mathcal{B})$. Thus, at best, if $\overline{r}$ exists, it must be carried as an internal

---

[2] A common assertion is that r is such that $R(r, \mathbf{x})$ stays constant as r becomes smaller, but contains a very large number of molecules. Given sufficient physical assumptions, it is possible to estimate the value of that r.

variable in the theory. Second, the differential calculus used as a tool in the study of mechanics is based on limits of the type (2), not (5).

A more satisfying approach to defining a useful concept of density in a corpuscular theory is as follows. For most purposes, the exact positions of the point masses are not known, and even possibly, not knowable. We can then view the body with the point masses located at certain positions as one *realization* in an *ensemble* of possible events, all of which are just as acceptable as any other as a description of the motion of the body $\mathcal{B}$. A useful variable corresponding to the density $\rho^*(\mathbf{x})$ is the *average*[3] of the ratio of mass to volume. Suppose μ denotes a specific realization. Then $M(b(r, \mathbf{x}), \mu)$ is the mass inside b in realization μ, and $V(b(r, \mathbf{x}))$ is its volume.[4] Consider

$$R(r, \mathbf{x}, \mu) = \frac{M(b(r, \mathbf{x}), \mu)}{V(b(r, \mathbf{x}))}. \tag{6}$$

The expected value of R can be computed by averaging, that is, by multiplying (6) by the probability of occurrence of realization μ, $dm(\mu)$ and "summing" over all possible realizations. Thus,

$$\overline{R}(r, \mathbf{x}) = \int_{\mathcal{E}} R(r, \mathbf{x}, \mu)\, dm(\mu). \tag{7}$$

The appropriate definition of "average density" is

$$\overline{\rho}(\mathbf{x}) = \lim_{r \to 0} \overline{R}(r, \mathbf{x}). \tag{8}$$

The average density so defined is not guaranteed to be smooth by any assumptions made so far. The smoothness depends on the ensemble and the relative probabilities of realizations in the ensemble. If the locations of the point masses are uniformly distributed, it seems reasonable to assume that the average density is also uniform. At the other extreme, if the ensemble $\mathcal{E}$ contains exactly one realization (indicating that the realization is quite special), then the "average density" obtained by this reasoning is that discussed after (5), and there is not a reasonable continuum theory with the corpuscular description behind it.

Suppose that $f(\mathbf{z}_1, \mathbf{z}_2, \ldots, \mathbf{z}_N)$ is the probability density function for the centers of N spheres of unit radius each having density ρ_s. Then, assuming no overlapping of the spheres the density is given by

$$\rho_s \left(H(1 - |\mathbf{x} - \mathbf{z}_1|) + H(1 - |\mathbf{x} - \mathbf{z}_2|) + \cdots + H(1 - |\mathbf{x} - \mathbf{z}_N|) \right), \tag{9}$$

where H is the Heaviside function, defined by

$$H(y) = \begin{cases} 1, & y < 0, \\ 0, & y \geq 0. \end{cases} \tag{10}$$

---

[3] Or expected value.
[4] The ball b and therefore its volume $V(b(r, \mathbf{x}))$ are independent of the particular arrangement of the point masses making up the realization μ.

The average density is given by

$$\bar{\rho}(\mathbf{x}) = \rho_s \int \cdots \int f(\mathbf{z}_1, \mathbf{z}_2, \ldots, \mathbf{z}_N) \left(\sum_{j=1}^{N} H(1 - |\mathbf{x} - \mathbf{z}_j|) \right) d\mathbf{z}_1 \ldots d\mathbf{z}_N, \quad (11)$$

where the multidimensional integral reflects the fact that no spheres overlap. If the average density is formally differentiated with respect to one component of the position vector x_i, and we assume that the point $\mathbf{x}$ is close to $\mathbf{z}_j$, then the derivative of ρ is given by

$$\frac{\partial}{\partial x_i} \bar{\rho}(\mathbf{x}) = \rho_s \int \cdots \int f(\mathbf{z}_1, \mathbf{z}_2, \ldots, \mathbf{z}_N) \left(\frac{\partial}{\partial x_i} H(1 - |\mathbf{x} - \mathbf{z}_j|) \right) d\mathbf{z}_1 \ldots d\mathbf{z}_N. \quad (12)$$

Care must be taken with this integral. The derivative of a Heaviside function must be treated as a generalized function. An appropriate treatment here is to use "integration by parts." This gives

$$\frac{\partial}{\partial x_i} \bar{\rho}(\mathbf{x}) = -\rho_s \int \cdots \int \frac{\partial}{\partial x_i} f(\mathbf{z}_1, \mathbf{z}_2, \ldots, \mathbf{z}_N) \left(H(1 - |\mathbf{x} - \mathbf{z}_j|) \right) d\mathbf{z}_1 \ldots d\mathbf{z}_N. \quad (13)$$

This demonstrates that the smoothness of the averaged density depends on the smoothness of the distribution function f. Indeed, by assuming that f has smooth derivatives of all orders, we can show that $\bar{\rho}(\mathbf{x})$ also possesses smooth derivatives of all orders, even though the exact density is discontinuous.

The average density $\bar{\rho}(\mathbf{x})$ has the property that

$$m(\mathcal{B}) = \int_{\chi(\mathcal{B})} \bar{\rho}(\mathbf{x}) \, dv(\mathcal{B}) \quad (14)$$

is the average mass of the "body" $\mathcal{B}$. Note that the concept of "body" refers to the sequence of bodies, each in different realizations, occupying the volume $\chi(\mathcal{B}) = b(r, \mathbf{x})$.

The continuum and corpuscular theories are alternative models for the same material bodies. To a large extent, the range of physical phenomena of interest to the users of such models do not overlap, so the existence of different models of approximately equal veracity presents no difficulty. Moreover, in the ranges of physical phenomena where there is overlap, it is possible to identify the variables in the continuum theory as average variables from the corpuscular theory. However, it is very difficult to derive useful[5] continuum theories from averaging a corpuscular theory. Indeed, the kinetic theory of gases is a corpuscular theory for a gas, and the Navier–Stokes equations are a continuum theory for a gas. These theories are among the simplest of each type, and they match quite well, but the correspondence is not perfect. Attempts to derive the viscosity of a gas from kinetic theory gives useful trends in the dependence of the viscosity on temperature and pressure, but agreement with experiment occurs only over a small range of conditions. However,

---

[5] In the context of multicomponent materials, "useful" usually implies "complicated."

both models are useful for understanding the phenomena observed in gases, and in predicting flows in complicated situations.

In the remainder of Part I, we present a summary of the continuum model, and the model of kinetic theory. We also present the theory of solutions. In so doing, we set many ideas that we use in later discussions, and provide motivation for the distinct models for multicomponent media.

2
Classical Continuum Theory

2.1. Kinematics

A *body* $\mathcal{B}$ is a set of particles X that can be mapped into a closed region of three-dimensional physical space at each time. Places in physical space are denoted by **x**, and times by t. Usually a preferred configuration of the body is chosen, and the location of the particles in that configuration is denoted by **X**. This configuration is called the *reference configuration*. It need not be one ever occupied by the body, but it may be. Here for convenience in visualizing results, we let the reference configuration be the configuration of the body at time 0. The symbols **X** and **x** do not denote coordinates. Rather, they are positions. Nonetheless they are often called, respectively, *material coordinates* and *spatial coordinates*. Popular usage also often assigns them the respective names *Lagrangian* and *Eulerian*. Though this terminology is not historically accurate [92], we do sometimes follow the popular usage.

The *motion* of $\mathcal{B}$ is the smooth mapping

$$\mathbf{x} = \chi(\mathbf{X}, t). \tag{1}$$

By our convention

$$\mathbf{X} = \chi(\mathbf{X}, 0). \tag{2}$$

The quantity given by

$$\mathbf{V}(\mathbf{X}, t) = \frac{\partial \chi(\mathbf{X}, t)}{\partial t}, \tag{3}$$

is the *velocity field*, or simply *velocity*, of the body $\mathcal{B}$. When written this way, the velocity is said to be in *material form*, or Lagrangian form. Assuming that (1) is smooth enough to be invertible almost everywhere, the velocity may also be written as a function of $\mathbf{x}$ and t. Then we have

$$\mathbf{V}(\mathbf{X}, t) = \mathbf{v}(\chi(\mathbf{X}, t), t) \tag{4}$$

and

$$\mathbf{v}(\mathbf{x}, t) = \mathbf{V}(\chi^{-1}(\mathbf{x}, t), t). \tag{5}$$

Then $\mathbf{v}(\mathbf{x}, t)$ is said to be in *spatial form*, or Eulerian form. The acceleration is given by a similar formula,

$$\mathbf{A} = \frac{\partial \mathbf{V}(\mathbf{X}, t)}{\partial t}. \tag{6}$$

It may also be written in spatial form. It is often convenient to use the same letter for a field, whether it is in spatial or material form. This leads to the cumbersome notation of having to write the arguments $\mathbf{x}$ or $\mathbf{X}$ to distinguish the two descriptions in spatial or material form. For derivatives with respect to time, the notation D/Dt, or alternatively a dot ($\cdot$), over the symbol for the function, is used to denote the time derivative of a function of material variables with $\mathbf{X}$ held constant, for example,

$$\mathbf{a} = \frac{D\mathbf{v}}{Dt} \equiv \dot{\mathbf{v}}. \tag{7}$$

If, for example, $\mathbf{v}$ is written in terms of spatial coordinates, then we have

$$\mathbf{a} = \frac{\partial \mathbf{v}}{\partial t} + \mathbf{v} \cdot \nabla \mathbf{v}, \tag{8}$$

where ∇ is the gradient with respect to spatial coordinates. For any function f, Df/Dt is called the *material derivative* of f.

The *velocity gradient* $\nabla \mathbf{v}$ is of interest in formulating theories of continua. Its symmetric part

$$\mathbf{D} = \frac{1}{2}[\nabla \mathbf{v} + (\nabla \mathbf{v})^T] \tag{9}$$

is the *deformation rate*. Its skew part

$$\mathbf{W} = \frac{1}{2}[\nabla \mathbf{v} - (\nabla \mathbf{v})^T] \tag{10}$$

is the *spin*, or *vorticity tensor*. The axial vector of the spin is the *vorticity* ζ. It is obvious that

$$\nabla \mathbf{v} = \mathbf{D} + \mathbf{W}. \tag{11}$$

Furthermore, this decomposition into symmetric and skew parts is unique.

2.2. Balance Equations

A desirable end product in formulating continuum theories of materials is a system of partial differential equations, valid pointwise in some region of space occupied by a body, and a set of boundary conditions, valid on the boundary of that body. This conflicts somewhat with our primitive ideas of mass, momentum, and energy, which are ideas about bodies as entities, rather than about points in bodies. There is a formal structure for reconciling these concepts. We sketch this structure in the next subsections.

2.2.1. Balance of Mass

Consider a material body $\mathcal{B}$. At a given time t, let the image of the body fill a portion of the region of three-dimensional Euclidean space, say $\chi(\mathcal{B}, t)$. One measure on the configuration of $\mathcal{B}$ is its *volume*, $v_\chi(\mathcal{B}, t)$. We wish this region of space to be smooth enough so that we may perform the usual analytic operations on it. A sufficient condition is that it be "regular in the sense of Kellogg" [54].

To make such a body interesting for mechanics, yet another measure is needed. This measure is the *mass*, $m(\mathcal{B})$. This quantity is assumed to be absolutely continuous with respect to volume; that is, the mass of a proper subbody of a body is strictly less than that of the body. Then the Radon–Nikodym theorem [75] proves the existence of a quantity ρ, defined pointwise on $\mathcal{B}$, with the property that for any subset $\overline{\mathcal{B}}$ of $\mathcal{B}$,

$$m(\chi(\overline{\mathcal{B}}, t)) = \int_{\chi(\overline{\mathcal{B}}, t)} \rho \, dv. \tag{12}$$

A quantity found thus from the Radon–Nikodym theorem is called a *density*. Most often, we call this particular density the *mass density*, though occasionally we simply call it *density*.

Our principal interest is in the dynamics of bodies, and thus it is interesting to let the shape and position of the body evolve in time. Recall that the region of space occupied by $\mathcal{B}$ at time t is $\chi(\mathcal{B}, t)$. We denote the reference configuration of the body by $\chi_0(\mathcal{B})$. Now we have sufficient concepts to formulate the statement of the *conservation of mass* of $\mathcal{B}$:

$$\frac{d}{dt} m(\chi(\mathcal{B}, t)) = 0. \tag{13}$$

That is, the mass of the body $\mathcal{B}$ is constant in time. This idea is intuitive. Here it is a *postulate*, that is, something we assume. Other postulates are possible, and indeed are appropriate for single constituents of the reacting media. By (12),

$$\frac{d}{dt} \int_{\chi(\mathcal{B}, t)} \rho \, dv = 0. \tag{14}$$

It is useful to render this global equation into local form. The transformation from the configuration χ_0 to the configuration χ is characterized by its Jacobian; that is, there is a quantity J with the property that

$$\int_{\chi(\mathcal{B},t)} \rho \, dv = \int_{\chi_0(\mathcal{B})} \rho J \, dv_0. \tag{15}$$

The Jacobian J expresses a local relation between points in the configuration $\chi(\mathcal{B}, t)$ and corresponding points in the configuration $\chi_0(\mathcal{B})$. Substituting (15) into (14),

$$\frac{d}{dt}\int_{\chi(\mathcal{B},t)} \rho \, dv = \frac{d}{dt}\int_{\chi_0(\mathcal{B})} \rho J \, dv_0 = \int_{\chi_0(\mathcal{B})} (\dot\rho J + \rho \dot J) \, dv_0 = 0. \tag{16}$$

By the Dubois–Reymond lemma it follows that the integrand in (16) must vanish,

$$\dot\rho J + \rho \dot J = 0. \tag{17}$$

This is the *material form* or *Lagrangian form of the equation of conservation of mass* [87]. Integrating yields the equation

$$\rho J = \text{constant}. \tag{18}$$

Euler's expansion formula [87] for the material derivative of J is

$$\dot J = J \nabla \cdot \mathbf{v}. \tag{19}$$

Substituting into (17) gives

$$\dot\rho + \rho \nabla \cdot \mathbf{v} = 0. \tag{20}$$

Next we note that the operation denoted by $\dot{}$ can be written as

$$\begin{aligned}
\dot\rho &= \frac{\partial \rho(\mathbf{X}, t)}{\partial t} \\
&= \frac{\partial \rho}{\partial t} + \frac{d\chi}{dt} \cdot \nabla \rho \\
&= \frac{\partial \rho}{\partial t} + \mathbf{v} \cdot \nabla \rho.
\end{aligned} \tag{21}$$

The last operation involves only fields as they depend on the spatial variable $\mathbf{x}$, and is called the *material derivative*. Substitution of this into (20) results in

$$\frac{\partial \rho}{\partial t} + \mathbf{v} \cdot \nabla \rho + \rho \nabla \cdot \mathbf{v} = \frac{\partial \rho}{\partial t} + \nabla \cdot \mathbf{v}\rho = 0. \tag{22}$$

This is the *spatial form* or *Eulerian form of the conservation of mass* [87].

2.2.2. Balance of Momentum

In addition to conservation of mass, it is appropriate to assume that certain quantities are balanced for material bodies $\mathcal{B}$. For example, our experience with mechanics of particles and rigid bodies is that the total force on a body is balanced by the rate of change of momentum of the body. This idea is appropriate for a continuous body $\mathcal{B}$ also, but because of the richness of the idea of "force" for such a body, formulating it is a much more subtle process. For bodies filling regions of space, two different realizations of the concept of force are appropriate. One is the concept of "body force." Such forces are absolutely continuous with respect to volume, so that the total body force on $\mathcal{B}$ is

$$\mathcal{F}_b = \int_{\chi(\mathcal{B},t)} \rho \mathbf{b}\, dv. \tag{23}$$

The quantity $\mathbf{b}$ is the body force per unit mass. We also allow $\chi(\mathcal{B}, t)$ to have surface tractions, that is, forces acting on the boundary of $\chi(\mathcal{B}, t)$,

$$\mathcal{F}_s = \oint_{\partial\chi(\mathcal{B},t)} \mathbf{t}\, ds. \tag{24}$$

The momentum of $\chi(\mathcal{B}, t)$ is given by

$$\mathcal{M} = \int_{\chi(\mathcal{B},t)} \rho \mathbf{v}\, dv. \tag{25}$$

We calculate the material derivative of the momentum. To do so, we transform the integral to the fixed configuration $\chi_0(\mathcal{B})$ using the equation of conservation of mass (22). The result is

$$\frac{d\mathcal{M}}{dt} = \int_{\chi(\mathcal{B},t)} \rho \dot{\mathbf{v}}\, dv(\mathcal{B}). \tag{26}$$

We postulate balance of momentum for $\chi(\mathcal{B}, t)$ in the form

$$\frac{d\mathcal{M}}{dt} = \mathcal{F}_b + \mathcal{F}_s, \tag{27}$$

that is, the rate of change of momentum of a body is balanced by the total force on the body. Under suitable conditions of smoothness of $\chi(\cdot, \cdot)$ and for suitably smooth subbodies $\overline{\mathcal{B}}$, it is possible to prove that almost everywhere in $\chi(\mathcal{B}, t)$ there is a tensor $\mathbf{T}$ with the property that

$$\mathbf{t} = \mathbf{n} \cdot \mathbf{T}, \tag{28}$$

where $\mathbf{n}$ is the unit exterior normal to $\partial\chi(\mathcal{B}, t)$.

The essence of the proof is summarized here. We apply the momentum equation (27) to a body with largest linear dimension l and let l be small. The volume integrals in (27) are of order l^3, while the surface integrals are order l^2. Consequently, the surface integrals must be zero for sufficiently small bodies. We take the body to be a "flake" with one dimension much smaller than the other two, oriented so that the smaller dimension is parallel to a given unit vector $\mathbf{n}$. We then have

$$\int_S \left[\mathbf{t}(\mathbf{n}^+) + \mathbf{t}(\mathbf{n}^-)\right] ds = 0,$$

where S denotes the larger face of the flake and $\mathbf{n}^+ = \mathbf{n}$ and $\mathbf{n}^- = -\mathbf{n}$. Since the face S of the flake is arbitrary, we see that

$$\mathbf{t}(\mathbf{n}) = -\mathbf{t}(-\mathbf{n}). \tag{29}$$

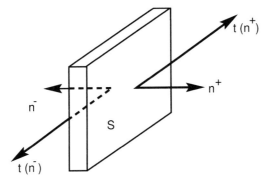

FIGURE 2.1. A "flake".

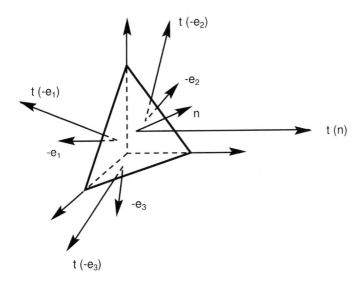

FIGURE 2.2. A tetrahedron.

We apply the momentum balance equation to a body shaped like a tetrahedron with one face having normal **n** and the other three faces having normals to the coordinate directions $\mathbf{e}_1$, $\mathbf{e}_2$, and $\mathbf{e}_3$. As the size of the tetrahedron decreases, the volume integrals become negligible compared to the surface integral. See Figure 2.2. We have

$$0 = \int_{S_n} \mathbf{t}(\mathbf{n})\,ds + \int_{S_1} \mathbf{t}(-\mathbf{e}_1)\,ds + \int_{S_2} \mathbf{t}(-\mathbf{e}_2)\,ds + \int_{S_3} \mathbf{t}(-\mathbf{e}_3)\,ds. \qquad (30)$$

Note that (29) gives

$$\mathbf{t}(-\mathbf{e}_1) = -\mathbf{t}(\mathbf{e}_1),$$
$$\mathbf{t}(-\mathbf{e}_2) = -\mathbf{t}(\mathbf{e}_2),$$
$$\mathbf{t}(-\mathbf{e}_3) = -\mathbf{t}(\mathbf{e}_3).$$

Also, the areas of S_1, S_2, and S_3 are related to the area of S_n by

$$S_1 = S_n(\mathbf{e}_1 \cdot \mathbf{n}),$$
$$S_2 = S_n(\mathbf{e}_2 \cdot \mathbf{n}),$$
$$S_3 = S_n(\mathbf{e}_3 \cdot \mathbf{n}).$$

If we use the integral mean value theorem on (30), we have

$$\mathbf{t}(\mathbf{n}) = (\mathbf{n} \cdot \mathbf{e}_1)\mathbf{t}(\mathbf{e}_1) + (\mathbf{n} \cdot \mathbf{e}_2)\mathbf{t}(\mathbf{e}_2) + (\mathbf{n} \cdot \mathbf{e}_3)\mathbf{t}(\mathbf{e}_3). \tag{31}$$

Thus, the traction vector is linear in the normal $\mathbf{n}$, and defines the stress tensor $\mathbf{T}$,

$$\mathbf{T} = \mathbf{e}_1 \mathbf{t}(\mathbf{e}_1) + \mathbf{e}_2 \mathbf{t}(\mathbf{e}_2) + \mathbf{e}_3 \mathbf{t}(\mathbf{e}_3), \tag{32}$$

so that

$$\mathbf{t} = \mathbf{n} \cdot \mathbf{T}.$$

The stress tensor represents the force per unit area across a surface element having unit normal in the appropriate direction.

The significance of the result (28) is partially lost because of the brevity of the notation. Since this result has a major role in discussions of multicomponent materials, let us examine it in more detail. A typical configuration of $\mathcal{B}$ is shown in Figure 2.3. The field $\mathbf{t}$ is defined at each point on the boundary $\partial \mathcal{B}$. Now consider a point P in the interior of the configuration of $\mathcal{B}$. Let that configuration be cut by the plane a. Then $\mathbf{n}$ and $\mathbf{t}$ at the point P are as shown in Figure 2.4. Likewise, the configuration cut by the plane b, and the corresponding $\mathbf{n}$ and $\mathbf{t}$ are shown in Figure 2.5. We see that $\mathbf{t}$ depends not only on P, but also on the direction of the cut through P. The result [87] is that, under reasonable conditions of smoothness, the dependence on the direction of the cut is a linear function of the unit normal $\mathbf{n}$. Thus, in more precise notation, (28) is

$$\mathbf{t}(\mathbf{x}, t; \mathbf{n}) = \mathbf{n} \cdot \mathbf{T}(\mathbf{x}, t). \tag{33}$$

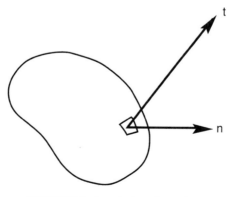

FIGURE 2.3. A typical configuration.

2.2. Balance Equations

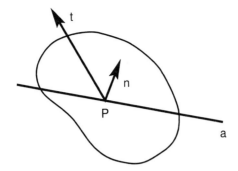

FIGURE 2.4. A typical configuration with a cut.

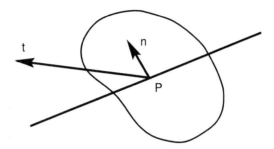

FIGURE 2.5. Another configuration with a cut.

It follows that the forces acting on the boundary of $\mathcal{B}$ are given by

$$\mathcal{F}_s = \oint_{\partial \chi(\mathcal{B},t)} \mathbf{n} \cdot \mathbf{T} \, ds. \tag{34}$$

An application of the divergence theorem then gives

$$\mathcal{F}_s = \int_{\chi(\mathcal{B},t)} \nabla \cdot \mathbf{T} \, dv. \tag{35}$$

By (23) and (24) the balance of momentum has the form

$$\int_{\chi(\mathcal{B},t)} \rho \dot{\mathbf{v}} \, dv = \int_{\chi(\mathcal{B},t)} (\rho \mathbf{b} + \nabla \cdot \mathbf{T}) \, dv. \tag{36}$$

As in the arguments for balance of mass, $\chi(\mathcal{B}, t)$ is arbitrary, and the *local form of the balance of momentum*, or *Cauchy's second law*,

$$\rho \dot{\mathbf{v}} = \rho \mathbf{b} + \nabla \cdot \mathbf{T}, \tag{37}$$

is obtained. This form of the momentum equation is an expression of the idea that "the product of mass and acceleration is balanced by applied forces" for a "particle." The derivative terms appearing on the left-hand side of this equation form the material derivative following a fluid particle. Thus, we shall call this equation the *material* (or "Lagrangian") form of the momentum equation. It should be noted, however, that the gradients are still "spatial" in that they are gradients with respect to the variable **x**. These gradients can be changed to gradients with respect to the variable **X** by using the relation

$$\frac{\partial}{\partial \mathbf{x}} = \frac{\partial \mathbf{X}}{\partial \mathbf{x}} \cdot \frac{\partial}{\partial \mathbf{X}}. \tag{38}$$

While Lagrangian forms are of use in solid mechanics, they have little advantage in fluid mechanics.

Changing $\dot{\mathbf{v}}$ into the material derivative gives the Eulerian, or spatial form, of the equation of balance of momentum

$$\rho \left(\frac{\partial \mathbf{v}}{\partial t} + \mathbf{v} \cdot \nabla \mathbf{v} \right) = \rho \mathbf{b} + \nabla \cdot \mathbf{T}. \tag{39}$$

2.2.3. Balance of Moment of Momentum

Moment of momentum is also a quantity that is balanced for material bodies $\mathcal{B}$. Experience with mechanics of particles and rigid bodies is that the total torque on a body is balanced by the rate of change of moment of momentum of the body. We apply this idea to a continuous body $\mathcal{B}$ also, where the torque is taken to have two different realizations. One is the concept of "body torque." Such torques are absolutely continuous with respect to volume, so that the total body torque on $\mathcal{B}$ is

$$\boldsymbol{T}_b = \int_{\chi(\mathcal{B},t)} \boldsymbol{\tau} \, dv. \tag{40}$$

We shall consider only materials which are nonpolar, so that the torque is the cross product of the position vector with the force. Thus,

$$\boldsymbol{T}_b = \int_{\chi(\mathcal{B},t)} \rho \mathbf{x} \wedge \mathbf{b} \, dv, \tag{41}$$

where the operator $\wedge$ represents the *cross product* operation, defined in terms of Cartesian coordinates by

$$(\mathbf{a} \wedge \mathbf{b})_i = \epsilon_{ijk} a_j b_k,$$

where ϵ_{ijk} is the alternating symbol, and the summation convention is used for repeating indices indicating coordinates. This operation is invariant under proper changes of frame (i.e., translations and rotations), but changes sign under reflections. We also allow $\chi(\mathcal{B}, t)$ to have surface torques, that is, torques acting on the boundary of $\chi(\mathcal{B}, t)$,

$$\boldsymbol{T}_s = \oint_{\partial \chi(\mathcal{B},t)} \mathbf{x} \wedge \mathbf{t} \, ds. \tag{42}$$

The moment of momentum of $\mathcal{B}$ is given by

$$\mathcal{N} = \int_{\chi(\mathcal{B},t)} \rho \mathbf{x} \wedge \mathbf{v} \, dv. \tag{43}$$

We calculate the material derivative of the moment of momentum. To do so, we transform the integral to the fixed configuration $\chi_0(B)$ using the equation of conservation of mass (22) and the fact that $\mathbf{v} \wedge \mathbf{v} = 0$. The result is

$$\frac{d\mathcal{N}}{dt} = \int_{\chi(B,t)} \rho \mathbf{x} \wedge \dot{\mathbf{v}}\, dv = \int_{\chi(B,t)} \mathbf{x} \wedge (\nabla \cdot \mathbf{T} + \rho \mathbf{b})\, dv. \tag{44}$$

We postulate the balance of moment of momentum for B in the form

$$\frac{d\mathcal{N}}{dt} = \mathbf{T}_b + \mathbf{T}_s, \tag{45}$$

that is, the rate of change of moment of momentum of a body is balanced by the total torque on the body.

The torques acting on the boundary of B are given by

$$\mathbf{T}_s = \oint_{\partial \chi(B,t)} \mathbf{n} \cdot \mathbf{x} \wedge \mathbf{T}\, ds. \tag{46}$$

An application of the divergence theorem then gives

$$\mathbf{T}_s = \int_{\chi(B,t)} (\text{vec } \mathbf{T} + \mathbf{x} \wedge \nabla \cdot \mathbf{T})\, dv, \tag{47}$$

where vec $\mathbf{T}$ is the axial vector of the tensor $\mathbf{T}$.

$$(\text{vec } \mathbf{T})_i = \epsilon_{ijk} T_{jk}.$$

It then follows that

$$\int_{\chi(B,t)} \text{vec } \mathbf{T}\, dv = 0. \tag{48}$$

Again, B is arbitrary, so the integrand vanishes. Thus the *local form of the balance of moment of momentum* is

$$\text{vec } \mathbf{T} = 0, \tag{49}$$

or

$$\mathbf{T} = \mathbf{T}^\mathsf{T}. \tag{50}$$

2.2.4. Balance of Energy

We also consider the balance of energy for a body B. Here, in addition to the concepts already introduced, we require u, the internal energy of B per unit mass, $\mathbf{q}$, the heat flux per unit area per unit time on the boundary of B, and r, a heat supply per unit mass per unit time[1] in B. We note that mathematical structures similar to those already introduced allow derivation of u from a statement that B

---

[1] Specifications such as this are rather lengthy. We commonly use the specification "heat supply" for r, when we actually mean "heat supply per unit mass per unit time", etc. Often the heat supply is due to radiation, and this term is often called the "radiation."

has a total internal energy that is absolutely continuous with respect to mass, and allows the derivation of **q** from the statement that at each point on the boundary of $\mathcal{B}$ there is a heat flux, which is dependent on the normal to $\partial\chi(\mathcal{B}, t)$. The postulate of balance of energy is

$$\frac{d}{dt}\int_{\chi(\mathcal{B},t)} \rho\left(u + \frac{1}{2}v^2\right) dv = \int_{\partial\chi(\mathcal{B},t)} \mathbf{n}\cdot(-\mathbf{q} + \mathbf{T}\cdot\mathbf{v})\,ds$$
$$+ \int_{\chi(\mathcal{B},t)} \rho(r + \mathbf{b}\cdot\mathbf{v})\,dv. \tag{51}$$

The usual interpretation of this equation is that the time rate of change of total energy of $\mathcal{B}$ is balanced by the working of the mechanical forces and the rate of energy added by heating. Manipulations of the type described for the momentum equation, and a substitution of the momentum equation itself, yields the local form of the energy equation in material coordinates

$$\rho\dot{u} = \mathbf{T}:\mathbf{D} - \nabla\cdot\mathbf{q} + \rho r. \tag{52}$$

Using the material derivative leads to the energy equation in spatial coordinates

$$\frac{\partial\rho u}{\partial t} + \nabla\cdot\rho u\mathbf{v} = -\nabla\cdot\mathbf{q} + \rho r + \mathbf{T}:\nabla\mathbf{v}. \tag{53}$$

A third form for the energy equation can be obtained by applying the equation for the decomposition of the stress tensor into a pressure p and a (viscous) shear stress τ, and by introducing the enthalpy h,

$$h = u + \frac{p}{\rho}. \tag{54}$$

This produces the enthalpy equation

$$\frac{\partial\rho h}{\partial t} + \nabla\cdot\rho h\mathbf{v} = -\nabla\cdot\mathbf{q} + \left(\frac{\partial p}{\partial t} + \mathbf{v}\cdot\nabla p\right) + \tau:\nabla\mathbf{v} + \rho r. \tag{55}$$

The classical kinetic theory of dilute monatomic gases indicates, in a sense that can be made precise in the context of that particular theory, that the three equations of balance—those for mass, momentum, and energy—are all that can be expected. This book is devoted to continuum theory, not kinetic theory, and to materials which cannot be interpreted as monatomic gases. Nonetheless, we use the kinetic theory as sufficient motivation so that we postulate no additional balance equations. We do note that for materials with microstructure there is motivation to consider additional balance equations [32], and a reasonable axiomatic structure has been built to deal with such additional balance equations in the case of mixtures. For example, Passman, Nunziato, and Walsh [68] sketch such an axiomatic structure. Much of the previous theory of mixtures with microstructure can be traced from the literature they cite. An excellent recent exposition on structured continua is that of Capriz [17].

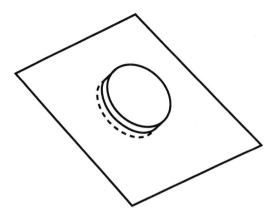

FIGURE 2.6. Cylindrical volume illustrating balances across interfaces.

2.3. Jump Conditions

The partial differential equations of balance of mass, momentum, and energy are not valid if the fields are not continuous; however, the statements of balance for bodies remain valid. Here we assume that the only interesting discontinuity is one across a smooth surface separating regions occupied by the body where the fields are smooth.

Consider the statement of balance of mass for a cylindrical volume containing a subset of the surface of discontinuity as shown in Figure 2.6.

$$\frac{d}{dt}\int_{\chi(B,t)} \rho\, dv = -\int_{\partial\chi(B,t)} \rho \mathbf{v}\cdot\mathbf{n}\, ds\,. \tag{56}$$

Applying standard arguments [87], we obtain the jump condition for mass

$$[\rho(\mathbf{v} - \mathbf{v}_i)] \cdot \mathbf{n} = 0\,. \tag{57}$$

Similar arguments give the jump condition for momentum

$$[(\rho\mathbf{v}(\mathbf{v} - \mathbf{v}_i) + \mathbf{T})] \cdot \mathbf{n} = \mathbf{m}_i^\sigma\,. \tag{58}$$

The jump condition for energy is

$$[\rho[u + \frac{1}{2}v^2(\mathbf{v} - \mathbf{v}_i)] + (\mathbf{T}\cdot\mathbf{v} - \mathbf{q})] \cdot \mathbf{n} = \epsilon_i^\sigma\,. \tag{59}$$

2.4. Frames, Frame Indifference, Objectivity

Processes occurring in mechanical systems occur at places and times. The answers to the questions "Where?" and "When?" require the concept of a "frame," or equivalently, "framing." A frame of reference is something with respect to which we can assign position (place) and instant (time) when and where events occur.

32   2. Classical Continuum Theory

Here, we give a short synopsis of ideas of frame indifference. The concept of "frame" is a primitive concept, and therefore does not have a definition. An intuitive explanation that gives the flavor of the underlying mathematics is given by Truesdell [90, p. 26]:

> For the purposes of classical mechanics, a frame of reference is interpreted by means of a rigid body and a clock. As far as the theory is concerned, the descriptive terms "observer" and "frame of reference of an observer" are synonymous with "as assigned by a framing." So as to further the desired interpretation, we shall occasionally say that **x** is the place and t is the instant of the event ε "as observed" in f and other things of a like kind, but the verb "observe," like the noun "observer," does not enter the mathematical structure.

Given a frame f, the change of frame from f to f^* is defined by

$$t^* = t + a, \tag{60}$$

$$\mathbf{x}^* = \mathbf{x}_0^*(t) + \mathbf{Q}(t)(\mathbf{x} - \mathbf{x}_0), \tag{61}$$

where a is a real number and **Q** is an orthogonal tensor,

$$\mathbf{Q} \cdot \mathbf{Q}^T = \mathbf{I}, \tag{62}$$

where **I** is the identity tensor. In rectangular Cartesian coordinates, this tensor has the same components as the Kronecker delta

$$\mathbf{I}_{ij} = \delta_{ij}.$$

The explanation of these equations is thus: Given a frame f with time measured by t, spatial origin $\mathbf{x}_0$, and general spatial point **x**, the same point in frame f^* is described by translation of time by a fixed amount a, a time-dependent change of origin in space, and a time-dependent rotation. The rotation **Q** need not be proper, that is, it may include a reflection also.

It is possible to compute the behavior of various kinematical quantities under a change of frame. We do a few such computations to illustrate how they are done. Note first that since t and t^* differ only by a translation, differentiation with respect to one is equivalent to differentiation with respect to the other. To shorten the notation, we leave out the arguments of functions. The velocity is given by

$$\dot{\mathbf{x}}^* = \dot{\mathbf{x}}_0^* + \dot{\mathbf{Q}} \cdot (\mathbf{x} - \mathbf{x}_0) + \mathbf{Q} \cdot (\mathbf{x} - \mathbf{x}_0)\dot{}. \tag{63}$$

The tensor **Q** has an interesting property. By (62)

$$\dot{\mathbf{Q}} \cdot \mathbf{Q}^T + \mathbf{Q} \cdot \dot{\mathbf{Q}}^T = 0, \tag{64}$$

so

$$\dot{\mathbf{Q}} \cdot \mathbf{Q}^T = -(\mathbf{Q} \cdot \dot{\mathbf{Q}}^T) = -(\dot{\mathbf{Q}} \cdot \mathbf{Q}^T)^T; \tag{65}$$

that is, $\dot{\mathbf{Q}} \cdot \mathbf{Q}^T$ is skew symmetric. Let

$$\mathbf{A} = \dot{\mathbf{Q}} \cdot \mathbf{Q}^T. \tag{66}$$

By (63)
$$(\mathbf{x}^* - \mathbf{x}_0^*)^{\cdot} = \mathbf{Q} \cdot (\mathbf{x} - \mathbf{x}_0)^{\cdot} + \mathbf{A} \cdot (\mathbf{x}^* - \mathbf{x}_0^*). \tag{67}$$

The quantity $\mathbf{x} - \mathbf{x}_0$ is the vector from the origin to a general point in f, and the quantity $\mathbf{x}^* - \mathbf{x}_0^*$ is the mapping of that vector into f^*. Equation (67) shows that the time rate of change of a radius vector in f^* is found by rotating the time rate of change of the radius vector in f by an amount $\mathbf{Q}$, and adding to that the rate of rotation $\mathbf{A}$ of f^* relative to f operating on the radius vector in f^*. The analogous result for accelerations is

$$(\mathbf{x}^* - \mathbf{x}_0^*)^{\cdot\cdot} = \mathbf{Q} \cdot (\mathbf{x} - \mathbf{x}_0)^{\cdot\cdot} + (\dot{\mathbf{A}} - \mathbf{A}^2) \cdot (\mathbf{x}^* - \mathbf{x}_0^*). \tag{68}$$

Here the term in addition to the rotation of f into f^* is $\dot{\mathbf{A}} - \mathbf{A}^2$, the time rate of change of $\mathbf{A}$, less the quantity $\mathbf{A}^2$, operating on the radius vector in f^*.

For functions obtained from the mapping of the position of a particle, it is possible to derive behavior under change of frame. The velocity field $\mathbf{v}(\mathbf{x}, t)$ is just the material derivative of the deformation. By (67), under a change of frame this becomes

$$\mathbf{v}^*(\mathbf{x}^*, t^*) = \mathbf{Q} \cdot \mathbf{v}(\mathbf{x}, t) + \mathbf{A} \cdot (\mathbf{x}^* - \mathbf{x}_0^*). \tag{69}$$

Let us take the gradient of (69). Of course the gradient with respect to $\mathbf{x}^*$ is different from the gradient with respect to $\mathbf{x}$. Again, we drop the explicit notation for functional dependences. We have

$$\nabla^* \mathbf{v}^* = \mathbf{Q} \cdot \nabla \mathbf{v} \cdot \mathbf{Q}^T + \mathbf{A}. \tag{70}$$

By the uniqueness of the additive decomposition of a tensor into symmetric and skew parts we have

$$\mathbf{D}^* = \mathbf{Q} \cdot \mathbf{D} \cdot \mathbf{Q}^T, \tag{71}$$

and

$$\mathbf{W}^* = \mathbf{Q} \cdot \mathbf{W} \cdot \mathbf{Q}^T + \mathbf{A}, \tag{72}$$

where $\mathbf{D}$ is the rate of deformation, and $\mathbf{W}$ is the rotation tensor. Finally, consider the deformation (1),

$$\mathbf{x} = \chi(\mathbf{X}, t). \tag{73}$$

The quantity

$$\mathbf{F} = \nabla_\mathbf{X} \chi, \tag{74}$$

where $\nabla_\mathbf{X}$ denotes the gradient with respect to $\mathbf{X}$, is shown [87] to transform according to

$$\mathbf{F}^* = \mathbf{Q} \cdot \mathbf{F}. \tag{75}$$

In some parts of mechanics, quantities called *strains* have an important role. Two tensors related to strain are the *right* and *left Cauchy–Green tensors* $\mathbf{C}$ and $\mathbf{B}$, given by

$$\mathbf{C} = \mathbf{F}^T \cdot \mathbf{F}, \tag{76}$$

and

$$\mathbf{B} = \mathbf{F} \cdot \mathbf{F}^T. \tag{77}$$

Let us investigate the transformations of these under changes of frame. By (76)

$$\begin{aligned}\mathbf{C}^* &= \mathbf{F}^{*T} \cdot \mathbf{F}^* \\ &= \mathbf{F}^T \cdot \mathbf{Q}^T \cdot \mathbf{Q} \cdot \mathbf{F} \\ &= \mathbf{F}^T \cdot \mathbf{F},\end{aligned} \tag{78}$$

so that

$$\mathbf{C}^* = \mathbf{C}. \tag{79}$$

A similar calculation gives

$$\mathbf{B}^* = \mathbf{Q} \cdot \mathbf{B} \cdot \mathbf{Q}^T. \tag{80}$$

The deformation rate or "strain rate" $\mathbf{D}$ cannot be derived from any single one of the strain measures $\mathbf{F}, \mathbf{B}, \mathbf{C}$ by any simple process of time differentiation.

The behavior of quantities other than those derived directly from the deformation function χ under changes of frame cannot usually be derived, rather they must be defined. A scalar $f(\mathbf{x}, t)$ is said to be *objective* if

$$f^* = f, \tag{81}$$

and a vector $\mathbf{u}$ is *objective* if

$$\mathbf{u}^* = \mathbf{Q} \cdot \mathbf{u}. \tag{82}$$

A tensor $\mathbf{T}$ is *objective* if it transforms objective vectors into objective vectors, that is

$$\mathbf{T}^* \cdot \mathbf{u}^* = \mathbf{Q} \cdot \mathbf{T} \cdot \mathbf{u} \tag{83}$$

for all $\mathbf{u}$. By (82) then

$$\mathbf{T}^* \cdot \mathbf{Q} \cdot \mathbf{u} = \mathbf{Q} \cdot \mathbf{T} \cdot \mathbf{u} \tag{84}$$

for all $\mathbf{u}$, and thus

$$\mathbf{T}^* = \mathbf{Q} \cdot \mathbf{T} \cdot \mathbf{Q}^T. \tag{85}$$

Thus $\mathbf{B}$ is objective, $\mathbf{C}$ is not.

2.5. Constitutive Equations, Well-Posedness of Boundary-Value Problems

The balance equations are seldom adequate to determine the evolution of points in a body in time and space. They must be supplemented by boundary conditions, and often by constitutive equations. Constitutive equations specify how the material distributes momentum and energy internally. Often they take the forms

$$\mathbf{T} = \mathcal{T}(\rho, \chi, u), \tag{86}$$

and

$$\mathbf{q} = \mathcal{Q}(\rho, \chi, u), \tag{87}$$

where $\mathcal{T}$ and $\mathcal{Q}$ are functionals of the motion and state variables. These functionals map the entire history of the motion and the state variables for all points in the body into the stress and heat flux at point $\mathbf{x}$ at time t. It is usually assumed that the future does not determine the stress or heat flux, so that the functionals are taken to be dependent on the past history of the motion and state variables. Many materials depend on the state of the material for parts of the body near the point $\mathbf{x}$. Assuming some smoothness in the functional, it can be shown that the dependence is

$$\mathbf{T}(\mathbf{x}, t) = \hat{\mathcal{T}}(\mathcal{S}) \tag{88}$$

and

$$\mathbf{q}(\mathbf{x}, t) = \hat{\mathcal{Q}}(\mathcal{S}) \tag{89}$$

where the state $\mathcal{S}$ is specified by

$$\mathcal{S} = \rho(\mathbf{x}, t), \chi(\mathbf{X}, t), u(\mathbf{x}, t), \rho_\mathbf{x}(\mathbf{x}, t), \chi_t(\mathbf{X}, t), \chi_\mathbf{X}(\mathbf{X}, t), u_\mathbf{x}(\mathbf{x}, t), \ldots, \tag{90}$$

where ... means that higher derivatives may be included.

The diversity of materials existing in nature and in technological applications implies that the set of possible constitutive equations is large. However, the set of constitutive equations must be subject to certain restrictions caused by invariance arguments. The most important of these restrictions is that imposed by *material frame indifference*. The principle of frame indifference asserts that the constitutive equations cannot depend on the frame of the observer. The motivation is that reference frames are inventions of the observer and are not intrinsic to the materials being described. Thus, the functional dependence of the constituted variable on the variables describing the mechanical state must be independent of frame, or to be frame indifferent; and the variables used to describe the mechanical state must also be independent of coordinate system, or to be objective.

2.6. Representation Theorems

In our presentation of multicomponent materials, scalar-, vector-, and tensor-valued functions of scalars, vectors, and tensors are also of interest. The study of such functions is sufficiently complicated so that we cannot do it justice in this book. Generally if we use such results we comment on each one specifically. There is a very large specialized literature on this subject. Much of it can be traced from Wang [95],[96], and the article by Smith [82] and the book by Boehler [12].

If **M** is a frame indifferent vector-valued function of the scalars $s_1, \ldots, s_j$, the objective vectors $\mathbf{V}_1, \ldots, \mathbf{V}_n$, and the objective symmetric tensors $\mathbf{T}_1, \ldots, \mathbf{T}_m$, then it must have the form

$$\mathbf{M} = \sum_{i=1}^{n} A_i \mathbf{V}_i + \sum_{i=1}^{n} \sum_{j=1}^{m} B_{ij} \mathbf{V}_i \cdot \mathbf{T}_j, \tag{91}$$

where A_i and B_{ij} are functions of the scalars $s_1, \ldots, s_j$, and the independent scalar invariants formed from $\mathbf{V}_1, \ldots, \mathbf{V}_n$ and $\mathbf{T}_1, \ldots, \mathbf{T}_m$. These scalar invariants are subsets of $\mathbf{V}_i \cdot \mathbf{V}_j$, $\mathbf{V}_i \cdot \mathbf{T}_j \cdot \mathbf{V}_k$, $\mathbf{V}_i \cdot \mathbf{T}_j \mathbf{T}_k \cdot \mathbf{V}_l$ and I_m, II_m, and III_m, where the scalar invariants of a tensor are given in Cartesian coordinate form by

$$I_m = T_{m_{ii}},$$

$$II_m = \frac{1}{2} \left(T_{m_{ii}} T_{m_{jj}} - T_{m_{ij}} T_{m_{ji}} \right),$$

and

$$III_m = \det |T_{m_{ij}}|.$$

Furthermore, if **R** is a frame indifferent symmetric tensor function of the same objective quantities, then it must have the form

$$\mathbf{R} = \sum_{i=1}^{m} C_i \mathbf{T}_i + \sum_{i=1}^{n} \sum_{j=1}^{m} D_{ij} \mathbf{V}_i \mathbf{V}_j$$
$$+ \sum_{i=1}^{n} \sum_{j=1}^{n} \sum_{k=1}^{m} \sum_{l=1}^{m} E_{ijkl} (\mathbf{V}_i \cdot \mathbf{T}_k \mathbf{V}_j \cdot \mathbf{T}_l + \mathbf{V}_j \cdot \mathbf{T}_l \mathbf{V}_i \cdot \mathbf{T}_k), \tag{92}$$

where C_i, D_{ij}, and E_{ijkl} are functions of the scalars and scalar invariants above.

2.7. Thermodynamic Processes

In addition to the balances of mass, energy, and momentum, another balance has an important role in the study of multicomponent materials, and that is the balance of entropy. This balance is sometimes called the "Second Law of Thermodynamics." We note that various other things are often given the same name, and they may or may not bear a quantifiable relation to our entropy inequality. There is a very large modern literature on this subject. Much of it may be traced from [18], [22], and [55]. The balance of entropy is often exploited in this paper, always in a particular way. In order to exemplify our methods, we work through a particular case of the exploitation of the entropy inequality for a single material in explicit detail. We follow almost exactly the presentation of Coleman and Noll [23].

2.7. Thermodynamic Processes

For each point in each configuration of a body $\mathcal{B}$, in addition to the fields we have discussed in the presentation of the other balance equations, we let $s(\mathbf{X}, t)$ be the entropy per unit mass, and let $\theta(\mathbf{X}, t)$ be the temperature. The entropy inequality is the balance principle

$$\frac{d}{dt} \int_{\chi(\mathcal{B},t)} \rho s \, dv \geq \int_{\chi(\mathcal{B},t)} \frac{\rho r}{\theta} \, dv - \oint_{\partial \chi(\mathcal{B},t)} \frac{\mathbf{q} \cdot \mathbf{n}}{\theta} \, ds. \tag{93}$$

Its local form is

$$\rho \dot{s} \geq \frac{\rho r}{\theta} - \nabla \cdot \left(\frac{\mathbf{q}}{\theta}\right). \tag{94}$$

We give a formal definition of the concept of thermodynamic process. Consider a mapping of a body $\mathcal{B}$ onto a region of space $\chi_0(\mathcal{B})$. Let $\mathbf{X}$ be a general point in space in that mapping. Consider the eight functions of $\mathbf{X}$ and t as follows:

(1) $\mathbf{x} = \chi(\mathbf{X}, t)$, the spatial position of $\mathbf{X}$;
(2) $\mathbf{T}(\mathbf{X}, t)$, the (symmetric) stress tensor;
(3) $\mathbf{b}(\mathbf{X}, t)$, the body force;
(4) $e(\mathbf{X}, t)$, the internal energy;
(5) $\mathbf{q}(\mathbf{X}, t)$, the heat flux vector;
(6) $r(\mathbf{X}, t)$, the heat supply;
(7) $s(\mathbf{X}, t)$, the entropy; and
(8) $\theta(\mathbf{X}, t)$, the temperature, subject to $\theta > 0$.

This set of eight functions is a *thermodynamic process* if it satisfies the balance of momentum

$$\rho \dot{\mathbf{v}} = \rho \mathbf{b} + \nabla \cdot \mathbf{T}, \tag{95}$$

and the balance of energy

$$\rho \dot{e} = \mathbf{T} : \mathbf{D} - \nabla \cdot \mathbf{q} + \rho r, \tag{96}$$

for every t and $\mathbf{X}$ in $\chi_0(\mathcal{B})$.

Coleman and Noll [23] note that in order to define a thermodynamic process it is sufficient to prescribe the six functions χ, $\mathbf{T}$, e, $\mathbf{q}$, s, and θ. The two remaining functions $\mathbf{b}$ and r are then uniquely determined by (95) and (96).

Thus far we have placed only minor restrictions on thermodynamic processes. A material is defined by a constitutive assumption, which is a restriction on the processes in a body consisting of the material. Here, we consider a very special constitutive assumption, but one which is of considerable physical interest. An elastic fluid with heat conduction and viscosity is defined by five response functions $\hat{e}$, $\hat{\theta}$, $\hat{p}$, $\hat{\mathbf{C}}$, and $\hat{\mathbf{q}}$. A process is said to be *admissible* in a body consisting of such a material if the following constitutive equations hold at each $\mathbf{X}$ and t:

$$e = \hat{e}(\rho, s), \tag{97}$$

$$\theta = \hat{\theta}(\rho, s), \tag{98}$$

$$\mathbf{T} = -\hat{p}(\rho, s)\mathbf{I} + \hat{\mathbf{C}}(\rho, s) : \mathbf{D}, \tag{99}$$

$$\mathbf{q} = \hat{\mathbf{q}}(\rho, s, \nabla \theta). \tag{100}$$

Here $\hat{\mathsf{C}}$ is a fourth-order tensor, subject to the obvious symmetries. In Cartesian component notation, the equation for stress is

$$T_{ij} = -p\delta_{ij} + \hat{C}_{ijkl} D_{kl}. \tag{101}$$

Now the crux of our use of thermodynamics is the postulate that for every process admissible in a body consisting of given material and for every part of this body the entropy inequality is satisfied.[2]

Equation (96) always holds. We substitute it into the entropy inequality, thus obtaining

$$\rho\theta\dot{s} - \rho\dot{e} + \mathbf{T}:\mathbf{D} - \frac{\mathbf{q}\cdot\nabla\theta}{\theta} \geq 0. \tag{102}$$

To use the postulate for the particular constitutive class of interest here, from the class of every admissible process, we choose only the processes that are defined by homogeneous densities and homogeneous entropy distributions. For such processes, ρ and s are independent of the material point and depend only on the time. By (98), the temperature is also homogeneous, that is, $\nabla\theta = 0$. The inequality then becomes, after substitution of (97) and (99),

$$\rho\left(\theta\dot{s} - \frac{\partial\hat{e}}{\partial s}\dot{s} - \frac{\partial\hat{e}}{\partial\rho}\dot{\rho}\right) - p\nabla\cdot\mathbf{v} + \mathbf{D}:\hat{\mathsf{C}}:\mathbf{D} \geq 0. \tag{103}$$

However, by conservation of mass, $\dot{\rho} = -\rho\nabla\cdot\mathbf{v}$, so

$$\rho\left(\theta - \frac{\partial\hat{e}}{\partial s}\right)\dot{s} + \rho^2\left(\frac{\partial\hat{e}}{\partial\rho} - \frac{p}{\rho^2}\right)\nabla\cdot\mathbf{v} + \mathbf{D}:\hat{\mathsf{C}}:\mathbf{D} \geq 0. \tag{104}$$

As functions of time, ρ and s can be chosen arbitrarily. It follows that for a particular time, the values of ρ, s, $\dot{\rho}$, and $\dot{s}$ can be assigned arbitrarily. Taking $\mathbf{D} = 0$ in (104) shows that $(\theta - \partial\hat{e}/\partial s)\dot{s} \geq 0$ must hold for every value of $\dot{s}$; this can be the case only when the temperature relation

$$\theta = \hat{\theta}(\rho, s) = \frac{\partial\hat{e}}{\partial s}(\rho, s) \tag{105}$$

holds.

Choosing $\dot{s} = 0$ and substituting $\alpha\mathbf{D}$ for $\mathbf{D}$ in (104), where α is an arbitrary number, we obtain

$$\alpha\rho^2\left(\frac{\partial\hat{e}}{\partial\rho} - \frac{p}{\rho^2}\right)\nabla\cdot\mathbf{v} + \alpha^2\mathbf{D}:\hat{\mathsf{C}}:\mathbf{D} \geq 0. \tag{106}$$

This polynomial inequality in α must be satisfied for all values of α when ρ, s, and $\mathbf{D}$ are kept fixed. This is the case if and only if the coefficient of α vanishes

---

[2]This postulate is quite different from the usage of the entropy inequality in many books on thermodynamics. There, a process is called *irreversible* if the entropy inequality is satisfied as an inequality, *reversible* if it is satisfied as an equality, and *impossible* if it is not satisfied. Then some of the relations we obtain by using Coleman and Noll's postulate must be obtained by arguments entailing additional physical and mathematical content.

2.7. Thermodynamic Processes

and the coefficient of α^2 is nonnegative. Thus we obtain the pressure relation

$$p = \rho^2 \frac{\partial \hat{e}}{\partial \rho}, \tag{107}$$

and the dissipation inequality

$$\mathbf{D} : \hat{\mathbf{C}} : \mathbf{D} \geq 0, \tag{108}$$

which must hold for all symmetric tensors $\mathbf{D}$.

Let us assume now that the temperature relation (98) can be solved for the entropy. This is not a trivial assumption. A sufficient condition is that for each fixed ρ, e is a convex function of s. Then

$$s = \check{s}(\rho, \theta). \tag{109}$$

Continuing to keep the homogeneous reference configuration fixed, let us consider deformations χ that are both homogeneous and constant in time. Each such deformation is characterized by a unique value of ρ, with $\mathbf{D} = 0$. Let us also consider time-independent temperature distributions $\theta = \theta(\mathbf{X})$. By (109), for each pair $\theta = \theta(\mathbf{X})$, $\rho = \text{constant}$, the corresponding time-independent entropy distribution $s = s(\mathbf{X})$ is uniquely determined. Thus a prescribed constant ρ and a prescribed function $\theta(\mathbf{X})$ uniquely determine an admissible process. For such a process, the heat conduction inequality (102) reduces to

$$-\hat{\mathbf{q}} \cdot \nabla \theta \geq 0. \tag{110}$$

Heretofore, we have left the functions $\hat{\mathbf{C}}$ and $\hat{\mathbf{q}}$ general. In fact, all fluids are isotropic, and all materials satisfy the principle of material frame indifference. This requires the transformation $\hat{\mathbf{C}}$ to be isotropic. Then it follows that

$$\hat{\mathbf{C}}(\rho, s) : \mathbf{D} = 2\mu \mathbf{D} + \lambda (\nabla \cdot \mathbf{v}) \mathbf{I}, \tag{111}$$

and

$$\hat{\mathbf{q}}(\rho, s, \nabla \theta) = -\kappa \nabla \theta, \tag{112}$$

where λ, μ, and κ may depend on ρ and s. Furthermore, the dissipation inequality and the heat conduction inequality require

$$\mu \geq 0, \tag{113}$$
$$3\lambda + 2\mu \geq 0, \tag{114}$$
$$\kappa \geq 0. \tag{115}$$

The restriction on the viscosities then require nontrivial arguments while the restrictions on the heat conductivity is obvious. Thus we have proved that, for a material of the class hypothesized,

$$\mathbf{T} = -p\mathbf{I} + 2\mu \mathbf{D} + \lambda (\nabla \cdot \mathbf{v}) \mathbf{I}, \tag{116}$$
$$\mathbf{q} = -\kappa \nabla \theta, \tag{117}$$
$$p = \rho^2 \frac{\partial \hat{e}}{\partial \rho}, \tag{118}$$

$$\theta = \frac{\partial \hat{e}}{\partial s}, \qquad (119)$$

$$\mu \geq 0, \qquad 3\lambda + 2\mu \geq 0, \qquad \kappa \geq 0. \qquad (120)$$

3
Viscous and Inviscid Fluids and Elastic Solids

In this chapter, we shall summarize the governing partial differential equations for inviscid fluids, Stokes flow, and elastic solids.

3.1. Fluids

In order to obtain the equations of motion for a viscous fluid, we assume that the density ρ is constant, and that the stress depends on the strain rate,

$$\text{sym}(\nabla \mathbf{v}) = \frac{1}{2}[\nabla \mathbf{v} + (\nabla \mathbf{v})^T]. \tag{1}$$

A Newtonian fluid is defined to have a linear relation between stress and the rate of strain. Thus [56],

$$\mathbf{T} = -p\mathbf{I} + \mu[\nabla \mathbf{v} + (\nabla \mathbf{v})^T] + \lambda \nabla \cdot \mathbf{v} \mathbf{I}, \tag{2}$$

where μ is the shear viscosity, and λ is the bulk viscosity. Substituting this in the momentum equation (39), and assuming that the density is constant, results in the Navier–Stokes equations

$$\nabla \cdot \mathbf{v} = 0, \tag{3}$$

$$\rho \left(\frac{\partial \mathbf{v}}{\partial t} + \mathbf{v} \cdot \nabla \mathbf{v} \right) = -\nabla p + \mu \nabla^2 \mathbf{v}, \tag{4}$$

where $\mathbf{v}$ is the velocity, and p is the pressure.

The boundary conditions for these equations are complicated. At solid walls, the fluid moves with the walls. At an inlet, the velocity of the incoming fluid can be prescribed. At an outlet, the pressure can be prescribed.

These equations have a long history, and it is considered difficult to obtain solutions. However, there are two limits that are amenable to analysis. To obtain these, we nondimensionalize the equations with a velocity scale U and a length scale L:

$$\begin{aligned} \mathbf{v} &= U\mathbf{v}', \\ \mathbf{x} &= L\mathbf{x}'. \end{aligned} \tag{5}$$

For the sake of convenience, we nondimensionalize time with

$$t = \frac{L}{U}t'. \tag{6}$$

3.1.1. Inviscid Fluid

Nondimensionalizing the pressure with

$$p = \rho U^2 p' \tag{7}$$

is appropriate for nearly inviscid flow. The equations of motion become

$$\nabla' \cdot \mathbf{v}' = 0, \tag{8}$$

$$\left(\frac{\partial \mathbf{v}'}{\partial t'} + \mathbf{v}' \cdot \nabla' \mathbf{v}' \right) = -\nabla' p' + \frac{1}{\text{Re}} \nabla'^2 \mathbf{v}', \tag{9}$$

where $\text{Re} = \rho U L/\mu$ is the Reynolds number. Formally letting $\text{Re} \to \infty$, and dropping the prime notation for dimensionless variables gives the equations of motion for an inviscid fluid

$$\nabla \cdot \mathbf{v} = 0, \tag{10}$$

$$\frac{\partial \mathbf{v}}{\partial t} + \mathbf{v} \cdot \nabla \mathbf{v} = -\nabla p. \tag{11}$$

The boundary condition for an inviscid fluid at a solid boundary is

$$\mathbf{n} \cdot \mathbf{v} = \mathbf{n} \cdot \mathbf{v}_s, \tag{12}$$

where $\mathbf{v}_s$ is the velocity of the solid.

If we use the identity

$$\mathbf{v} \wedge (\nabla \wedge \mathbf{v}) = \frac{1}{2} \nabla v^2 - \mathbf{v} \cdot \nabla \mathbf{v}, \tag{13}$$

we have

$$\frac{\partial \mathbf{v}}{\partial t} - \mathbf{v} \wedge \nabla \wedge \mathbf{v} = -\nabla \left(p + \frac{1}{2} v^2 \right). \tag{14}$$

Vorticity

If we take the curl of (14), define the vorticity $\boldsymbol{\zeta} = \nabla \wedge \mathbf{v}$, and use the identities that $\nabla \wedge \nabla s = 0$ for a smooth scalar field s, and

$$\nabla \wedge (\mathbf{v} \wedge \boldsymbol{\zeta}) = \boldsymbol{\zeta} \cdot \nabla \mathbf{v} - \mathbf{v} \cdot \nabla \boldsymbol{\zeta}, \tag{15}$$

we have [86]

$$\frac{\partial \zeta}{\partial t} + \mathbf{v} \cdot \nabla \zeta = \zeta \cdot \nabla \mathbf{v}. \tag{16}$$

The vorticity can be obtained from its initial value by noting that the vorticity equation (16) can be put into material coordinates as

$$\frac{D\zeta}{Dt} = \frac{\partial \zeta(\mathbf{X}, t)}{\partial t} = \zeta \frac{\partial}{\partial t} \nabla_\mathbf{X} \chi, \tag{17}$$

so that $\zeta = \zeta(\mathbf{X}, 0) \exp(\nabla_\mathbf{X} \chi)$. It should be noted that $\nabla_\mathbf{X} \chi$ is also not easy to obtain.

Any sufficiently smooth vector field can be written as the sum of the gradient of a scalar field and the curl of a vector field. Without loss of generality, that vector field can be chosen so that its divergence vanishes. This is called the *Helmholtz representation* [87]. For the velocity field considered here, that representation is

$$\mathbf{v} = \nabla \phi + \nabla \wedge \mathbf{a} = \nabla \phi + \mathbf{v}', \tag{18}$$

where $\nabla \cdot \mathbf{a} = 0$. The relation between the vorticity ζ and the Helmholtz potential is given by

$$\zeta = -\nabla^2 \mathbf{a}. \tag{19}$$

If $G(\mathbf{x}, \mathbf{y})$ is a Green's function for Laplace's equation for the pertinent geometry, then the Helmholtz potential $\mathbf{a}$ is given by

$$\mathbf{a}(\mathbf{x}) = -\int G(\mathbf{x}, \mathbf{y}) \zeta(\mathbf{y}) \, d\mathbf{y}, \tag{20}$$

and the velocity $\mathbf{v}'$ is given by

$$\mathbf{v}' = \nabla \wedge \mathbf{a} = -\int \nabla G(\mathbf{x}, \mathbf{y}) \wedge \zeta(\mathbf{y}) \, d\mathbf{y}. \tag{21}$$

The boundary condition for G is taken to be $G = 0$ on a solid surface. Thus,

$$\mathbf{n} \cdot \nabla G = 0$$

on solid surfaces, and consequently $\mathbf{n} \cdot \mathbf{v}' = 0$.

Potential Flow

The continuity equation (10) is

$$\nabla \cdot \mathbf{v} = \nabla^2 \phi + \nabla \cdot (\nabla \wedge \mathbf{a}) = \nabla^2 \phi = 0. \tag{22}$$

Thus, part of the velocity field is a potential flow. The boundary condition on a solid surface is that the normal component of the velocity of the fluid is equal to the normal component of the velocity of the solid surface. Thus,

$$\mathbf{n} \cdot \mathbf{v} = \mathbf{n} \cdot \mathbf{v}_s, \tag{23}$$

where $\mathbf{n}$ is the unit exterior normal to the surface. From (18) and (12),

$$\mathbf{n} \cdot \mathbf{v} = \mathbf{n} \cdot \nabla \phi + \mathbf{n} \cdot \nabla \wedge \mathbf{a} = \mathbf{n} \cdot \mathbf{v}_s. \tag{24}$$

3.1.2. *Stokes Flow*

If we nondimensionalize the pressure with

$$p = \frac{\mu U}{L} p', \tag{25}$$

the scaling is appropriate for slow flow. The equations of motion become

$$\nabla' \cdot \mathbf{v}' = 0, \tag{26}$$

$$\mathrm{Re}\left(\frac{\partial \mathbf{v}'}{\partial t'} + \mathbf{v}' \cdot \nabla' \mathbf{v}'\right) = -\nabla' p' + \nabla'^2 \mathbf{v}'. \tag{27}$$

Formally letting Re $\to$ 0 gives the Stokes equations

$$\nabla \cdot \mathbf{v} = 0, \tag{28}$$

$$-\nabla' p' + \nabla'^2 \mathbf{v} = 0. \tag{29}$$

3.2. Solids

Classically, solids are materials with a "preferred" state, and for which the stress depends on the strain,

$$\mathbf{T} = \mathbf{T}(\nabla_X \chi). \tag{30}$$

Invariance requires that the stress depends on the symmetric part of $\nabla_X \chi$. If we further write $\chi = \mathbf{X} + \mathbf{U}$, and linearize for small $\mathbf{U}$, we have

$$\rho_0 \frac{\partial^2 \mathbf{u}}{\partial t^2} = \nabla \cdot \mathbf{T}(\gamma), \tag{31}$$

where

$$\gamma = \mathrm{sym}\,\nabla \mathbf{u} \tag{32}$$

is the linear strain. The stress–deformation relation for linear elasticity is given by Love [59] as

$$\mathbf{T} = 2\mu_s\,\mathrm{sym}\,\nabla \mathbf{u} + \lambda_s(\nabla \cdot \mathbf{u})\mathbf{I}, \tag{33}$$

where μ_s and λ_s are the Lamé constants for the elastic solid material. This yields the momentum equation for linear elasticity

$$\rho_0 \frac{\partial^2 \mathbf{u}}{\partial t} = \mu_s \nabla^2 \mathbf{u} + (\lambda_s + \mu_s)\nabla(\nabla \cdot \mathbf{u}). \tag{34}$$

The boundary conditions for an elastic solid include the possibility of imposing the stress:

$$\mathbf{t} = \mathbf{n} \cdot \mathbf{T}. \tag{35}$$

If we use the Helmholtz representation for $\mathbf{u}$, we have

$$\mathbf{u} = \nabla \Phi + \nabla \wedge \mathbf{A}, \tag{36}$$

where $\nabla \cdot \mathbf{A} = 0$. Then we see that there are solutions where

$$\frac{\partial^2 \Phi}{\partial t^2} = c_s^2 \nabla^2 \Phi, \tag{37}$$

and

$$\frac{\partial^2 \mathbf{A}}{\partial t^2} = c_p^2 \nabla^2 \mathbf{A}. \tag{38}$$

Here

$$c_p = \left(\frac{\mu_s + 2\lambda_s}{\rho_0}\right)^{1/2}, \tag{39}$$

$$c_s = \left(\frac{\mu_s}{\rho_0}\right)^{1/2}, \tag{40}$$

are the so-called primary and secondary elastic wave speeds.

Quasi-Equilibrium

If we scale with

$$\mathbf{x} = L\mathbf{x}', \tag{41}$$
$$t = Tt', \tag{42}$$

we have

$$\epsilon_s^2 \frac{\partial^2 \Phi}{\partial (t')^2} = \nabla'^2 \Phi, \tag{43}$$

and

$$\epsilon_p^2 \frac{\partial^2 \mathbf{A}}{\partial t^2} = c_p^2 \nabla^2 \mathbf{A}, \tag{44}$$

where $\epsilon_s = L/c_s T$ and $\epsilon_p = L/c_p T$. If $T \gg L/c_s$, then the deformation in the solid is sufficiently slow that

$$\nabla \cdot \mathbf{T} = 0. \tag{45}$$

3.2.1. Rigid Solids

Consider an elastic solid that is in quasi-equilibrium. We scale the displacement by

$$\mathbf{u} = \delta \mathbf{u}'. \tag{46}$$

The stress on the boundary of the solid satisfies

$$\mathbf{t} = \frac{2\delta \mu_s}{L} \mathbf{n} \cdot \left(\text{sym } \nabla' \mathbf{u}' + \frac{\lambda_s}{\mu_s} \nabla' \cdot \mathbf{u}' \mathbf{I}\right). \tag{47}$$

If the scale of the stress on the boundary is small compared to $\delta \mu_s / L$, the displacement will be small, and the solid will behave as if it were rigid.

46   3. Viscous and Inviscid Fluids and Elastic Solids

For a rigid solid, the deformation inside the solid body is zero. Consequently, the motion inside such a body satisfies

$$\text{sym}\,\nabla \mathbf{v} = 0. \tag{48}$$

It follows that inside each sphere

$$\nabla \cdot \mathbf{v} = 0. \tag{49}$$

The equation of motion in the solid is

$$\nabla \cdot \mathbf{T} = 0. \tag{50}$$

Since the solid material is assumed rigid, there is no constitutive equation for the stress. The solution for the motion of rigid solids is

$$\mathbf{x} = \mathbf{X} + \int \mathbf{V}\,dt + \int \mathbf{x} \wedge \boldsymbol{\omega}\,dt, \tag{51}$$

where $\mathbf{V}$ is the velocity of some point inside the body, and $\boldsymbol{\omega}$ is the rotation rate with respect to this point.

For linearly elastic solids, we have

$$\boldsymbol{\chi} = \mathbf{X} + \mathbf{U}, \tag{52}$$

where $\mathbf{U}$ is the displacement. The momentum equation can be written as

$$\rho \frac{\partial^2 \mathbf{U}}{\partial t^2} = \nabla \cdot \mathbf{T}, \tag{53}$$

where the change to material coordinates is implied on the right-hand side of (53). For small deformation, we have $\mathbf{X} \approx \mathbf{x}$, and $\mathbf{U}(\mathbf{X}, t) \approx \mathbf{u}(\mathbf{x}, t)$. Then the momentum equation is approximated by

$$\rho \frac{\partial^2 \mathbf{u}}{\partial t^2} = \nabla \cdot \mathbf{T}, \tag{54}$$

and the difference between material and spatial coordinates on the stress divergence is small.

We note that the solutions of the equations permit elastic waves. If we are interested in time scales much longer than the time to propagate elastic waves across the body, we approximate these solutions by omitting the inertia term, leaving the equation

$$\nabla \cdot \mathbf{T} = 0. \tag{55}$$

Substituting the stress-deformation law (33), we obtain

$$(\lambda_s + \mu_s)\nabla(\nabla \cdot \mathbf{u}) + \mu_s \nabla^2 \mathbf{u} = 0. \tag{56}$$

This can be written as

$$\mathbf{T} = \boldsymbol{\sigma} + \Theta \mathbf{I}, \tag{57}$$

where Θ is the trace of the stress

$$\Theta = (2\mu_s + 3\lambda)\nabla \cdot \mathbf{u}, \tag{58}$$

and the trace of $\boldsymbol{\sigma}$ vanishes,

$$\boldsymbol{\sigma} = \mu_s \operatorname{sym} \nabla \mathbf{u} - \frac{2\mu_s}{3}(\nabla \cdot \mathbf{u})\mathbf{I}. \tag{59}$$

By taking the divergence of (56), we see that the spherical part of the stress satisfies [59]

$$\nabla^2 \Theta = 0. \tag{60}$$

If Θ is known, the displacements can be found from (56).

4
Kinetic Theory

A popular model for gases is based on a corpuscular theory where the gas molecules are modeled as rigid spheres. Random motions of the particle around the mean motion are caused by collisions, which in turn cause transfers of momentum and energy. We also note that the model can be applied to granular materials. However, for practical models of granular materials, the effects of inelastic collisions are important [50].

In this chapter, we summarize a kinetic theory model [20] for the velocity fluctuations that leads to expressions for the viscosity and thermal conductivity for a gas consisting of identical hard spheres.

Let $f(\mathbf{z}, \mathbf{v}, t)$ be the number density of identical spheres in phase space. That is, in a small volume $d\mathbf{z}$, there are $f(\mathbf{z}, \mathbf{v}, t)\, d\mathbf{z}\, d\mathbf{v}$ spheres having velocity within $d\mathbf{v}$ of velocity $\mathbf{v}$. In the absence of external forces on the spheres, f satisfies Boltzmann's equation in the form

$$\frac{\partial f}{\partial t} + \nabla \cdot (\mathbf{v} f) = \mathcal{C}(f), \tag{1}$$

where $\mathcal{C}(f)$ is the collision operator. The number density is defined as

$$n(\mathbf{x}, t) = \int f(\mathbf{v}, \mathbf{x}, t)\, d\mathbf{v}, \tag{2}$$

where the integral is over all velocities. The average velocity of the particles is

$$\bar{\mathbf{v}} = \frac{1}{n} \int \mathbf{v} f\, d\mathbf{v}. \tag{3}$$

4.1. Collision Operator

Consider two particles that collide. Suppose the velocity of the first, before collision, is $\mathbf{v}_1$, and the velocity of the second, before collision, is $\mathbf{v}_2$. Also, the velocity of the first particle, after collision, is $\mathbf{v}'_1$, and the velocity of the second, after collision, is $\mathbf{v}'_2$. If the collision conserves momentum, then the center of mass velocity $\mathbf{V}$ does not change, so that

$$\mathbf{V} = \mathbf{v}_1 + \mathbf{v}_2 = \mathbf{v}'_1 + \mathbf{v}'_2. \tag{4}$$

Then the relative velocities are

$$\mathbf{v}_{12} = \mathbf{v}_1 - \mathbf{v}_2 = -\mathbf{v}_{21}, \tag{5}$$

and

$$\mathbf{v}'_{12} = \mathbf{v}'_1 - \mathbf{v}'_2 = -\mathbf{v}'_{21}. \tag{6}$$

Note that

$$\mathbf{v}_1 = \mathbf{V} - \mathbf{v}_{21}, \tag{7}$$

$$\mathbf{v}_2 = \mathbf{V} + \mathbf{v}_{21}, \tag{8}$$

$$\mathbf{v}'_1 = \mathbf{V} - \mathbf{v}'_{21}, \tag{9}$$

and

$$\mathbf{v}'_2 = \mathbf{V} + \mathbf{v}'_{21}. \tag{10}$$

If energy is also conserved in collision, we have

$$v_1^2 + v_2^2 = (v'_1)^2 + (v'_2)^2. \tag{11}$$

Substituting (7)–(10) in (11) leads to

$$V^2 + (v_{21})^2 = V^2 + (v'_{21})^2. \tag{12}$$

Thus, conservation of energy gives

$$(v_{21})^2 = (v'_{21})^2. \tag{13}$$

The dynamics of a collision depends on $\mathbf{v}_1$ and $\mathbf{v}_2$, and on the collision geometry. Let $\mathbf{k}$ be the unit vector defining the direction from the center of particle 1 through the center of particle 2 at the instant of collision (see Figure 4.1). The components of $\mathbf{v}_{21}$ and $\mathbf{v}'_{21}$ in the direction of $\mathbf{k}$ are equal in magnitude, but opposite in sign. Thus, $\mathbf{v}_{21} \cdot \mathbf{k} = -\mathbf{v}'_{21} \cdot \mathbf{k}$, and the components perpendicular to $\mathbf{k}$ are equal. Therefore,

$$\mathbf{v}_{21} - \mathbf{v}'_{21} = 2(\mathbf{v}_{21} \cdot \mathbf{k})\mathbf{k} = -2(\mathbf{v}'_{21} \cdot \mathbf{k})\mathbf{k}. \tag{14}$$

Then

$$\mathbf{v}'_1 - \mathbf{v}_1 = 2(\mathbf{v}_{21} \cdot \mathbf{k})\mathbf{k} = -2(\mathbf{v}'_{21} \cdot \mathbf{k})\mathbf{k}, \tag{15}$$

50   4. Kinetic Theory

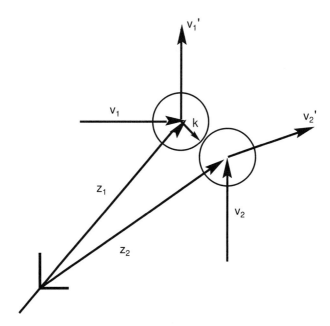

FIGURE 4.1. Collision geometry.

and
$$\mathbf{v}_2' - \mathbf{v}_2 = -2(\mathbf{v}_{21} \cdot \mathbf{k})\mathbf{k} = 2(\mathbf{v}_{21}' \cdot \mathbf{k})\mathbf{k}. \tag{16}$$

Consider particle 1 to be within $d\mathbf{x}$ of the point $\mathbf{x}$, and to have velocity within $d\mathbf{v}_1$ of some velocity $\mathbf{v}_1$. Then, in order for particle 2, having velocity within $d\mathbf{v}_2$ of $\mathbf{v}_2$, to collide with it in time interval dt, so that the collision point is within the element $d\mathbf{k}$ of the solid angle centered about the direction $\mathbf{k}$, particle 2 must lie in a cylindrical volume that is the projection of the length $\mathbf{v}_{21}\, dt$ on the area $(2a)^2\, d\mathbf{k}$. See Figure 4.2. Thus, the number of collisions occurring in time dt is

$$f^{(2)}(\mathbf{x}, \mathbf{v}_1, \mathbf{x} + 2\mathbf{k}a, \mathbf{v}_2, t)\, d\mathbf{x}\, d\mathbf{v}_1\, dt\, \mathbf{v}_{21} \cdot d\mathbf{k}\, (2a)^2\, d\mathbf{v}_2, \tag{17}$$

where $f^{(2)}(\mathbf{x}, \mathbf{v}_1, \mathbf{x}_2, \mathbf{v}_2, t)$ is the joint particle distribution function.

The two-particle distribution is equal to the product of the one-particle distributions, if the positions and velocities are independently distributed. Thus, we assume that [20]

$$f^{(2)}(\mathbf{x}_1, \mathbf{v}_1, \mathbf{x}_2, \mathbf{v}_2, t) = f(\mathbf{x}_1, \mathbf{v}_1, t) f(\mathbf{x}_2, \mathbf{v}_2, t). \tag{18}$$

The rate of loss of particles from velocity $\mathbf{v}$ can be obtained by letting $\mathbf{v}_1 = \mathbf{v}$ and integrating over all the other collision parameters, $\mathbf{k}$ and $\mathbf{v}_2$. If we assume that particle collisions conserve energy, the number gained at velocity $\mathbf{v}$ in time dt can be found by simply reversing the roles of the primed and unprimed quantities. The

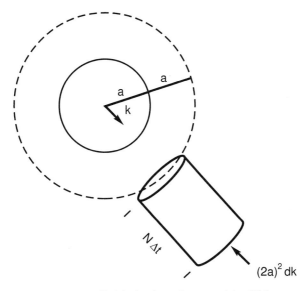

FIGURE 4.2. Cylindrical volume for potential collisions.

result is that the collision operator is

$$\mathcal{C}(f) = \iint \left(f(\mathbf{x}, \mathbf{v}, t) f(\mathbf{x} + 2a\mathbf{k}, \mathbf{v}_2, t) - f(\mathbf{x}, \mathbf{v}', t) f(\mathbf{x} - 2a\mathbf{k}, \mathbf{v}'_2, t) \right)$$
$$\times (2a)^2 \mathbf{v}_{21} \cdot \mathbf{k} \, d\mathbf{v}_2 \, d\mathbf{k}. \qquad (19)$$

Within the kinetic theory model, the stress[1] is given by

$$\mathbf{T} = -m_p \int f(\mathbf{x}, \mathbf{v}, t)(\mathbf{v} - \bar{\mathbf{v}})(\mathbf{v} - \bar{\mathbf{v}}) \, d\mathbf{v}, \qquad (20)$$

and the heat (fluctuation energy) flux is

$$\mathbf{q} = m_p \int f(\mathbf{x}, \mathbf{v}, t) \frac{1}{2}(v - \bar{v})^2 (\mathbf{v} - \bar{\mathbf{v}}) \, d\mathbf{v}. \qquad (21)$$

In order to compute the stress and heat flux, we shall find an approximate solution for f. We shall follow the reasoning of Enskog, as discussed by Chapman and Cowling [20]. The approximation is based on the assumption that the velocity distribution is nearly Maxwellian.

4.2. The H-Theorem

For convenience of notation, we shall let

$$f = f(\mathbf{x}, \mathbf{v}, t), \qquad f' = f(\mathbf{x}, \mathbf{v}', t),$$

---

[1] Actually, this quantity is the negative of the momentum flux due to velocity fluctuations.

4. Kinetic Theory

$$f_2 = f(\mathbf{x} + 2a\mathbf{k}, \mathbf{v}_2, t), \qquad f_2' = f(\mathbf{x} + 2a\mathbf{k}, \mathbf{v}_2', t). \tag{22}$$

Let $H = \int f \ln f \, d\mathbf{v}$. Then H is a measure of order in the velocity distribution, in the sense that

$$\begin{aligned}
\frac{\partial H}{\partial t} &= \int \frac{\partial f \ln f}{\partial t} d\mathbf{v} \\
&= \int (1 + \ln f) \frac{\partial f}{\partial t} d\mathbf{v} \\
&= \iiint (1 + \ln f)(f_2' f' - f_2 f) \mathbf{v}_{21} \cdot \mathbf{k} \, d\mathbf{k} \, d\mathbf{v}_2 \, d\mathbf{v}. \tag{23}
\end{aligned}$$

By the assumption of reversibility in the collisions, we see that

$$\begin{aligned}
\frac{\partial H}{\partial t} &= \iiint (1 + \ln f)(f_2' f' - f_2 f) \mathbf{v}_{21} \cdot \mathbf{k} \, d\mathbf{k} \, d\mathbf{v}_2 \, d\mathbf{v} \\
&= \iiint (1 + \ln f_2)(f_2' f' - f_2 f) \mathbf{v}_{21} \cdot \mathbf{k} \, d\mathbf{k} \, d\mathbf{v}_2 \, d\mathbf{v} \\
&= -\iiint (1 + \ln f')(f_2' f' - f_2 f) \mathbf{v}_{21} \cdot \mathbf{k} \, d\mathbf{k} \, d\mathbf{v}_2 \, d\mathbf{v} \\
&= -\iiint (1 + \ln f_2')(f_2' f' - f_2 f) \mathbf{v}_{21} \cdot \mathbf{k} \, d\mathbf{k} \, d\mathbf{v}_2 \, d\mathbf{v}. \tag{24}
\end{aligned}$$

Thus, adding these expressions, and noting that

$$\ln\left(\frac{f f_2}{f' f_2'}\right) = \ln f + \ln f_2 - \ln f' - \ln f_2', \tag{25}$$

we see that

$$\frac{\partial H}{\partial t} = \frac{1}{4} \iiint \ln\left(\frac{f f_2}{f' f_2'}\right)(f_2' f' - f_2 f) \mathbf{v}_{21} \cdot \mathbf{k} \, d\mathbf{k} \, d\mathbf{v}_2 \, d\mathbf{v}. \tag{26}$$

Then, since

$$\ln\left(\frac{f f_2}{f' f_2'}\right) < 0, \tag{27}$$

when

$$f f_2 < f' f_2', \tag{28}$$

and

$$\ln\left(\frac{f f_2}{f' f_2'}\right) > 0, \tag{29}$$

when

$$f f_2 > f' f_2', \tag{30}$$

it is obvious that

$$\frac{\partial H}{\partial t} < 0. \tag{31}$$

Thus, even though each individual collision is reversible, the order of the system, as measured by H, decreases. This remains one of the strongest arguments for the existence of an entropy inequality.

4.3. Maxwellian Distribution

At steady state, $H = 0$, so that $ff_2 = f'f_2'$, or, equivalently, $\ln f + \ln f_2 = \ln f' + \ln f_2'$. Thus, $\ln f$ is conserved in collisions in the steady state. Since number, momentum, and energy are the physical parameters that are conserved, we expect that

$$\ln f = \alpha_0 + \boldsymbol{\alpha}_1 \cdot \mathbf{v} + \alpha_2 v^2, \tag{32}$$

where α_0, $\boldsymbol{\alpha}_1$, and α_2 are constants. Thus,

$$f = f_0 \exp(\boldsymbol{\alpha}_1 \cdot \mathbf{v} + \alpha_2 v^2), \tag{33}$$

where $f_0 = \exp(\alpha_0)$. By symmetry, $\boldsymbol{\alpha}_1 = 0$. If we note that the particle kinetic energy[2] u^{Re} is equal to

$$u^{Re} = \frac{1}{2n} \int v^2 f \, d\mathbf{v}, \tag{34}$$

then we see that

$$f = f_M(\mathbf{v}) = n \left(\frac{3}{\pi u^{Re}}\right)^{3/2} \exp\left(\frac{-v^2}{3u^{Re}}\right). \tag{35}$$

This distribution is called the Maxwellian.

4.4. Slight Disequilibrium

We seek a solution to Boltzmann's equation that is close to the Maxwellian distribution. We assume that $f = f_M(\mathbf{v}')(1 + \Phi)$, where $\mathbf{v} = \bar{\mathbf{v}} + \mathbf{v}'$, with $\mathbf{v}'$ being the velocity fluctuation. Thus,

$$f_M(\mathbf{v}') = n \left(\frac{3}{\pi u^{Re}}\right)^{3/2} \exp(-\mathcal{V}^2), \tag{36}$$

---

[2] The temperature of the gas is proportional to the kinetic energy, so that $T = 3m_p u^{Re}/2k$, where k is Boltzmann's constant. We retain the fluctuation kinetic energy here for comparison to multicomponent flow dynamics.

54   4. Kinetic Theory

where

$$\mathcal{V} = \frac{(\mathbf{v} - \bar{\mathbf{v}})}{\sqrt{3u^{Re}}}. \tag{37}$$

We assume that Φ is small, and we allow the parameters n, $\bar{\mathbf{v}}$, and u^{Re} to be functions of position and time. We expand the collision operator for slight disequilibrium. The equation becomes

$$\delta_\Phi \mathcal{C} = \mathcal{D} f_M, \tag{38}$$

where the following operators are defined:

$$\mathcal{D} f_M = \frac{\partial f_M}{\partial t} + \mathbf{v} \cdot \frac{\partial}{\partial \mathbf{z}} f_M, \tag{39}$$

$$\delta_\Phi \mathcal{C} = \iint (\Phi + \Phi_2 - \Phi' - \Phi'_2) f_M f_{M2} (2a)^2 \mathbf{v}_{21} \cdot \mathbf{k} \, d\mathbf{v}_2 \, d\mathbf{k}. \tag{40}$$

We recognize that

$$\frac{\partial}{\partial t} f_{M2} = f_{M2} \left(\frac{1}{n} \frac{\partial n}{\partial t} + \frac{2}{\sqrt{3u^{Re}}} \mathcal{V} \cdot \frac{\partial}{\partial t} \bar{\mathbf{v}} + 3\mathcal{V}^2 \frac{1}{u^{Re}} \frac{\partial u^{Re}}{\partial t} \right), \tag{41}$$

and

$$\nabla f_{M2} = f_{M2} \left(\frac{\nabla n}{n} + \frac{2}{\sqrt{3u^{Re}}} \mathcal{V} \cdot \nabla \bar{\mathbf{v}} + 3\mathcal{V}^2 \frac{\nabla u^{Re}}{u^{Re}} \right). \tag{42}$$

Furthermore, we have

$$\mathcal{D} f_M = f_M \left\{ \frac{1}{n} \left[\frac{Dn}{Dt} + (\mathbf{v} - \bar{\mathbf{v}}) \cdot \nabla n \right] \right.$$
$$+ \frac{3\mathcal{V}^2}{u^{Re}} \left[\frac{Du^{Re}}{Dt} + (\mathbf{v} - \bar{\mathbf{v}}) \cdot \nabla u^{Re} \right]$$
$$\left. + \frac{2}{\sqrt{3u^{Re}}} \mathcal{V} \cdot \left(\frac{D\bar{\mathbf{v}}}{Dt} + (\mathbf{v} - \bar{\mathbf{v}}) \cdot \nabla \bar{\mathbf{v}} \right) \right\}, \tag{43}$$

where D/Dt is the material derivative.

The condition for the solvability of (38) is that the right-hand side of (38) must be orthogonal to the solutions of the homogeneous equation. In terms of Φ, the solutions of the homogeneous equation are the conserved quantities 1, $\mathbf{v}$, and v^2. Thus, the conditions for solvability are

$$\int \mathcal{D} f_M \, d\mathbf{v} = 0, \tag{44}$$

$$\int \mathbf{v} \mathcal{D} f_M \, d\mathbf{v} = 0, \tag{45}$$

$$\int v^2 \mathcal{D} f_M \, d\mathbf{v} = 0. \tag{46}$$

4.4. Slight Disequilibrium

The conditions for solubility, (44), (45), and (46), give the equations for balance of mass, momentum, and energy, to lowest order in a, within the context of the Boltzmann model. They are

$$\frac{\partial n}{\partial t} + \nabla \cdot (n\bar{\mathbf{v}}) = 0, \tag{47}$$

$$\rho \left(\frac{\partial \bar{\mathbf{v}}}{\partial t} + \bar{\mathbf{v}} \cdot \nabla \bar{\mathbf{v}} \right) = -\nabla P, \tag{48}$$

$$\alpha \rho_p \left(\frac{\partial u^{Re}}{\partial t} + \bar{\mathbf{v}} \cdot \nabla u^{Re} \right) = -\frac{2}{3} P \nabla \cdot \bar{\mathbf{v}}, \tag{49}$$

where $\rho = n m_p$, m_p is the mass of a particle, and

$$P = \rho \frac{2}{3} u^{Re}. \tag{50}$$

Substituting in (43) leads to the equation

$$\iint (\Phi + \Phi_2 - \Phi' - \Phi_2') f_M f_{M2} (2a)^2 \mathbf{v}_{21} \cdot \mathbf{k} \, d\mathbf{v}_2 \, d\mathbf{k}$$
$$= -f_M \{ \sqrt{3u^{Re}} (\mathcal{V}^2 - \frac{5}{2}) \mathcal{V} \cdot \nabla \ln u^{Re} + 2 (\mathcal{V}\mathcal{V} - \mathcal{V}^2 \mathbf{I}) : \nabla \bar{\mathbf{v}} \}. \tag{51}$$

Thus we can write

$$\Phi = -\{ \sqrt{3u^{Re}} \mathbf{A} \cdot \nabla \ln u^{Re} + 2\mathbf{B} : \nabla \bar{\mathbf{v}} \}, \tag{52}$$

where $\mathbf{A}$ and $\mathbf{B}$ are the vector and tensor solutions of

$$\iint (\mathbf{A} + \mathbf{A}_2 - \mathbf{A}' - \mathbf{A}_2') f_{M2} \mathbf{v}_{21} \cdot \mathbf{k} \, d\mathbf{v}_2 \, d\mathbf{k} = \mathcal{V}(\mathcal{V}^2 - \frac{5}{2}), \tag{53}$$

$$\iint (\mathbf{B} + \mathbf{B}_2 - \mathbf{B}' - \mathbf{B}_2') f_{M2} \mathbf{v}_{21} \cdot \mathbf{k} \, d\mathbf{v}_2 \, d\mathbf{k} = (\mathcal{V}\mathcal{V} - \mathcal{V}^2 \mathbf{I}). \tag{54}$$

We note that, by exploiting the symmetry of $\mathbf{A}$ and $\mathbf{B}$,

$$\mathbf{A} = \mathcal{V} A(\mathcal{V}), \tag{55}$$
$$\mathbf{B} = (\mathcal{V}\mathcal{V} - \mathcal{V}^2 \mathbf{I}) B(\mathcal{V}). \tag{56}$$

The stress and heat flux are given by

$$\mathbf{T}_d^{Re} = -P\mathbf{I} + m_p (3u^{Re})^{3/2} \int \mathcal{V}\mathcal{V} f_M \Phi \, d\mathcal{V}, \tag{57}$$

and

$$\mathbf{q}_d^{Re} = m_p (3u^{Re})^2 \int \frac{1}{2} \mathcal{V} \mathcal{V}^2 f_M \Phi \, d\mathcal{V}. \tag{58}$$

Substituting (55) and (56), and using the symmetry of f_M gives

$$\mathbf{T} + P\mathbf{I} = \mu \left[\nabla \bar{\mathbf{v}} + (\nabla \bar{\mathbf{v}})^T - \frac{1}{3} (\nabla \cdot \bar{\mathbf{v}}) \mathbf{I} \right], \tag{59}$$

$$\mathbf{q} = -\lambda \nabla u^{Re}, \tag{60}$$

where

$$\mu = \frac{4}{15}m_p(3u^{Re})^{3/2}\int f_M \mathcal{V}^4 B(\mathcal{V})\,d\mathcal{V}, \qquad (61)$$

$$\lambda = 3m_p u^{Re}\int f_M \mathcal{V}^4 A(\mathcal{V})\,d\mathcal{V}. \qquad (62)$$

Expanding A and B in powers of $\mathcal{V}$ gives, to first approximation

$$A(\mathcal{V}) = A_0 + A_2\left(\frac{5}{2} - \mathcal{V}^2\right) + \cdots, \qquad (63)$$

$$B(\mathcal{V}) = B_0 + \cdots. \qquad (64)$$

The solutions to (53) and (54) require $A_0 = 0$, and A_2 and B_0 to be given by

$$A_2 = \frac{1}{\displaystyle\int_0^\infty f_M \mathcal{V}^3 \left(\frac{5}{2} - \mathcal{V}^2\right)^2 d\mathcal{V}}, \qquad (65)$$

$$B_0 = \frac{1}{\displaystyle\int_0^\infty f_M \mathcal{V}\,d\mathcal{V}}. \qquad (66)$$

The results are

$$\mu = \frac{5\sqrt{3}m_p}{64\sqrt{2\pi}a^2}\sqrt{u^{Re}}, \qquad (67)$$

$$\lambda = \frac{225\sqrt{3}m_p}{512\sqrt{2\pi}a^2}\sqrt{u^{Re}}. \qquad (68)$$

4.5. Dense Gases

In dense gases, the finite volume of particles, and the interference of a third particle in collisions can have a significant effect. Chapman and Cowling [20] assume that

$$f^{(2)}(\mathbf{x}_1, \mathbf{v}_1, \mathbf{x}_2, \mathbf{v}_2, t) = G(\alpha)f(\mathbf{x}_1, \mathbf{v}_1, t)f(\mathbf{x}_2, \mathbf{v}_2, t), \qquad (69)$$

where $G(\alpha)$ represents the change in the collision rate due to the finite volume of the particles.

For relatively dilute gases, the effect can be computed by assuming that it can be broken up into two parts, an effect of the "excluded volume," and a shielding effect. First, the volume excluded by a sphere of radius a is a sphere concentric with it, of radius $2a$. Thus, the volume excluded by the finite size of the spheres, per unit volume, is

$$n\frac{4}{3}\pi(2a)^3 = 8\alpha.$$

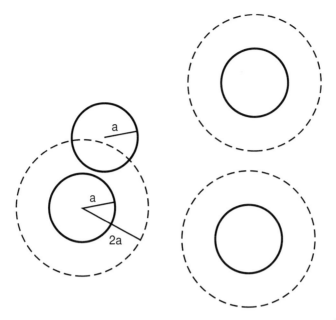

FIGURE 4.3. Excluded volume: The potential colliding sphere can occupy less volume than $(1 - \alpha)$.

Thus, the probability of a collision between spheres is increased by a factor

$$\frac{1}{1 - 8\alpha}.$$

See Figure 4.3.

To calculate the shielding effect, consider the spherical annulus centered about one of the spheres (target sphere), such that a sphere (shielding sphere) whose center is in this annulus will be capable of shielding the target sphere from collision with a third sphere (colliding sphere). Suppose the center of the shielding sphere lies on a surface of radius x about the target sphere. See Figure 4.4. The shielding sphere will interfere with collisions if $2a < x < 4a$. Now consider the spherical shell of area $4\pi(2a)^2$ where the center of the colliding sphere would lie at the instant of collision. The colliding sphere will be shielded from a spherical cap on this sphere, of area $2\pi ax$. The probability of a collision being shielded by any sphere is the probability of the shielding sphere lying within dx of x, times the probability that the shielding sphere interferes with the collision. This probability is

$$4\pi x^2 \, dx \, n \frac{4\pi(2a)^2 - 2\pi ax}{4\pi(2a)^2}.$$

Integrating from $x = 2a$ to $x = 4a$ gives

$$\frac{11}{2}\alpha, \tag{70}$$

4. Kinetic Theory

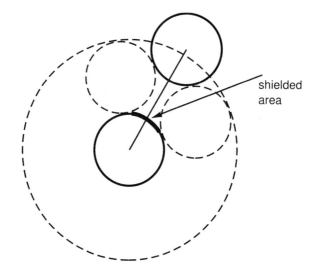

FIGURE 4.4. Shielding effect: A nearby sphere keeps others from reaching the shielded area.

as the probability of shielding. The probability of collision is changed by a factor $1 - 11/2\alpha$.

The combined effect of the excluded volume and the shielding is that the factor G is given by

$$G(\alpha) = \frac{1 - \frac{11}{2}\alpha}{1 - 8\alpha} \approx 1 + \frac{5}{2}\alpha. \tag{71}$$

The collision operator is given by

$$C(f) = \iint [G(\alpha(\mathbf{x} + a\mathbf{k}) f(\mathbf{x}, \mathbf{v}, t) f(\mathbf{x} + 2a\mathbf{k}, \mathbf{v}_2, t) \\ - G(\alpha(\mathbf{x} - a\mathbf{k})) f(\mathbf{x}, \mathbf{v}', t) f(\mathbf{x} - 2a\mathbf{k}, \mathbf{v}'_2, t)] \times (2a)^2 \mathbf{v}_{21} \cdot \mathbf{k} \, d\mathbf{v}_2 \, d\mathbf{k} \tag{72}$$

The theory of the preceding section can be carried through, and the results are that the dense gas viscosity μ_δ and conductivity λ_δ are given by

$$\mu_\delta = \frac{\left[1 + \frac{8}{5}\alpha G(\alpha)\right]}{G(\alpha)} \mu, \tag{73}$$

$$\lambda_\delta = \frac{\left[1 + \frac{12}{5}\alpha G(\alpha)\right]}{G(\alpha)} \lambda. \tag{74}$$

5
Classical Theory of Solutions

The major focus of this book is multicomponent materials. The mathematical theory appropriate to such mixtures of materials is still in a stage of development. A simpler type of material is the *classical mixture*, or *solution*. In such a material, the components are not physically distinct, that is, the mixing of the materials is at a molecular level. A kinetic theory for such materials was given by Maxwell [62], and was transposed into a form appropriate to continuum theory by Truesdell [91].[1]

A *solution* is a sequence of bodies $\mathcal{B}_k, k = 1, \ldots, n$, all of which are supposed to be able to occupy the same region of space at the same time. If $\mathbf{X}_k$ is the place occupied by a particle of $\mathcal{B}_k$ in some reference configuration, then the motion of $\mathcal{B}_k$ is the smooth mapping

$$\mathbf{x} = \chi_k(\mathbf{X}_k, t). \tag{1}$$

Define

$$\grave{\mathbf{x}}_k = \frac{\partial \chi_k(\mathbf{X}_k, t)}{\partial t}. \tag{2}$$

Thus

$$\mathbf{v}_k = \frac{\partial \chi_k(\mathbf{X}_k, t)}{\partial t} = \grave{\mathbf{x}}_k, \tag{3}$$

---

[1] The work is given an elegant synopsis in Lecture 5 of C. Truesdell, *Rational Thermodynamics*, Springer-Verlag, New York, 1985. We express our appreciation to Professor Truesdell and Springer-Verlag for permission to reproduce much of the material in that lecture here.

is the velocity of constituent k. By (1) this velocity may also be written as a function of $\mathbf{x}$ and t. The acceleration is given by a similar formula,

$$\mathbf{a}_k = \frac{\partial \mathbf{v}_k(\mathbf{X}_k, t)}{\partial t} = \grave{\mathbf{v}}_k. \tag{4}$$

It may also be written as a function of $\mathbf{x}$ and t.

Each body has its own mass, and thus its own mass density ρ_k, generally a function of $\mathbf{x}$ and t. Balance principles for the constituents resemble those for single continua, except that the constituents are allowed to interact with one another. The balance principles for mass, momentum, and energy are

$$\int c_k \, dv = \frac{d_k}{dt}\left(\int \rho_k \, dv\right), \tag{5}$$

$$\int \mathbf{m}_k \, dv = \frac{d_k}{dt}\left(\int \rho_k \mathbf{v}_k \, dv\right) - \oint \mathbf{n} \cdot \mathbf{T}_k \, ds - \int \rho_k \mathbf{b}_k \, dv, \tag{6}$$

$$\int \epsilon_k \, dv = \frac{d_k}{dt}\left(\int \rho_k \left(e_k + \frac{v_k^2}{2}\, dv\right)\right) - \oint \mathbf{n} \cdot \mathbf{T}_k \cdot \mathbf{v}_k \, ds$$

$$ - \oint \mathbf{q} \cdot \mathbf{n} \, ds - \int \rho_k r_k \, dv. \tag{7}$$

In these equations $\int$ indicates integration over any region of space occupied by the mixture and $\oint$ indicates integration over the surface of that region. If we assume the appropriate fields to be adequately smooth, the balance principles become

$$c_k = \grave{\rho}_k + \rho_k \nabla \cdot \mathbf{v}_k, \tag{8}$$

$$\mathbf{m}_k = c_k \mathbf{v}_k + \rho_k \mathbf{a}_k - \nabla \cdot \mathbf{T}_k - \rho_k \mathbf{b}_k, \tag{9}$$

$$\epsilon_k = \mathbf{m}_k \cdot \mathbf{v}_k + c_k \left(e_k - \frac{v_k^2}{2}\right) + \rho_k \grave{e}_k - \mathbf{T}_k \cdot \nabla \mathbf{v}_k - \nabla \cdot \mathbf{q}_k - \rho_k r_k. \tag{10}$$

We consider only places in space that are occupied simultaneously by one particle of each of the constituent bodies. Then the total mass density ρ is defined as the sum of the densities of the constituents,

$$\rho = \sum \rho_k. \tag{11}$$

Here and henceforth, $\sum$ stands for summation from $k = 1$ to $k = n$. The mixture may be regarded as being in motion also as a single body. Quantities associated with it are of two kinds. First, there are densities of additive set functions. These are simply the sums, weighted by the concentrations, of the densities of the constituents. For example, for the density of the momentum, we define

$$\rho \mathbf{v} = \sum \rho_k \mathbf{v}_k. \tag{12}$$

A similar definition is used for the total body force and the heating source

$$\rho \mathbf{b} = \sum \rho_k \mathbf{b}_k, \tag{13}$$

$$\rho r = \sum \rho_k r_k. \tag{14}$$

This type of definition is not appropriate for many other quantities having to do with the mixture. For the stress, the definition is

$$\mathbf{T} - \rho \mathbf{vv} = \sum (\mathbf{T}_k - \rho_k \mathbf{v}_k \mathbf{v}_k), \qquad (15)$$

and for the internal energy it is

$$\rho \left(e + \frac{v^2}{2} \right) = \sum \rho_k \left(e_k + \frac{v_k^2}{2} \right), \qquad (16)$$

while for the heat flux it is

$$\mathbf{q} - \mathbf{T} \cdot \mathbf{v} - \rho \mathbf{v} \left(e + \frac{v^2}{2} \right) = \sum \left(\mathbf{q}_k - \mathbf{T}_k \cdot \mathbf{v}_k - \rho_k \mathbf{v}_k \left(e_k + \frac{v_k^2}{2} \right) \right). \qquad (17)$$

Generally, it is not possible to prove from balance principles alone that the stresses in the constituents are symmetric. In fact, (15) shows that some constituents may have unsymmetric stresses, yet the stress in the mixture may still be symmetric.

It is plausible that the body consisting of the sum of all of the constituents have the same equations of balance as an ordinary single continuum. We *postulate* that for this body, the sums of the interactions of mass, energy, and momentum vanish, that is,

$$\sum c_k = 0, \qquad (18)$$

$$\sum \mathbf{m}_k = 0, \qquad (19)$$

$$\sum e_k = 0, \qquad (20)$$

It then follows that

$$\dot{\rho} + \rho \nabla \cdot \mathbf{v} = 0, \qquad (21)$$

$$\rho \mathbf{a} = \nabla \cdot \mathbf{T} + \rho \mathbf{b}, \qquad (22)$$

$$\rho \dot{e} = \mathbf{T} : \nabla \mathbf{v} + \nabla \cdot \mathbf{q} + \rho r. \qquad (23)$$

That is, the equations of balance for a mixture are the same as the equations of balance for a single continuum.

For a mixture consisting of a single continuum, (18)–(20) force (8)–(10) to reduce to the balance equations (21)–(23) of a single continuum.

Part II
Continuum Theory

6
Continuum Balance Equations for Multicomponent Fluids

6.1. Multicomponent Mixtures

The mixtures described in Chapter 5 are based on the concept that a mixture may be represented by "a sequence of bodies $\mathcal{B}_k$, all of which ... occupy regions of space ... simultaneously" [89, p. 81]. Examples of such materials are air (a mixture of nitrogen, oxygen, and other materials in small amounts), and whisky (a mixture of water, alcohol, and other materials in small amounts). However, it was commonly recognized at an early stage that so strong an assumption of intermiscibility was not appropriate to all physical situations. For example, soils, porous rock, suspensions of coal particles in water, packed powders, granular propellants, etc., consist of identifiable solid particles surrounded by one or more continuous media, or an identifiable porous matrix through which one or more of the continua are dispersed. Motions of the individual components are possible and, as long as there are no chemical reactions, each constituent retains its integrity.[1] We call such materials *multicomponent mixtures*. They are more complicated than classical mixtures in the sense that they have geometrical structure, but less complicated in the sense that the constituents are not intimately intermixed. A theory describing them should reflect these facts.

For example, a suspension of elastic spherical particles in a liquid is a very simple two-component mixture. It is possible to interpret an initial-boundary-value problem for such a mixture as simply a complicated problem in mechanics. In order

---

[1] Chemical reactions or other interchanges of mass among constituents are not part of the structure presented here. However, an extension to this case is possible.

to formulate such a problem we would need to know the geometry of the container, the nature of the driving forces, the constitutive properties of the liquid and the particles, and the initial position and velocity of every single particle, and of each point in the fluid. The solution of the problem would be the evolution of position and deformation of every particle, and the complete history of the flow field of the liquid. We know of no general solution of this type.[2] Moreover, to find such solutions, we require an unrealistic amount of information about initial conditions, and the solutions would give much more detail about the flow fields than could possibly be used. What is needed is a model, one which requires only reasonable and accessible initial and boundary conditions, and which yields a reasonable amount of information. Multicomponent mixture theories provide models of this sort.

Generally, in order to use a continuum description for multicomponent materials, we attempt to argue that the smallest dimension of the boundaries of a boundary-value problem is larger than a typical particle or pore. When this is the case, the idea of a continuum mixture pertains when each particle of the continuum is given a mathematical structure to account for the gross phenomena associated with the heterogeneous nature of the mixture. However, this is not the only way to justify a continuum description. In Chapter 8, we present averaging arguments to substantiate this idea. It should be noted that a continuum model is incapable of describing the motions of individual particles or the flow in individual pores. Just as the continuum is an idealization of nature, appropriate to the description of materials on some level of resolution and not to others, so is the multicomponent mixture theory an idealization, giving an appropriate description on some level of detail, but not on others.

Many multicomponent mixtures have considerable material microstructure. Particles may be spherical, prolate, or oblate rods, biconcave disks, or have arbitrary shape. Rotation of particles may have a significant role in the physics of the boundary-value problems of interest. Some work on relatively uncomplicated materials of this type has been done. Even for such materials, the complication of the theory is almost overwhelming, so much so that we do not discuss such materials here. Rather, we model the presence of more than one component in the mixture by the concept of volume fractions[3] $\alpha_k(\mathbf{x}, t)$ for each constituent. If $b(r, \mathbf{x})$ is the ball of radius r centered at $\mathbf{x}$, and $V_k(b(r, \mathbf{x}))$ is the volume of component k inside b, then, for example,

$$A_k(r, \mathbf{x}) = \frac{V_k(b(r, \mathbf{x}))}{V(b(r, \mathbf{x}))} \tag{1}$$

is the ratio of volume of component k to the total volume in the ball b. We again have all the difficulties of defining the appropriate continuum variable $\alpha_k(\mathbf{x}, t)$. The most pleasing resolution of the difficulties lies in the interpretation of $A_k(r, \mathbf{x})$

---

[2]There is a very extensive body of work of approximate computer solutions of problems of this type.

[3]This idea is old. A formal mathematical basis is given by Goodman and Cowin [36].

in terms of averages. This approach is the subject of Part III of this book. In this Part II, we assume that the limiting processes make sense, and then proceed.

Thus, let $V_k(b(r, \mathbf{x}; \mu))$ be the volume of component k inside b. Then

$$\alpha_k(\mathbf{x}, t) = \lim_{r \to 0} A_k(r, \mathbf{x}) = \lim_{r \to 0} \overline{\frac{V_k(b(r, \mathbf{x}))}{V(b(r, \mathbf{x}))}}, \tag{2}$$

is the average volume fraction of component k inside b, In terms of this variable alone, it is impossible, for example, to tell the difference between two bodies with a uniform distribution of grains, one with large grains, and one with small grains. Take, for instance, a cubic body of side L, with spherical solid grains in a uniform cubic lattice. Call this constituent 1. Let there be a fluid filling the interstices. Call it constituent 2. Then, whatever the size of the grains, $\alpha_1 = \pi/6$ and $\alpha_2 = 1 - \pi/6$. This fact of course does not preclude the appearance of some other geometric parameters, for example, a characteristic particle size, elsewhere in the theory.

6.2. Kinematics

Our presentation of the kinematics and balance equations for multicomponent mixtures follows our presentation of the kinematics and balance equations for a single continuum, so much so that often we omit the definitions of certain quantities when the definitions for constituents of a multicomponent mixture[4] are obvious from those for a single continuum. We consider a sequence of bodies $\mathcal{B}_k$, $k = 1, ..., N$, the configurations of which are all able to occupy the same region of space simultaneously. The motions of the configurations of $\mathcal{B}_k$ are the smooth mappings

$$\mathbf{x} = \chi_k(\mathbf{X}_k, t), \tag{3}$$

where $\mathbf{X}_k$ are the positions of material points in a fixed reference configuration, and t denotes time. This is depicted in Figure 6.1. We assume that (3) is sufficiently smooth almost everywhere on $\mathcal{B}_k$ so that it is invertible. Thus

$$\mathbf{X} = \chi_k^{-1}(\mathbf{x}_k, t), \tag{4}$$

where

$$\mathbf{X} = \chi_k^{-1}(\chi_k(\mathbf{X}, t), t). \tag{5}$$

If $\chi_k(\mathbf{X}_k, t) = \chi_l(\mathbf{X}_k, t)$ for $k \neq l$, then material k has no relative motion with respect to material l. Then we say that k does not slip[5] with respect to l.

---

[4] Our presentation follows closely that of Bowen [9] and Müller [63], which in turn follows closely that of Truesdell [89]. Motivation in terms of kinetic theory of a mixture of monatomic gases, the simplest corpuscular theory applicable to this situation, is given by Maxwell [62].

[5] Another term is "diffuse."

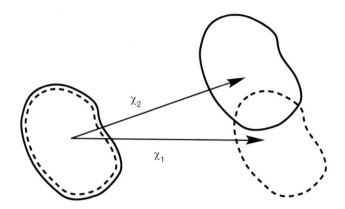

FIGURE 6.1. Configurations and mappings.

Define

$$\dot{\mathbf{x}}_k = \frac{\partial \chi_k(\mathbf{X}_k, t)}{\partial t}. \tag{6}$$

The quantity $\mathbf{V}_k = \dot{\mathbf{x}}_k$ is the velocity of $\mathbf{X}_k$ at time t. By the assumption that (3) is invertible, then $\mathbf{v}_k(\mathbf{x}, t) = \mathbf{V}_k(\chi_k^{-1}(\mathbf{x}, t), t)$ is the velocity of the material point written in terms of $\mathbf{x}$ and t. This spatial ("Eulerian") description is the convenient one for most of our development and is used unless we explicitly state otherwise. If $\phi_k(\mathbf{x}, t)$ is the spatial description of some function, and $\Phi_k(\mathbf{X}_k, t)$ is the corresponding material ("Lagrangian") description of the same function, then

$$\begin{aligned}
\dot{\phi}_k(\mathbf{x}, t) &= \frac{\partial \Phi_k(\mathbf{X}_k, t)}{\partial t} \\
&= \frac{\partial \phi_k(\mathbf{x}, t)}{\partial t} + \frac{\partial \mathbf{x}}{\partial t} \cdot \nabla \phi_k(\mathbf{x}, t) \\
&= \frac{\partial \phi_k(\mathbf{x}, t)}{\partial t} + \mathbf{v}_k(\mathbf{x}, t) \cdot \nabla \phi_k(\mathbf{x}, t).
\end{aligned} \tag{7}$$

Acceleration, deformation rate, and spin of constituent k are defined in the obvious ways.

6.3. Balance Equations

Mass density of configurations of $\mathcal{B}_k$ must be treated with some care, because there are two obvious definitions, and both of them are useful. Consider a body $\mathcal{B}_k$. One definition of density results in the mass of component k in body $\mathcal{B}_k$ being given by

$$m_k(\mathcal{B}) = \int_{\chi(\mathcal{B}_k)} \underline{\rho}_k \, dv, \tag{8}$$

where $\mathcal{B}_k$ is the body of component k that coincides with body $\mathcal{B}$ at the instant t. Then the mass density $\underline{\rho}_k$ is the mass of $\chi_k(\mathcal{B})$ per unit *total* volume. This definition is analogous to our discussion of mass of a single body. Of course, in fact, the structure underlying B_k does not occupy all of space, it only occupies a portion α_k. The function ρ_k, defined by

$$\alpha_k \rho_k = \underline{\rho}_k, \quad (9)$$

is the mass of constituent k per unit volume of constituent k.

For each constituent body B_k we define equations of balance of mass, momentum, moment of momentum, and energy. These equations are similar in form to those for the classical theory of solutions.

For each of our equations of balance, integrals are taken over $V = \chi(B)$ or $\partial V = \partial \chi(B)$ as appropriate. In order to shorten the notation, we omit explicit specification of the regions.

6.3.1. Balance of Mass

The postulate of the balance of mass of component k is

$$\frac{d}{dt} \int \underline{\rho}_k \, dv = \int \Gamma_k \, dv,$$

where Γ_k is the rate of production of mass of component k per unit volume, due to phase change or chemical reaction. This is the rate at which material of the other components change to material of component k. An interpretation for such loss or gain is that it is a result of "phase change," or "chemical reactions," and in fact there is an extensive and elegant body of theory for such reactions.[6] However, that is not the only interpretation, and often exchanges of mass at corpuscular levels greater (or less) in scale than those associated with classical chemistry may be of interest. For example, the simultaneous flow and dissolution of solid particles in a fluid medium is of interest for such phenomena as ablation.

Changing the integration to the reference configuration, and using

$$\dot{J}_k = J_k \nabla \cdot \mathbf{v}_k,$$

we obtain

$$\int (\dot{\underline{\rho}}_k + \underline{\rho}_k \nabla \cdot \mathbf{v}_k - \Gamma_k) \, dv = 0. \quad (10)$$

If we assume that the integrand in (10) is continuous, then by the Dubois–Reymond lemma, the integrand vanishes. Thus, we have the equation of balance of mass:

$$\dot{\underline{\rho}}_k + \underline{\rho}_k \nabla \cdot \mathbf{v}_k = \Gamma_k. \quad (11)$$

This equation can be written as

$$\frac{\partial \alpha_k \rho_k}{\partial t} + \nabla \cdot (\alpha_k \rho_k \mathbf{v}_k) = \Gamma_k. \quad (12)$$

---

[6] Much of the modern literature can be traced from (R. Aris, 1969 [3]).

Another equivalent integral form of the equation of balance of mass can be obtained from (12) by integrating over a volume fixed in space.

$$\frac{d}{dt}\int \alpha_k \rho_k \, dv + \int \nabla \cdot (\alpha_k \rho_k \mathbf{v}_k) \, dv = \int \Gamma_k \, dv. \tag{13}$$

Use of the divergence theorem results in the equation of balance of mass of component k in the form

$$\frac{d}{dt}\int \alpha_k \rho_k \, dv + \oint \alpha_k \rho_k \mathbf{v}_k \cdot \mathbf{n} \, ds = \int \Gamma_k \, dv. \tag{14}$$

This equation has a useful interpretation. The mass of component k inside a fixed volume V can change due to two different effects. First, mass can move into or out of V across the boundary ∂V. The flux of mass of component k, the mass per unit area per unit time, across an element of surface with unit normal $\mathbf{n}$, is given by $\alpha_k \rho_k \mathbf{v}_k \cdot \mathbf{n}$. This mass flux defines the *center of mass velocity* or *barycentric velocity* $\mathbf{v}_k$. Also, the mass of component k can change due to phase change or chemical production of component k.

6.3.2. Balance of Momentum

The momentum of component k in body $\mathcal{B}_k$ is defined to be

$$\int \underline{\rho}_k \mathbf{v}_k \, dv.$$

Balance of momentum is

$$\frac{d}{dt}\int \underline{\rho}_k \mathbf{v}_k \, dv = \mathcal{F}_k^b + \mathcal{F}_k^s, \tag{15}$$

where $\mathcal{F}_k^b$ is the body source of momentum to component k, and $\mathcal{F}_k^s$ is the surface source. The body source is

$$\mathcal{F}_k^b = \int (\underline{\rho}_k \mathbf{b}_k + \mathbf{M}_k + \Gamma_k \mathbf{v}_{ki}) \, dv. \tag{16}$$

The quantity $\mathbf{b}_k$ is the body force per unit mass, and is usually attributed to "external" sources. The quantity $\mathbf{M}_k$ is the force on component k due to interaction with other components inside the body $\mathcal{B}$. The remaining source term, $\Gamma_k \mathbf{v}_{ki}$ is the source of momentum due to the source of mass. The surface source is

$$\mathcal{F}_k^s = \oint \underline{\mathbf{t}}_k \, ds, \tag{17}$$

where $\underline{\mathbf{t}}_k$ is the traction, which is the source per unit area of momentum to component k. By analogy to the definition of density of component k, the quantity $\underline{\mathbf{t}}_k$ is the source of momentum per unit total area. Since the structure underlying $\mathcal{B}_k$ does not occupy all of space, we define the *traction* of component k, $\mathbf{t}_k$ by

$$\alpha_k \mathbf{t}_k = \underline{\mathbf{t}}_k. \tag{18}$$

6.3. Balance Equations

The traction exerted depends on the orientation of the surface element ds, and the dependence is similar to that noted in Chapter 2. This dependence is denoted by

$$\underline{\mathbf{t}}_k = \underline{\mathbf{t}}_k(\mathbf{x}, t; \mathbf{n}).$$

Note that $\underline{\mathbf{t}}_k$ is the force per unit area of component k acting through ∂V. It seems reasonable to assume that the force per unit (mixture) area is equal to α_k times the force per unit area of component k. An argument similar to that given in Chapter 2 shows the existence of a stress tensor $\underline{\mathbf{T}}_k$ such that

$$\underline{\mathbf{t}}_k = \mathbf{n} \cdot \underline{\mathbf{T}}_k, \tag{19}$$

or, equivalently, there exists a component stress tensor $\mathbf{T}_k$ such that

$$\alpha_k \mathbf{t}_k = \alpha_k \mathbf{n} \cdot \mathbf{T}_k. \tag{20}$$

Balance of momentum then becomes

$$\frac{d}{dt}\int \underline{\rho}_k \mathbf{v}_k \, dv = \oint \underline{\mathbf{t}}_k \, ds + \int (\underline{\rho}_k \mathbf{b}_k + \mathbf{M}_k + \Gamma_k \mathbf{v}_{ki}) \, dv, \tag{21}$$

We transform the integral to the fixed configuration $\chi_0(\mathcal{B}_0)$, use the equation of balance of mass (11) and the divergence theorem. The resulting equation for balance of momentum has the form

$$\int (\underline{\rho}_k \dot{\mathbf{v}}_k + \Gamma_k(\mathbf{v}_k - \mathbf{v}_{ki}) - \nabla \cdot \underline{\mathbf{T}}_k - \underline{\rho}_k \mathbf{b}_k - \mathbf{M}_k) \, dv = 0. \tag{22}$$

Again, assuming sufficient continuity, the Dubois–Reymond lemma shows that the integrand vanishes. We then obtain the equation of balance of momentum of component k

$$\underline{\rho}_k \dot{\mathbf{v}}_k = \Gamma_k(\mathbf{v}_{ki} - \mathbf{v}_k) + \nabla \cdot \underline{\mathbf{T}}_k + \underline{\rho}_k \mathbf{b}_k + \mathbf{M}_k. \tag{23}$$

This equation can be written as

$$\alpha_k \rho_k \left(\frac{\partial \mathbf{v}_k}{\partial t} + \mathbf{v}_k \cdot \nabla \mathbf{v}_k \right) = \Gamma_k(\mathbf{v}_{ki} - \mathbf{v}_k) + \nabla \cdot (\alpha_k \mathbf{T}_k) + \alpha_k \rho_k \mathbf{b}_k + \mathbf{M}_k, \tag{24}$$

or, in the equivalent form,

$$\frac{\partial \alpha_k \rho_k \mathbf{v}_k}{\partial t} + \nabla \cdot \alpha_k \rho_k \mathbf{v}_k \mathbf{v}_k = \Gamma_k \mathbf{v}_{ki} + \nabla \cdot (\alpha_k \mathbf{T}_k) + \alpha_k \rho_k \mathbf{b}_k + \mathbf{M}_k. \tag{25}$$

Another equivalent integral form of the equation of balance of momentum can be obtained from (25) by integrating over a volume fixed in space.

$$\frac{d}{dt}\int \alpha_k \rho_k \mathbf{v}_k \, dv + \int \nabla \cdot (\alpha_k \rho_k \mathbf{v}_k \mathbf{v}_k) \, dv$$
$$= \int (\nabla \cdot (\alpha \mathbf{T}_k) + \Gamma_k \mathbf{v}_{ki} + \mathbf{M}_k + \alpha_k \rho_k \mathbf{b}_k) \, dv. \tag{26}$$

72   6. Continuum Balance Equations for Multicomponent Fluids

Use of the divergence theorem results in the form

$$\frac{d}{dt}\int \alpha_k \rho_k \mathbf{v}_k \, dv + \oint \alpha_k \rho_k \mathbf{v}_k \mathbf{v}_k \cdot \mathbf{n} \, ds$$
$$= \oint \alpha_k \mathbf{t}_k \, ds + \int (\alpha_k \rho_k \mathbf{b}_k + \mathbf{M}_k + \Gamma_k \mathbf{v}_{ki}) \, dv. \qquad (27)$$

This equation has a useful interpretation. The momentum of component k inside V can change due to several different effects. First, momentum can move into or out of V with the mass that moves into or out of V. This is expressed as

$$\oint \alpha_k \rho_k \mathbf{v}_k \mathbf{v}_k \cdot \mathbf{n} \, ds.$$

In addition, the material outside V exerts a force on the material inside V through the boundary ∂V of V. This momentum flux is given as

$$\int \alpha_k \mathbf{t}_k \, ds.$$

Also, the momentum of component k can be increased inside V due to body sources. The body source has three parts, the external body source, the interfacial force, and the rate of momentum gain due to phase change or chemical reactions.

6.3.3.   Balance of Moment of Momentum

The moment of momentum of component k inside V is defined to be

$$\int \underline{\rho}_k \mathbf{x} \wedge \mathbf{v}_k \, dv,$$

balance of moment of momentum is

$$\frac{d}{dt}\int \underline{\rho}_k \mathbf{x} \wedge \mathbf{v}_k \, dv = \boldsymbol{T}^b_k + \boldsymbol{T}^s_k, \qquad (28)$$

where $\boldsymbol{T}^b_k$ is the body source of momentum to component k, and $\boldsymbol{T}^s_k$ is the surface source. The body source is

$$\boldsymbol{T}^b_k = \int \underline{\rho}_k(\tau_k + \mathbf{x} \wedge \mathbf{b}_k) + (\psi_k + \mathbf{x} \wedge \mathbf{M}_k) + \Gamma_k(\alpha_{ki} + \mathbf{x} \wedge \mathbf{v}_{ki}) \, dv. \qquad (29)$$

The quantity τ_k is the body torque per unit mass, and is usually attributed to "external" torques. The quantity ψ_k is the torque on component k due to interaction with other components inside the body $\mathcal{B}$. The remaining source term, $\Gamma_k \alpha_{ki}$ is the source of moment of momentum due to the source of mass. The surface source is

$$\boldsymbol{T}^s_k = \oint (\underline{\mathbf{l}}_k + \mathbf{x} \wedge \underline{\mathbf{t}}_k) \, ds, \qquad (30)$$

where $\underline{\mathbf{l}}_k$ is the couple stress, which is the source per unit area of moment of momentum to component k. By analogy to similar definitions, the quantity $\underline{\mathbf{l}}_k =$

$\alpha_k \mathbf{l}_k$, where $\mathbf{l}_k$ is the source of moment of momentum per unit area of component k. An argument similar to those given previously shows the existence of a tensor $\underline{\mathbf{L}}_k$ such that

$$\mathbf{l}_k = \mathbf{n} \cdot \underline{\mathbf{L}}_k, \qquad (31)$$

or, equivalently, there exists a tensor $\mathbf{L}_k$ such that

$$\alpha_k \mathbf{l}_k = \alpha_k \mathbf{n} \cdot \mathbf{L}_k. \qquad (32)$$

The local form of the balance of moment of momentum is

$$\nabla \cdot \alpha_k \mathbf{L}_k + \alpha_k (\mathbf{T}_k - \mathbf{T}_k^T) + \boldsymbol{\tau}_k + \Gamma_k \alpha_{ki} = 0. \qquad (33)$$

In all of this book, we assume that the multicomponent mixtures are nonpolar $\mathbf{L}_k = 0, \boldsymbol{\tau}_k = 0,$ and $\alpha_{ki} = 0$, so that the stresses of the constituents are symmetric.

6.3.4. Balance of Energy

The energy of component k for a body $\mathcal{B}$ inside V is defined to be

$$\int \underline{\rho}_k \left(u_k + \frac{1}{2} v_k^2 \right) dv,$$

where u_k is the internal energy per unit mass. Note that the term $\frac{1}{2} v_k^2$ represents the kinetic energy of the mean motion; if there are velocity fluctuations due to the relative motions of the materials, that kinetic energy will be included in the internal energy u_k. Balance of energy is

$$\frac{d}{dt} \int \underline{\rho}_k \left(u_k + \frac{1}{2} v_k^2 \right) dv = \mathcal{W}_k^b + \mathcal{W}_k^s, \qquad (34)$$

where $\mathcal{W}_k^b$ is the body source of energy to component k, and $\mathcal{W}_k^s$ is the surface source. The body source is

$$\mathcal{W}_k^b = \int \underline{\rho}_k (r_k + \mathbf{v}_k \cdot \mathbf{b}_k) + (E_k + \mathbf{v}_k \cdot \mathbf{M}_k) + \Gamma_k u_{ki} \, dv. \qquad (35)$$

The quantity r_k is the body force per unit mass, and is usually attributed to "external" sources. The quantity E_k is the source to component k due to interaction with other components inside the body $\mathcal{B}$. The remaining source term, $\Gamma_k u_{ki}$ is the source of energy due to the source of mass. The surface source is

$$\mathcal{W}_k^s = \oint (\mathbf{v}_k \cdot \mathbf{t}_k - \underline{Q}_k) \, ds, \qquad (36)$$

where $\underline{Q}_k$ is the source per unit area of energy to component k. By analogy to the definition of density of component k, the quantity $\mathbf{Q}_k$ is the source of energy per unit total area. Since the structure underlying $\mathcal{B}_k$ does not occupy all of the space, we define the source of energy per unit area of component k, $\mathbf{Q}_k$, by

$$\alpha_k \mathbf{Q}_k = \underline{Q}_k. \qquad (37)$$

An argument similar to that used previously shows the existence of a vector field $\underline{\mathbf{q}}_k$ such that

$$\mathbf{Q}_k = \mathbf{n} \cdot \underline{\mathbf{q}}_k, \qquad (38)$$

or, equivalently, there exists a vector $\mathbf{q}_k$ such that

$$\alpha_k \mathbf{Q}_k = \alpha_k \mathbf{n} \cdot \mathbf{q}_k. \qquad (39)$$

Applying arguments similar to those used to derive local forms of the equations of balance of momentum and moment of momentum, we obtain the local form of balance of energy

$$\alpha_k \rho_k \dot{u}_k = \alpha_k \mathbf{T}_k : \nabla \mathbf{v}_k - \nabla \cdot \alpha_k \mathbf{q}_k$$
$$+ \alpha_k \rho_k r_k + E_k + \Gamma_k \left(u_{ki} - u_k - \frac{1}{2} v_k^2 + \mathbf{v}_{ki} \cdot \mathbf{v}_k \right). \qquad (40)$$

This equation can also be written as

$$\alpha_k \rho_k \left(\frac{\partial u_k}{\partial t} + \mathbf{v}_k \cdot \nabla u_k \right) = \alpha_k \mathbf{T}_k : \nabla \mathbf{v}_k - \nabla \cdot \alpha_k \mathbf{q}_k + \alpha_k \rho_k r_k$$
$$+ E_k + \Gamma_k \left(u_{ki} - u_k - \frac{1}{2} v_k^2 + \mathbf{v}_{ki} \cdot \mathbf{v}_k \right). \qquad (41)$$

By analogy with the momentum, the internal energy of component k inside a fixed volume V can change due to surface sources and due to body sources. The stress does work throughout the volume, in the form

$$\int \alpha_k \mathbf{T}_k : \nabla \mathbf{v}_k \, dv$$

and the body force does not contribute to the change of internal energy.

Exchange of momentum and energy has a significant role in much of the theory. In the momentum equation (25), the exchange of momentum has the same form as a body force, and in the energy equation (40), the exchange of energy has the same form as the body heating ("radiation"). In our treatment of thermodynamics we assume that the body force and body heating are specified externally, while the exchanges of momentum and energy are specified by constitutive equations. In fact, motivating the forms of such equations has a central role in the theory of multicomponent mixtures.

6.4. Multicomponent Entropy Inequalities

The concept of irreversibility in multicomponent mixtures is suggested by two different views of continuum mechanics and their interpretations in terms of continuum modeling of multicomponent mixtures. First, continuum mechanical theories of single-component materials use the concept of entropy to quantify the idea of "disorder" at the molecular scale. Internal energy, then, depends on the amount of "disorder" in the molecular motions. "Temperature" becomes a measure of the disordered energy. The rest, loosely called "free" energy, can be converted back into mechanical energy. This idea of entropy motivates the definition of molecular scale, or microscale, entropy.

If we consider the kinetic model for a gas, where the molecules are thought of as rigid spheres having a distribution of velocities, there is a concept of "information content," in the velocity distribution function. This approach is sometimes used to motivate the definition of entropy in the continuum model, but the connection is tenuous. However, it is precisely this connection that motivates us to introduce a second entropy. This entropy quantifies the disorder[7] associated with the motions of the components, and is distinct from the concept of microscale entropy discussed previously. In what follows, we introduce some concepts about the structure of the multicomponent medium as to the origin of stress, heat flux, and internal energy. These ideas have some further justification in terms of the model derived from averaging.

6.4.1. Microscale and Mesoscale Energy Equations

Let the stress $\mathbf{T}_k$ be written as

$$\mathbf{T}_k = \mathbf{T}_k^m + \mathbf{T}_k^M, \tag{42}$$

where $\mathbf{T}_k^m$ is the microscale stress, and corresponds to the force exerted on an element of area by the material of component k; and $\mathbf{T}_k^M$ is the mesoscale stress, corresponding to the momentum transferred by fluctuations of the motions of the components around their mean motion. Similarly, let the heat flux $\mathbf{q}_k$ be written as

$$\mathbf{q}_k = \mathbf{q}_k^m + \mathbf{q}_k^M, \tag{43}$$

where $\mathbf{q}_k^m$ is the microscale heat flux, and corresponds to the energy flux through an element of area by conduction through the material of component k; and $\mathbf{q}_k^M$ is the mesoscale heat flux, corresponding to the energy transferred by the fluctuations. Also let the internal energy density u_k be written as

$$u_k = u_k^m + u_k^M, \tag{44}$$

where u_k^m is the microscale internal energy density, and u_k^M is the mesoscale internal energy density; and the interfacial internal energy source E_k can be written as

$$E_k = E_k^m + E_k^M, \tag{45}$$

where E_k^m is the interfacial source of microscale energy, and E_k^M is the interfacial source of mesocale energy. Assume that these two internal energies separately satisfy balance equations of the form

$$\frac{\partial \alpha_k \rho_k u_k^m}{\partial t} + \nabla \cdot \alpha_k \rho_k u_k^m \mathbf{v}_k = \alpha_k \mathbf{T}_k^m : \nabla \mathbf{v}_k - \nabla \cdot \alpha_k \mathbf{q}_k^m + u_{ki}^m \Gamma_k$$

---

[7] Or "departure from equilibrium."

$$+ \alpha_k \rho_k r_k + E_k^m + \alpha_k \mathcal{D}_k, \tag{46}$$

$$\frac{\partial \alpha_k \rho_k u_k^M}{\partial t} + \nabla \cdot \alpha_k \rho_k u_k^M \mathbf{v}_k = \alpha_k \mathbf{T}_k^M : \nabla \mathbf{v}_k - \nabla \cdot \alpha_k \mathbf{q}_k^M + u_{ki}^M \Gamma_k$$
$$+ E_k^M - \alpha_k \mathcal{D}_k, \tag{47}$$

where $\mathcal{D}_k$ is the rate of energy loss from the mesoscale energy to the microscale energy. Note that we have assumed that external sources of energy are added to the microscale energy, as, for example, radiant heat.

6.4.2. Microscale and Mesoscale Entropy Equations

The microscale and mesoscale entropies of component k inside V are defined to be

$$\int \alpha_k \rho_k s_k^m \, dv, \qquad \int \alpha_k \rho_k s_k^M \, dv,$$

respectively. The entropy of component k inside V is assumed to change at least as fast as entropy flows in, and is produced inside. The rates of flux of microscale and mesoscale entropies through the surface of V due to mass moving through ∂V are given by

$$\oint \alpha_k \rho_k s_k^m \mathbf{v}_k \cdot \mathbf{n} \, ds, \qquad \oint \alpha_k \rho_k s_k^M \mathbf{v}_k \cdot \mathbf{n} \, ds.$$

In addition, microscale and mesoscale entropies cross the surface ∂V of V due to the corresponding energy crossing ∂V. These terms are given by

$$\oint \alpha_k \Phi_k^m \, ds, \qquad \oint \alpha_k \Phi_k^M \, ds,$$

where Φ_k^m is the microscale flux of entropy per unit area of component k and Φ_k^M is the mesoscale flux of entropy per unit area of component k. The external sources of microscale and mesoscale entropies of component k are given by

$$\int \alpha_k \rho_k \sigma_k^m \, dv, \qquad \int \alpha_k \rho_k \sigma_k^M \, dv. \tag{48}$$

The second body source of entropy for the microscale and the mesoscale is associated with the rate of gain of entropy to component k from the other components. These are given by

$$\int S_k^m \, dv, \qquad \int S_k^M \, dv. \tag{49}$$

where S_k^m and S_k^M are the rates at which component k gains microscale and mesoscale entropies from the other components, per unit volume. Finally, the sources of microscale and mesoscale entropies due to phase change are given by

$$\int \Gamma_k s_{ki}^m \, dv, \qquad \int \Gamma_k s_{ki}^M \, dv, \tag{50}$$

6.4. Multicomponent Entropy Inequalities 77

where s_{ki}^m and s_{ki}^M are the microscale and mesoscale interfacial entropy densities. Thus, for the balance of microscale and mesoscale entropies, we have

$$\frac{d}{dt}\int \alpha_k \rho_k s_k^m \, dv + \oint \alpha_k \rho_k s_k^m \mathbf{v}_k \cdot \mathbf{n}\, ds$$
$$\geq \oint \alpha_k \Phi_k^m \, ds + \int \left[\alpha_k \rho_k \sigma_k^m + S_k^m + \Gamma_k s_{ki}^m\right] dv, \quad (51)$$

$$\frac{d}{dt}\int \alpha_k \rho_k s_k^M \, dv + \oint \alpha_k \rho_k s_k^M \mathbf{v}_k \cdot \mathbf{n}\, ds$$
$$\geq \oint \alpha_k \Phi_k^M \, ds + \int \left[\alpha_k \rho_k \sigma_k^M + S_k^M + \Gamma_k s_{ki}^M\right] dv. \quad (52)$$

A tetrahedron argument, similar to the ones given before, leads to the conclusion that the entropy fluxes Φ_k^m and Φ_k^M are given by $\mathbf{n} \cdot \boldsymbol{\phi}_k^m$ and $\mathbf{n} \cdot \boldsymbol{\phi}_k^M$. This leads to the entropy inequalities

$$\frac{\partial \alpha_k \rho_k s_k^m}{\partial t} + \nabla \cdot (\alpha_k \rho_k \mathbf{v}_k s_k^m) \geq \nabla \cdot \alpha_k \boldsymbol{\phi}_k^m + \alpha_k \rho_k \sigma_k^m + S_k^m + \Gamma_k s_{ki}^m, \quad (53)$$

$$\frac{\partial \alpha_k \rho_k s_k^M}{\partial t} + \nabla \cdot (\alpha_k \rho_k \mathbf{v}_k s_k^M) \geq \nabla \cdot \alpha_k \boldsymbol{\phi}_k^M + \alpha_k \rho_k \sigma_k^M + S_k^M + \Gamma_k s_{ki}^M. \quad (54)$$

6.4.3. Temperatures

We postulate that the entropy sources are absolutely continuous with respect to the corresponding energy sources. That is, we assume that

$$\mathcal{S}_k^m(\mathcal{B}) = \int \alpha_k \rho_k \sigma_k^m \, dv, \quad (55)$$

$$\mathcal{S}_k^m(\partial \mathcal{B}) = -\oint \alpha_k \boldsymbol{\phi}_k^m \cdot \mathbf{n}\, ds, \quad (56)$$

$$\mathcal{S}_k^{mi}(\mathcal{B}) = \int \Sigma_k^m \, dv, \quad (57)$$

for the microscale; and

$$\mathcal{S}_k^M(\mathcal{B}) = \int \alpha_k \rho_k \sigma_k^M \, dv, \quad (58)$$

$$\mathcal{S}_k^M(\partial \mathcal{B}) = -\oint \alpha_k \boldsymbol{\phi}_k^M \cdot \mathbf{n}\, ds, \quad (59)$$

$$\mathcal{S}_k^{Mi}(\mathcal{B}) = \int \Sigma_k^M \, dv, \quad (60)$$

for the mesoscale, are additive with respect to the corresponding energy sources; namely,

$$\mathcal{E}_k^m(\mathcal{B}) = \int \alpha_k \rho_k R_k^m \, dv, \quad (61)$$

$$\mathcal{E}_k^m(\partial \mathcal{B}) = -\oint \alpha_k \mathbf{Q}_k^m \cdot \mathbf{n}\, ds,\qquad(62)$$

$$\mathcal{E}_k^{mi}(\mathcal{B}) = \int \mathcal{E}_k^m\, dv,\qquad(63)$$

for the microscale; and

$$\mathcal{E}_k^M(\mathcal{B}) = \int \alpha_k \rho_k R_k^M\, dv,\qquad(64)$$

$$\mathcal{E}_k^M(\partial \mathcal{B}) = -\oint \alpha_k \mathbf{Q}_k^M \cdot \mathbf{n}\, ds,\qquad(65)$$

$$\mathcal{E}_k^{Mi}(\mathcal{B}) = \int \mathcal{E}_k^M\, dv,\qquad(66)$$

for the mesoscale. Here the energy source terms are the flux terms $\mathbf{Q}_k^m$ and $\mathbf{Q}_k^M$, the body source terms R_k^m and R_k^M, and the interfacial source terms $\mathcal{E}_k^m$ and $\mathcal{E}_k^M$. These terms remain to be identified. By the Radon–Nikodym theorem, there exist densities ϑ_k^{mc}, ϑ_k^{mr}, ϑ_k^{Mc}, ϑ_k^{Mr}, ϑ_{ki}^m, and ϑ_{ki}^M, such that

$$S_k^m(\mathcal{B}) = \int \vartheta_k^{mr} \alpha_k \rho_k R_k^m\, dv,\qquad(67)$$

$$S_k^m(\partial \mathcal{B}) = -\oint \vartheta_k^{mc} \alpha_k \mathbf{Q}_k^m \cdot \mathbf{n}\, ds,\qquad(68)$$

$$S_k^{mi}(\mathcal{B}) = \int \vartheta_{ki}^m \mathcal{E}_k^m\, dv,\qquad(69)$$

$$S_k^M(\mathcal{B}) = \int \vartheta_k^{Mr} \alpha_k \rho_k R_k^M\, dv,\qquad(70)$$

$$S_k^M(\partial \mathcal{B}) = -\oint \vartheta_k^{Mc} \alpha_k \mathbf{Q}_k^M \cdot \mathbf{n}\, ds,\qquad(71)$$

$$S_k^{Mi}(\mathcal{B}) = \int \vartheta_{ki}^M \mathcal{E}_k^M\, dv.\qquad(72)$$

By analogy with the classical theory for single-component materials, the Radon–Nikodym densities can be written as

$$\vartheta_k^{mc} = \frac{1}{\theta_k^{mc}}\quad \vartheta_k^{mr} = \frac{1}{\theta_k^{mr}}\quad \vartheta_k^{Mc} = \frac{1}{\theta_k^{Mc}},$$

$$\vartheta_k^{Mr} = \frac{1}{\theta_k^{Mr}}\quad \vartheta_{ki}^m = \frac{1}{\theta_{ki}^m}\quad \vartheta_k^{Mi} = \frac{1}{\theta_{ki}^M}.\qquad(73)$$

These six quantities are analogous to the temperature defined for homogeneous materials and, consequently, we shall refer to them as temperatures.

The entropy inequalities are then

$$\frac{\partial \alpha_k \rho_k s_k^m}{\partial t} + \nabla \cdot \alpha_k \rho_k s_k^m \mathbf{v}_k \geq -\nabla \cdot \alpha_k \frac{\mathbf{Q}_k^m}{\theta_k^{mc}} + \alpha_k \rho_k \frac{R_k^m}{\theta_k^{mr}} + \frac{\mathcal{E}_k^{mi}}{\theta_{ki}^m} + s_{ki}^m \Gamma_k,\qquad(74)$$

$$\frac{\partial \alpha_k \rho_k s_k^M}{\partial t} + \nabla \cdot \alpha_k \rho_k s_k^M \mathbf{v}_k \geq -\nabla \cdot \alpha_k \frac{\mathbf{Q}_k^M}{\theta_k^{Mc}} + \alpha_k \rho_k \frac{R_k^M}{\theta_k^{Mr}} + \frac{\mathcal{E}_k^{Mi}}{\theta_{ki}^M} + s_{ki}^M \Gamma_k. \quad (75)$$

This approach introduces six temperatures per component; the microscale conduction temperature θ_k^{mc}, the mesoscale conduction temperature θ_k^{Mc}, the microscale radiative temperature θ_k^{mr}, the mesoscale radiative temperature θ_k^{Mr}, and the interfacial temperatures θ_k^{mi} and θ_k^{Mi}.

With so many temperatures, the theory is unwieldy. We shall now make some plausible assumptions regarding these temperatures, in order to proceed to a theory that indicates the usefulness of the concepts introduced here.

First, in analogy with the classical theory for single-component materials, we shall assume that, for each component, the conduction temperature and the radiation temperature are equal:

$$\theta_k^{mc} = \theta_k^{mr} = \theta_k^m. \quad (76)$$

We also assume that there is only one temperature of the mesoscale motions for each component. Thus,

$$\theta_k^{Mc} = \theta_k^{Mr} = \theta_k^{Mi} = \theta_k^M. \quad (77)$$

We shall also assume thermodynamic equilibrium of the interface. This implies that the interfacial temperature is the same for both components:

$$\theta_{1i}^m = \theta_{2i}^m = \theta_i^m. \quad (78)$$

Using these temperatures, and the equations of balance of mass, the entropy inequalities become

$$\alpha_k \rho_k \left(\frac{\partial s_k^m}{\partial t} + \mathbf{v}_k \cdot \nabla s_k^m \right) \geq -\nabla \cdot \alpha_k \frac{\mathbf{Q}_k^m}{\theta_k^m} + \alpha_k \rho_k \frac{R_k^m}{\theta_k^m}$$
$$+ \frac{\mathcal{E}_k^{mi}}{\theta_i^m} + \left(s_{ki}^m - s_k^m \right) \Gamma_k, \quad (79)$$

$$\alpha_k \rho_k \left(\frac{\partial s_k^M}{\partial t} + \mathbf{v}_k \cdot \nabla s_k^M \right) \geq -\nabla \cdot \alpha_k \frac{\mathbf{Q}_k^M}{\theta_k^M} + \alpha_k \rho_k \frac{R_k^M}{\theta_k^M}$$
$$+ \frac{\mathcal{E}_k^{Mi}}{\theta_i^M} + \left(s_{ki}^M - s_k^M \right) \Gamma_k. \quad (80)$$

6.4.4. Energy and Entropy Source Assumptions

Let us now partition the energy source terms appearing in (46) and (47). We assume that

$$\mathbf{Q}_k^m = \mathbf{q}_k^m + \tilde{\mathbf{Q}}_k^m, \quad (81)$$

$$\mathbf{Q}_k^M = \mathbf{q}_k^M + \tilde{\mathbf{Q}}_k^M. \quad (82)$$

6. Continuum Balance Equations for Multicomponent Fluids

The energy source term for the microscale is

$$R_k^m = r_k. \tag{83}$$

We further assume that the source of energy to the mesoscale is due to interactions between the mesoscale and the microscale. Thus,

$$R_k^M = -\frac{\alpha_k \mathcal{D}_k}{\rho_k}. \tag{84}$$

For the interfacial energy transfer terms, we assume that

$$E_k^m = \mathcal{E}_k^m + \tilde{E}_k^m, \tag{85}$$

$$E_k^M = -\mathbf{M}_k \cdot (\mathbf{v}_k - \mathbf{v}_i) + \tilde{E}_k^M, \tag{86}$$

Using the equations of balance of the microscale and mesoscale energy in the entropy inequality leads to the equations

$$\alpha_k \rho_k \left(\theta_k^m \frac{D_k s_k^m}{Dt} - \frac{D_k u_k^m}{Dt} \right) \geq \alpha_k \frac{1}{\theta_k^m} (\mathbf{q}_k^m + \tilde{\mathbf{Q}}_k^m) \cdot \nabla \theta_k^m - \nabla \cdot \alpha_k \tilde{\mathbf{Q}}_k^m$$

$$- \alpha_k \mathbf{T}_k^m : \nabla \mathbf{v}_k + \mathcal{E}_k^{mi} \left(\frac{\theta_k^m}{\theta_i^m} - 1 \right) - \tilde{E}_k^{mi} \tag{87}$$

$$- \alpha_k \mathcal{D}_k + \left[\theta_k^m (s_{ki}^m - s_k^m) + u_k^m \right] \Gamma_k,$$

$$\alpha_k \rho_k \left(\theta_k^M \frac{D_k s_k^M}{Dt} - \frac{D_k u_k^M}{Dt} \right) \geq \alpha_k \frac{1}{\theta_k^M} (\mathbf{q}_k^M + \tilde{\mathbf{Q}}_k^M) \cdot \nabla \theta_k^M - \nabla \cdot \alpha_k \tilde{\mathbf{Q}}_k^M$$

$$+ \mathbf{M}_k \cdot (\mathbf{v}_k - \mathbf{v}_i) - \alpha_k \mathbf{T}_k^M : \nabla \mathbf{v}_k \tag{88}$$

$$- \tilde{E}_k^M + \left[\theta_k^M (s_{ki}^M - s_k^M) + u_k^M \right] \Gamma_k.$$

While the theory seems cumbersome with two equations of the balance of the energy and two entropy inequalities for each component, it allows the separation of energy and entropy into thermal (called here microscale), and turbulent (actually, fluctuation, called here mesoscale). Further, it is complicated because the system for which the statements of irreversibility are postulated are not isolated, and can exchange energy across the interfaces. These exchange rates are incorporated into the theory, but must be constitued with care. The utility of this theory will be seen in Section 14.4.

7
Mixture Equations

At each point **x** and at each time t, the agglomeration of configurations of $\mathcal{B}_k$ forms a composite body $\mathcal{B}$. However, it is useful to construct the body from the viewpoint of physical motivation, because it is the entity most easily observed and measured with conventional instruments. For example, much of the older work on the rheometry of suspensions [24] is based on measurements of $\mathcal{B}$ alone. It is also convenient to construct $\mathcal{B}$ from the viewpoint of mathematics, principally because boundary conditions are often more easily stated or understood in terms of $\mathcal{B}$ than in terms of $\mathcal{B}_k$.

The definitions of the properties of $\mathcal{B}$ in terms of the properties of $\mathcal{B}_k$ are almost entirely arbitrary. Here we use a set of definitions motivated by elementary measure-theoretic considerations and by the work of Maxwell [62] on a corpuscular theory intended to describe a similar physical situation to that described here with a continuum theory. We do not claim that this is the only set of definitions possible. It is true that the set constitutes a complete and self-consistent structure in a sense that will become clear later in this chapter.

To motivate the definition of the mass density of $\mathcal{B}$ in terms of $\mathcal{B}_k$, consider a set of bodies $\mathcal{B}_k$, each occupying the same region of space, say V. Assume each body is homogeneous, so that its mass M_k is uniformly distributed over V. Let its mass density be ρ_k. The mass of $\mathcal{B}$ is the sum of the masses of the constituent bodies. Since each of those bodies has the same volume, the mass density of $\mathcal{B}$ is

$$\rho = \sum_k \rho_k. \tag{1}$$

7. Mixture Equations

This motivational argument is exact for homogeneous bodies only. We take the definition given by (1) to be *general*, that is, each of the fields described in that equation is taken to be a function of position in space and time.

The definition of the volume fraction of B is similar to that of the mass density. In a multicomponent material, each component is physically distinct from every other component. We idealize this idea by saying that in a given region of space, each component takes up a portion of the volume, or *volume fraction* α_k. The total portion of the volume taken up by the composite body B is

$$\alpha = \sum_k \alpha_k. \tag{2}$$

We assume this equation holds pointwise, and generalize it to fields in the same way as mass density. The restrictions

$$\alpha_k \leq 1, \qquad \alpha \leq 1, \tag{3}$$

are obvious. If $\alpha < 1$ at a material point, the body is called *unsaturated* at that point, while if $\alpha = 1$, the body is called *saturated*. In this work, we always assume that $\alpha = 1$.

For definitions of other quantities for B, we use the set given by Truesdell [89]. These are based on a corpuscular theory of Maxwell [62]. The velocity $\mathbf{v}$ in B is given by

$$\rho \mathbf{v} = \sum_k \alpha_k \rho_k \mathbf{v}_k. \tag{4}$$

A different definition of velocity is

$$\mathbf{v}_c = \sum_k \alpha_k \mathbf{v}_k. \tag{5}$$

The velocity defined by (4) is the center-of-mass velocity, and is what appears in the equations of balance of mass and momentum. The velocity defined by (5) is called the center-of-volume velocity.

Generally it is hard to make statements about B on the basis of statements about B_k. However, it is essential to have reasonable balance laws for B. To obtain them, we define the stress in B by

$$\mathbf{T} - \rho \mathbf{v}\mathbf{v} = \sum_k (\alpha_k \mathbf{T}_k - \alpha_k \rho_k \mathbf{v}_k \mathbf{v}_k). \tag{6}$$

We define the *relative velocity* or *slip* of component k by

$$\mathbf{v}'_k = \mathbf{v}_k - \mathbf{v}. \tag{7}$$

then

$$\sum_k \alpha_k \rho_k \mathbf{v}'_k = 0 \tag{8}$$

and (6) can be put into the form

$$\mathbf{T} = \sum_k (\alpha_k \mathbf{T}_k - \alpha_k \rho_k \mathbf{v}'_k \mathbf{v}'_k). \tag{9}$$

Other definitions are

$$\rho \mathbf{b} = \sum_k \alpha_k \rho_k \mathbf{b}_k, \quad (10)$$

$$\mathbf{q} - \mathbf{T} \cdot \mathbf{v} - \rho(e + \frac{1}{2}v^2)\mathbf{v} = \sum_k \left[\alpha_k \mathbf{q}_k - \alpha_k \mathbf{T}_k \cdot \mathbf{v}_k - \alpha_k \rho_k (e_k + \frac{1}{2}v_k^2)\mathbf{v}_k \right]. \quad (11)$$

Now (6.3.12), (6.3.25), and (6.3.34) are statements of balance among the $\mathcal{B}_k$. Each of them allows loss or gain of the appropriate quantity to the other bodies. Here we postulate that the exchanges are exactly that—exchanges among the constituents, so that their sums vanish. Thus

$$\sum_k \Gamma_k = 0, \quad (12)$$

$$\sum_k (\mathbf{M}_k + \Gamma_k \mathbf{v}_{ki}) = 0, \quad (13)$$

$$\sum_k (E_k + W_k + \Gamma_k u_{ki}) = 0. \quad (14)$$

Given that the balances for the constituents (6.3.12), (6.3.25), (6.3.40) hold, and the definitions (1), (4), (6), (10), and (11), these equations are equivalent to

$$\frac{\partial \rho}{\partial t} + \nabla \cdot \rho \mathbf{v} = 0, \quad (15)$$

$$\frac{\partial \rho \mathbf{v}}{\partial t} + \nabla \cdot \rho \mathbf{v}\mathbf{v} = \nabla \cdot \mathbf{T} + \rho \mathbf{b}, \quad (16)$$

$$\frac{\partial \rho(e + \frac{1}{2}v^2)}{\partial t} + \nabla \cdot \rho \left(e + \frac{1}{2}v^2 \right) \mathbf{v} = \nabla \cdot (-\mathbf{q} + \mathbf{v} \cdot \mathbf{T}) + \rho(r + \mathbf{v} \cdot \mathbf{b}), \quad (17)$$

that is, the classical equations of the balance of mass, momentum, and energy for a single body.

The results above, though requiring substantial calculation, are conceptually simple. The principal hypothesis is that, for balance equations only, the obvious generalization of the classical equations to the constituents of a mixture lead in an obvious way to the classical equations of balance. The only intermediaries are the definitions (1), based on the measure-theoretic idea of mass (4), a definition of the velocity of the mixture, and (6), (10), (11), which are motivated by the classic kinetic theory of mixtures of monatomic gases. Moreover, the whole system of equations is quite underdetermined in the sense described in [69] and [25]. Thus an infinite number of substitutes for the equations expressing properties of the mixture in terms of properties of the constituents lead from the balance equations for the constituents (6.3.12), (6.3.25), (6.3.40), to the balance equations for the mixture (15), (16), (17).[1] This is not a statement about constitutive equations, only a statement about balance equations. For materials with microstructure, there are

---

[1] An explicit formulation of these ideas has been given by Samohýl and Šilhavý [77].

some where limited aspects of the constitutive information for the microstructure are included in the balance equations. In fact, balance equations in addition to those for mass, momentum, and energy have been proposed and found to be useful for the descriptions of some natural phenomena.[2] Theories have been built for mixtures of such materials,[3] and the extreme algebraic difficulty of obtaining results such as those we have obtained has caused some authors to state that the classical equations of balance of mass, momentum, and energy do not hold for mixtures with microstructure. We see no reason to accept such arguments.

Not all kinematical variables for the mixture are obtained from the corresponding kinematical variables of the constituents in any "obvious way." As a consequence of the definition (7), we have

$$\rho \mathbf{D} = \sum \alpha_k \rho_k \mathbf{D}_k + \sum \left[\operatorname{sym} \nabla(\alpha_k \rho_k) \mathbf{v}_k \right] . \tag{18}$$

We note that a reasonable definition of "rigid" is that $\mathbf{D} = 0$. Thus, if each component is rigid, (18) shows that the mixture may not be rigid, if $\nabla(\alpha_k \rho_k)\mathbf{v}_k \neq 0$. This is a result which appears to violate intuition: A mixture of two rigid constituents is not itself always rigid.

---

[2] The first explicit statement of this idea is that of Ericksen [32].
[3] See [68] for a clean treatment of a special case.

Part III
Averaging Theory

8
Introduction

A prime characteristic of many flows of multicomponent materials is that there is uncertainty in the exact locations of the particular constituents at any particular time. For some predictions, this is not important. Often, we are concerned with more gross features of the motion. This means that, for equivalent macroscopic flows, there will be uncertainty in the locations of particular constituents for *all* times. For instance, consider a suspension of small particles in a liquid. If such a suspension is to be used in, for example, a falling ball viscometer, it might be mixed outside the viscometer, then placed in that device prior to the conduction of an experiment with a falling ball. In a properly conducted experiment, the particles would be approximately uniformly distributed in the fluid. However, there would be no assurance that a particular point in the fluid would contain a particle or not. Experiments conducted with nominally identical fluids in identical viscometers would yield identical gross results, even though the exact initial locations of the suspended particles in the two experiments were noticeably different. As another example, in a sedimenting suspension, the exact distribution of the locations of the particles is immaterial as long as they are reasonably "spread out." We would not allow the particles to be lumped in some way that is not consistent with the initial conditions appropriate for the flow; neither would we allow the particles to be "paired" so that very near each particle is exactly one neighbor. A more interesting case is the set of all experiments with the same boundary conditions, and initial conditions with some (undefined) properties that we would like to associate with the mean and distribution of the particles and their velocities. We call this set an *ensemble*. Such ensembles are reasonable sets over which to perform averages because variations in the details of the flows are assured in all situations, while at the same time variations in the gross flows cannot occur.

We take the approach here that we wish to predict averages over ensembles of multicomponent flows. In the flows we consider, it does not matter that we allow the particles or bubbles or droplets to be rearranged in space within reason. The averaging procedure we choose is the ensemble average. We shall refer to a "process" as the set of possible multicomponent flows that could occur, given appropriate initial and boundary conditions. A "realization" of the flow is a possible motion that could have happened. This is what is usually meant as a process, but when we wish to use this terminology, we shall say "microscopic process." Generally, we expect an infinite number of realizations of the flow, consisting of variations of position, attitude, and velocities of the discrete units and the fluid between them.

The ensemble average allows the interpretation of phenomena in terms of the repeatability of multicomponent flows. Any particular exact experiment or realization will not be repeatable; however, any repetition of the experiment will lead to another member of the ensemble. Usually in multicomponent materials, we think of the ensemble corresponding to a given motion as a very large set of realizations, with variations that we can observe.

In Part III, we derive the balance equations by applying an averaging operation to the equations of motion for two continua separated by an interface across which the densities, velocities, etc., may jump. We then define appropriate averaged variables in Section 11.3, and write the averaged equations in terms of them in Section 11. In Section 12, we compare the equations resulting from the averaging approach of this chapter, and the equations derived from the balance principles of Chapter 6.

8.1. Local Conservation Equations

Under a reasonably broad range of conditions, many common substances, including air and water, and water and its vapor, behave as compressible Newtonian fluids. The interface between them supports surface tension. Also, many solids suspended in fluids are in the range of deformations where linear elasticity is valid. Often, solid materials behave as rigid units. In general, it is necessary to consider the equations governing the three-dimensional, unsteady motion of these materials.

Here the components are assumed to be chemically inert, nonpolar, and not under the influence of electromagnetic fields. It is this system, together with appropriate initial and boundary conditions, from which we wish to extract the necessary averaged model.

The exact equations of motion, valid inside each material, are given in Chapter 2, and are:

Mass

$$\frac{\partial \rho}{\partial t} + \nabla \cdot \rho \mathbf{v} = 0 ; \tag{1}$$

Momentum

$$\frac{\partial \rho \mathbf{v}}{\partial t} + \nabla \cdot \rho \mathbf{v}\mathbf{v} = \nabla \cdot \mathbf{T} + \rho \mathbf{b}; \qquad (2)$$

Energy

$$\frac{\partial}{\partial t}\rho(u + \frac{1}{2}v^2) + \nabla \cdot \rho \mathbf{v}(u + \frac{1}{2}v^2) = \nabla \cdot (\mathbf{T} \cdot \mathbf{v} - \mathbf{q}) + \rho(r + \mathbf{b} \cdot \mathbf{v}); \qquad (3)$$

Entropy

$$\frac{\partial \rho s}{\partial t} + \nabla \cdot \rho \mathbf{v} s \geq \nabla \cdot \frac{\mathbf{q}}{\theta} + \frac{r}{\theta}; \qquad (4)$$

where ρ is the density, $\mathbf{v}$ is the velocity, $\mathbf{T}$ is the stress tensor, $\mathbf{b}$ is the body force, u is the specific internal energy or energy per unit mass, $\mathbf{q}$ is the heat flux, r is the heating source per unit mass, s is the specific entropy or entropy per unit mass, $\mathbf{q}/\theta$ is the entropy flux, and r/θ is the entropy supply.

Ordinary materials such as water and air, under ordinary conditions of stress and strain, do not exhibit the strange features of polar materials. These features are often due to the long-chain nature of molecules, and result, at this level of description, in a nonsymmetric stress tensor. For nonpolar materials, conservation of moment of momentum implies

$$\mathbf{T} = \mathbf{T}^T. \qquad (5)$$

These equations are valid in the interiors of each material involved in the flow. The canonical form of the balance equations for each component is

$$\frac{\partial \rho \Psi}{\partial t} + \nabla \cdot \rho \Psi \mathbf{v} = \nabla \cdot \mathbf{J} + \rho f, \qquad (6)$$

where Ψ is the quantity conserved, $\mathbf{J}$ is its (molecular or diffusive) flux, and f is its source density. When linear constitutive equations of the usual type are written, the fluxes $\mathbf{J}$ involve transport properties. For example, the heat flux vector involves thermal conductivity; the momentum flux vector, viscosity.

8.2. Jump Conditions

Multicomponent materials are characterized by the presence and motion of interfaces separating the components. At an interface between components, properties are discontinuous, although mass, momentum, and energy must be conserved. The canonical form of the jump conditions is

$$[\rho \Psi(\mathbf{v} - \mathbf{v}_i) + \mathbf{J}] \cdot \mathbf{n} = m. \qquad (7)$$

The jump conditions valid across the interface are given in Section 2.3, and are:

Mass

$$[\rho(\mathbf{v} - \mathbf{v}_i)] \cdot \mathbf{n} = 0 ; \tag{8}$$

Momentum

$$[\rho \mathbf{v}(\mathbf{v} - \mathbf{v}_i) + \mathbf{T}] \cdot \mathbf{n} = \mathbf{m}_i^\sigma ; \tag{9}$$

Energy

$$\left[\rho\left[u + \frac{1}{2}v^2(\mathbf{v} - \mathbf{v}_i)\right] + \mathbf{T} \cdot \mathbf{v} - \mathbf{q}\right] \cdot \mathbf{n} = \epsilon_i^\sigma ; \tag{10}$$

Entropy

$$\left[\rho s(\mathbf{v} - \mathbf{v}_i) + \frac{\mathbf{q}}{\theta}\right] \cdot \mathbf{n} = \frac{\epsilon_i^\sigma}{\theta_i} ; \tag{11}$$

where $\mathbf{v}_i$ is the velocity of the interface, $\mathbf{n}$ is the unit normal, and the boldface brackets denote jumps across the interfaces. Here $\mathbf{m}_i^\sigma$ is the traction associated with surface tension, ϵ_i^σ is the surface energy associated with the interface, and θ_i is the temperature of the interface. Implicit in (11) is the condition of thermal equilibrium at the interface, which forces the temperature to be continuous at the interface.

The surface traction, which has the dimensions of stress, is defined in the notation of generalized tensors in the interface as [2]

$$\mathbf{m}_i^\sigma = (H\sigma \mathbf{n} + \nabla_i \sigma), \tag{12}$$

where σ is the surface tension, H is the mean curvature of the interface, $\mathbf{n}$ is the unit normal, and ∇_i denotes the gradient in surface coordinates. The surface energy source term is given by

$$\epsilon_i^\sigma = H\sigma \mathbf{n} \cdot \mathbf{v}_i + \nabla \cdot (\sigma \mathbf{v}_i) - \frac{d_i u_i}{dt} - u_i \nabla_i \cdot \mathbf{v}_i , \tag{13}$$

where u_i is the surface internal energy density, and d_i/dt is the material derivative in the interface. [2], [45].

The canonical conservation principle for an interface is expressed by the following jump condition:

$$[(\rho \Psi(\mathbf{v} - \mathbf{v}_i) + \mathbf{J})] \cdot \mathbf{n} = m , \tag{14}$$

where $\mathbf{v}_i$ is the velocity of the interface, $\mathbf{n}$ is the unit normal, and m is the interfacial source of Ψ.

TABLE 8.1. Variables in generic conservation and jump equations.

Conservation Principle	Ψ	**J**	f	m
Mass	1	0	0	0
Momentum	**v**	**T**	**b**	m_i^σ
Energy	$u + \frac{1}{2}v^2$	$\mathbf{T}\cdot\mathbf{v} - \mathbf{q}$	$\mathbf{b}\cdot\mathbf{v} + r$	ϵ_i^σ
Entropy	s	$\mathbf{q}/\theta$	r/θ	q_i/θ_i

8.3. Summary of the Exact Equations

The canonical form of the equations of motion for the exact motions of the materials involved are

$$\frac{\partial \rho \Psi}{\partial t} + \nabla \cdot \rho \Psi \mathbf{v} = \nabla \cdot \mathbf{J} + \rho f, \qquad (15)$$

with jump conditions

$$[\rho\Psi(\mathbf{v} - \mathbf{v}_i) +]\mathbf{J} \cdot \mathbf{n} = m. \qquad (16)$$

The usual values for Ψ, **J**, f, and m are given in Table 8.2.

9
Ensemble Averaging

Here, we define the ensemble average, and give some results pertaining to its application to multicomponent flows.

Let us consider a specific type of physical situation in order to set our ideas. Suppose we have a saturated mixture of rigid spheres in an incompressible liquid. Suppose that at some initial instant, the spheres are approximately equally likely to be found at any point in the liquid. Clearly there are a large number of possible positions that the spheres could occupy; the concept of "large number" is purposely left undefined here. Let us cause the mixture to flow, and express an interest in its properties, say the velocity and pressure fields over a period of time. Setting up an experiment of this type is reasonably easy in a certain sense, for it is easy to mix spheres into a liquid and place the mixture into a machine. It is also easy to cause subsequent motion, and to follow the gross motion of the particles and the fluid. What is difficult to establish is the exact initial and subsequent locations of each of the particles. Furthermore, the exact initial and subsequent locations of each of the particles is not of particular interest in many applications. For example, consider the initial condition consisting of an array of 10^6 spheres of 1 cm in diameter evenly distributed in a cubic array in a box 2 m on one side. Consider an identical array, except with one of the spheres displaced 1 mm in a specific direction. The behavior of the two systems under identical forces should be very nearly the same. For purposes of, for example, the solution of engineering problems, these two initial conditions, and very many others like them, lead to motions that are indistinguishable. These motions are realizations *of the same system*. Thus, *in some averaged sense,* the systems are identical. In this case and for this purpose, it is not possible to keep an exact accounting of the exact evolution of the details of the system, *nor is it desirable.*

Indeed, it is a necessity for most useful predictions that the features that are of interest for a particular motion must be insensitive to small changes in the initial conditions, boundary conditions, and the source terms. Were it otherwise, the system might not be repeatable, so that other researchers cannot verify the results.

The ensemble average is an average that allows the interpretation of the phenomena in terms of the repeatability of multicomponent flows. Any particular exact experiment or realization will not be repeatable; however, any repetition of the experiment will lead to another member of the ensemble.

Usually in multicomponent materials, we think of the ensemble corresponding to a given motion as a very large set of realizations, with variations that we can observe. Indeed, the set can (theoretically) contain rearrangements of the atoms within the components; however, these different realizations are not different within the context of continuum mechanics, and we do not consider them further here. Thus, we shall start from the equations of continuum mechanics.

Consider the flow of spheres and fluid in a device. This situation is represented schematically in Figure 9.1. For most purposes, the differences between the different realizations are of no consequence. However, average properties of the distributions of the spheres, or of the motions, are important. The different realizations offer the possibility of samples to obtain the probability that the spheres appear at a given point, and values of mechanical variables. These samples can be averaged to obtain averaged fields.

To be consistent with the averaging procedure, or to interpret the fields in terms of the averaged variables, the initial conditions should be thought of as giving the average values of the fields so specified. Similarly, we should also think of inlet conditions as specifying average properties of the incoming mixture. Under most conditions, however, the boundaries of the device, and the boundary conditions imposed thereon, can be taken to be deterministic.

Using the ensemble average as the basis for the theory of multicomponent flow has several interesting implications.

First, we note that in situations where time and/or volume averaging are appropriate, these averages can be used as "samples" of the ensemble. Data acquisition based on time averaging is much easier than that implied by ensemble averaging.

Second, ensemble averaging does not require that a control volume contain a large number of particles in any given realization. Consider the following situation

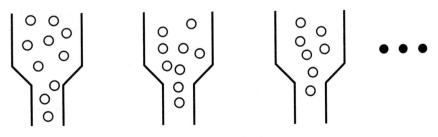

FIGURE 9.1. Realizations.

where the average of a particle–fluid mixture is of interest. Gas turbines are eroded by particulate matter passing through the inlet and impacting on the various parts of the machine. The trajectories of the particles moving through the device depend on where and when the particles enter the inlet of the device. One prediction of interest to the designer is the *expected* value of the particle flux near parts in the device which are susceptible to erosion. Thus, average, or *expected* values of the concentration and velocities of the components are of interest. Since suppression of erosion is crucial to the continued operation of the turbine, sufficient precautions will be taken so that there will be only a few particles in the device at any one time. Indeed, if control of the intake of particles is sufficiently good, there may be no particles in the device most of the time, and the situation where there is only one particle is in the device during some time interval will not be uncommon. The role of the prediction of average flow and particle concentration variables then takes the following meaning. The concentration of particles is proportional to the probability that the particle will be at the various points in the device at various times. The particle velocity field will be the mean velocity that the particle will have if it is at that position in the device. With this information, the design engineer will be able to assess the places where erosion due to particle impact will occur. Note that there will be no time when there will be many particles in some representative control volume. Thus, averaging schemes that depend on the concept of many representative particles will fail in this situation. Ensemble averaging, on the other hand, is appropriate. In this case, the ensemble is the set of motions of a single particle through the device, given that it started at a random point at the inlet (with some distribution of position associated with the dynamics of the particle moving through the inlet flow) at a random time during the transient flow through the device. It is clear that the solution for the average concentration and average velocity gives little information about the behavior of a single particle in the device; however, the information is quite appropriate for assessing the probability of damage to the device.

Third, a very elementary concept of averaging involves simply adding the observed values and dividing by the number of observations. From the point of view of ensemble averaging, this represents sampling the distribution implicit in the description of the ensemble a finite number of times. The law of large numbers, which is a general statement of ergodicity, suggests that a sufficiently large number of independent observations can be used to approximate expected values of the distribution. The difficulty is in describing the set of observations. In any flow, there are variations which are considered incidental and others which are fundamental. For example, in a monodisperse bubbly flow, if the initial positions of the bubbles are varied within reason, the properties of the flow should not change. If the initial positions are arranged so that the bubbles coalesce, this flow may be a fundamental variation and may be either excluded from the allowed set or retained as an event having only a low probability of occurrence.

Fourth, the ensemble average view of a physical event allows for an interpretation in terms of the view that all realizations are only approximations of the ideal.

Consider a system where the variability is contained in the locations of the particles or bubbles at the initial instant. If the observer were able to control the initial locations more precisely, the ensemble could then be viewed as smaller, or as having large variations assigned smaller probability. Indeed, if the observer could control the variability precisely, the ensemble would consist of one event. Consider the flow of a single particle released from a position that is known exactly, in a fluid flow that is controlled so as to have no fluctuations. In this case, the ensemble contains *one* realization, and the "averaging" leads to the exact equations and jump conditions. The averaged field equations that we shall derive here allow this as a special case.

To make these ideas precise, they must be made exact mathematically. This requires mathematics at a certain level of sophistication, for example, that of Royden [75]. The emphasis here is on the mechanics of multicomponent mixtures, not on the details of mathematical analysis, so we use the ideas of [75] without further comment.

Our first task is to sort out what is amenable to treatment by analysis and what is not. The essence of the ideas above is the comparison of things from certain sets and averaging over those sets. In the example with the 10^6 spheres, several sets of initial conditions were assumed to lead to motions that were, for practical purposes, identical. We find it difficult to formalize the concept of "practical purpose," so we leave it undefined. Rather, we assume a set, called an *ensemble*, and focus attention on it. An example of an ensemble might be a set motions resulting from a set of similar initial conditions, for example, those cited above. It might be difficult to place a specific particle exactly at a particular place at a particular time. Then a reasonable ensemble might be the motions resulting from initial conditions lying in open spheres of initial positions with given centers. Notice in this case, it would be most likely that the initial condition were a particular position, and less likely that it be a position far from that one. It might be totally unlikely that positions some given distance from a particular point would occur. This suggests that we must assign probabilities to the motions. A different example is to assume a turbulent flow containing one solid particle whose initial conditions are known exactly. If a number of experiments are done in which the particle is tracked exactly, the set of positions of the particle as a function of time might also be an ensemble. An example of a different extreme occurs when exact initial and subsequent information about a whole flow field might be useful. Then, the only useful ensemble would have exactly one member.

Formally, then, *an* ensemble *is a set of motions "possible" in the system*. There may be other restrictions. To give the theory enough structure that we may assign probabilities to the motions, the ensemble is given certain properties. This requires that we deal with subsets of the ensemble that are *measurable*. Clearly, some sets could be so mathematically pathological that they would not be worthy of consideration as ensembles. Thus not all sets of motions can be ensembles.

9. Ensemble Averaging

Formalizing these ideas is straightforward: Choose a set X. Let A be a subset of X, and let A^c be the complement of A, that is, the set of elements of X that is not in A. Thus $A \cup A^c = X$.

Definition. An *algebra* $\mathcal{A}$ is a collection of subsets $A, B, \ldots$ with three properties:
(1) if $A \in \mathcal{A}$ and $B \in \mathcal{A}$, then $A \cup B \in \mathcal{A}$;
(2) if $A \in \mathcal{A}$, then $A^c \in \mathcal{A}$; and
(3) if $A \in \mathcal{A}$ and $B \in \mathcal{A}$, then $A \cap B \in \mathcal{A}$.

It is straightforward to prove that finite unions are also in $\mathcal{A}$. It is impossible to prove that countably infinite unions are in $\mathcal{A}$. However, it is possible, given X, to find an algebra $\mathcal{A}$ of subsets of X with the property that if each of $A_i \in \mathcal{A}$, then $\bigcup_i^\infty A_i \in \mathcal{A}$. This is called a σ-*algebra*.

Definition. A σ-*algebra* $\mathcal{A}$ is an algebra such that if the sequence of subsets $A_i \in \mathcal{A}$, then $\bigcup_i^\infty A_i \in \mathcal{A}$.

A significant result showing that this formalism is not empty and is interesting, is that, given any collection of subsets of X, there is a smallest σ-algebra containing that collection. The elements of this σ-algebra are called *Borel sets*.

Definition. *Borel sets* are elements of the smallest σ-algebra that contains the open sets.

Since if a σ-algebra contains a set, it also contains the complement of the set, the σ-algebra contains closed sets also.

The elements chosen for inclusion in realizations are thus largely at the discretion of the user. For example, in multicomponent fluids, a realization could consist of the values of the velocity, the stress (pressure or elastic stress), the density, the temperature, and the position of the interfaces for the duration of the motion, over all spatial points that are occupied by the material during its motion. Suppose f is one of these fields. We denote the realization by μ, and denote the dependence of the field on the realization by $f(\mathbf{x}, t; \mu)$. We refer to the set of all realizations μ, denoted by $\mathcal{E}$, as the *ensemble*. Any subset of an ensemble is called a *subensemble*, and is an ensemble. The totality of all possible realizations for the same macroscopic process is quite large, and requires some further discussions of large sets.

Roughly speaking, Borel sets are sets to which it is possible to assign probabilities. For multicomponent flow, it is possible to prove that the Borel sets of interest can be described in terms of elementary sets of realizations that have the form

$$\overline{\mathcal{E}}(\mathbf{x}, t; F) = \{\mu \,|\, f(\mathbf{x}, t; \mu) \leq F\}, \tag{1}$$

for all real F. A Borel set can be formed from countably many intersections, unions, and complements of these sets of realizations. For example, if $f(\mathbf{x}, t; \mu)$ is some field, and F and dF are some numbers, then the Borel set of processes for which f takes values between F and $F + dF$ is

$$d\overline{\mathcal{E}}(\mathbf{x}, t; F, F + dF) = \overline{\mathcal{E}}(\mathbf{x}, t; F + dF) \cap \overline{\mathcal{E}}^c(\mathbf{x}, t; F). \tag{2}$$

Intrinsic to defining the ensemble averaging process is the choice of a measure. For the Borel set of realizations, the measure of interest is the *probability*. It must satisfy:

- $m(\emptyset) = 0$;
- $m(\mathcal{E}) = 1$;
- if $\overline{\mathcal{E}}_1 \subset \mathcal{E}$ and $\overline{\mathcal{E}}_2 \subset \mathcal{E}$ are Borel sets, with $\overline{\mathcal{E}}_1 \subset \overline{\mathcal{E}}_2$, then $m(\overline{\mathcal{E}}_1) \leq m(\overline{\mathcal{E}}_2)$; and
- if $\overline{\mathcal{E}}_i \subset \mathcal{E}, i = 1, 2, 3, \ldots$, with $\overline{\mathcal{E}}_i \bigcup \overline{\mathcal{E}}_j = \emptyset$, then $m(\cup_{i=1}^{\infty} \overline{\mathcal{E}}_i) = \sum_{i=1}^{\infty} m(\overline{\mathcal{E}}_i)$.

The probabilities associated with realizations, or more precisely, the probabilities associated with Borel sets of realizations, will be different for each different macroscopic process, even though the same realizations might be in both sets of possible outcomes. For example, if we wish to study sedimentation of spheres in a given apparatus, we may wish to start with a uniform dispersion of spheres for one set of experiments, and a clump of spheres for another. Depending on how the initial conditions are achieved, a realization that starts from some clump may lead to a possible outcome for the uniform dispersion experiment. We should hope that it would not, but we might be prepared to include that outcome as a realization in the ensemble for the case of uniform dispersion, but assign to it a low probability. That is, we should assign the probability to the Borel sets, that are centered on that outcome, a small probability for the uniform dispersion case. Thus, we see that the assignment of measure is specific to the macroscopic physical phenomenon being studied. We shall see that the assignment of a specific measure will not affect the balance equations, but some descriptions of the outcomes and the measures are needed to use the ideas of ensemble averaging to derive relations that suggest constitutive equations.

Implicit in the definition of the measure defined on the ensemble or, more precisely, on the Borel subsets of the ensemble, is the idea of what the important features of the motion are. The measure $m(\overline{\mathcal{E}})$ of a Borel subset $\overline{\mathcal{E}}$ is the probability of the processes in that subset occurring in the motion. If the Borel set is one of the elementary ones defined in (1), then this measure is the probability that the value of f is less than F. The probability of achieving process μ^* is meaningless, but we can discuss the probability of being within dF of a given function $f^*(\mathbf{x}, t)$ by considering the Borel set defined by

$$d\overline{\mathcal{E}}(f^*, dF) = \left\{ \mu \,\middle|\, \max_{(\mathbf{x}, t)} |f(\mathbf{x}, t; \mu) - f^*(\mathbf{x}, t)| \leq dF \right\}. \tag{3}$$

The function f^* need not be a field corresponding to any process, but it could be, with $f^*(\mathbf{x}, t) = f(\mathbf{x}, t; \mu^*)$. Roughly speaking, dF is the tolerance. The probability of being within dF of f^*, then, is the measure of the Borel set

$$m(d\overline{\mathcal{E}}(f^*, dF)). \tag{4}$$

9. Ensemble Averaging

The probability of achieving a particular process or, more precisely, being near a particular process, depends on what conditions are being specified, and how well they can be achieved in the given apparatus. Indeed, if we consider setting up and running two different flows in the same apparatus, many different microscopic processes might be common to both. However, the probabilities will be different. Thus, what is meant by a "process" includes information about both the ensemble of microscopic processes, and the probabilities of the occurrence of each. For the purposes of discussion, we shall refer to the ensemble plus the measure as the "macroscopic process," $\mathcal{M}$,

$$\mathcal{M} = (\mathcal{E}, m) \tag{5}$$

or, when the context is clear, just "process."

If f is some field, then the average of f is defined in terms of Borel sets as follows. Partition the interval $(-\infty, \infty)$ into

$$-\infty < F_1 < F_2 < \cdots < F_N < \infty. \tag{6}$$

Let $dF_1 = F_2 - F_1, dF_2 = F_3 - F_2$, etc. Then the average of f is

$$\overline{f}(\mathbf{x}, t) = \lim_{N \to \infty} \sum_{i=1}^{N-1} F_i \, m(d\overline{\mathcal{E}}(\mathbf{x}, t; F_i, dF_i)), \tag{7}$$

where, in the limit, $\max dF_i \to 0$. This procedure leads to the Lebesgue integral over Borel sets.

It is also characteristic of multicomponent materials that there do not exist special realizations that occur with some nonzero probability. Thus, for a Borel set of the form $\overline{\mathcal{E}}(\mathbf{x}, t; F, F + dF)$, we assume that its measure is absolutely continuous with respect to dF, for each F. Then the measure of this set, $m(\overline{\mathcal{E}}(\mathbf{x}, t; F, F + dF))$, is given by a density $dm(\mu)$ [75], and we define the *average* of f by

Definition. The *ensemble average* of f is

$$\overline{f}(\mathbf{x}, t) = \int_{\mathcal{E}} f(\mathbf{x}, t; \mu) \, dm(\mu), \tag{8}$$

where $dm(\cdot)$ is the density for the measure (probability) on the set of all processes $\mathcal{E}$.

The notation

$$\int_{\mathcal{E}} f(\mathbf{x}, t; \mu) \, dm(\mu) \tag{9}$$

is suggestive, and we wish to exploit the properties associated with the linear integration operation, specifically with respect to interchange of the order of integration and differentiation operators.

9.1. Results for Ensemble Averaging

In order to apply the averaging procedure to the equations of motion, we shall need some results about the averaging procedure. We shall also give analogous results for time and volume averaging.

9.1.1. Reynolds Rules

Equation (8) has several properties that may be called "linearity." Let c_1 and c_2 be constants, and let f_1 and f_2 be fields from realizations in the ensemble $\mathcal{E}$. Then trivially from the definition

$$\int_{\mathcal{E}} [c_1 f_1(\mathbf{x}, t; \mu) + c_2 f_2(\mathbf{x}, t; \mu)] \, dm(\mu)$$
$$= c_1 \int_{\mathcal{E}} f_1(\mathbf{x}, t; \mu) \, dm(\mu) + c_2 \int_{\mathcal{E}} f_2(\mathbf{x}, t; \mu) \, dm(\mu), \quad (10)$$

that is, in this sense,

$$\overline{c_1 f_1 + c_2 f_2} = c_1 \overline{f_1} + c_2 \overline{f_2}. \quad (11)$$

It is also possible to break the ensemble up into disjoint parts, just as it is possible to break a line interval up into disjoint parts. Thus let $\mathcal{E} = \mathcal{E}_1 \cup \mathcal{E}_2$ with $\mathcal{E}_1 \cap \mathcal{E}_2 = 0$. Then

$$\int_{\mathcal{E}} f(\mathbf{x}, t; \mu) \, dm(\mu) = \int_{\mathcal{E}_1} f(\mathbf{x}, t; \mu) \, dm(\mu) + \int_{\mathcal{E}_2} f(\mathbf{x}, t; \mu) \, dm(\mu). \quad (12)$$

There is yet another linearity property with respect to ensembles, analogous to constructing a two-dimensional Euclidean space from two one-dimensional spaces.

Let $\overline{f}$ denote an average. Let c_1 and c_2 be constants, and let f_1 and f_2 be fields from realizations in some ensembles $\mathcal{E}_1$ and $\mathcal{E}_2$, respectively. Then the field $c_1 f_1 + c_2 f_2$ is a field in the product ensemble $\mathcal{E}_1 \times \mathcal{E}_2$, and we have the result of linearity:

$$\overline{c_1 f_1 + c_2 f_2} = c_1 \overline{f_1} + c_2 \overline{f_2}. \quad (13)$$

The proof is as follows:

$$\overline{c_1 f_1 + c_2 f_2} = \int_{\mathcal{E}_1 \times \mathcal{E}_2} (c_1 f_1 + c_2 f_2) \, dm(\mu_1 \times \mu_2)$$
$$= c_1 \int_{\mathcal{E}_1 \times \mathcal{E}_2} f_1 \, dm(\mu_1 \times \mu_2) + c_2 \int_{\mathcal{E}_1 \times \mathcal{E}_2} f_2 \, dm(\mu_1 \times \mu_2)$$
$$= c_1 \int_{\mathcal{E}_1} f_1 \, dm(\mu_1) \int_{\mathcal{E}_2} dm(\mu_2) + c_2 \int_{\mathcal{E}_2} f_2 \, dm(\mu_2) \int_{\mathcal{E}_1} dm(\mu_1)$$
$$= c_1 \int_{\mathcal{E}_1} f_1 \, dm(\mu_1) + c_2 \int_{\mathcal{E}_2} f_2 \, dm(\mu_2)$$
$$= c_1 \overline{f_1} + c_2 \overline{f_2}. \quad (14)$$

If f_1 and f_2 are fields in the realization in $\mathcal{E}$, then the field $\overline{f_1 f_2}$ is also a field in the realization in $\mathcal{E}$, and its average is given by

$$\overline{\overline{f_1} f_2} = \int_{\mathcal{E}} \overline{f_1} f_2 \, dm(\mu)$$
$$= \overline{f_1} \int_{\mathcal{E}} f_2 \, dm(\mu)$$
$$= \overline{f_1}\, \overline{f_2}. \qquad (15)$$

In particular, if $f_2 \equiv 1$, then $\overline{\overline{f_1}} = \overline{f_1}$.

9.1.2. Treating Generalized Functions

Multicomponent mixtures are characterized by the presence of discontinuities in the fields. In many of these mixtures, the material properties such as viscosity or elasticity are different in different components. Thus, the fields $\partial f/\partial t$ and ∇f are generalized functions. We shall assume that within each material, the fields of interest are smooth, and that they can change discontinuously across reasonably complicated surfaces dividing the components. A full treatment requires a description with test functions. We outline such a treatment here.

Consider the set of test functions, Φ, on space and time. That is, $\phi(\mathbf{x}, t) \in \Phi$ if ϕ has compact support, and has derivatives to all orders. Then if $f(\mathbf{x}, t)$ is a generalized function, the derivatives of f are defined by

$$\int_\Omega \phi(\mathbf{x}, t) \frac{\partial f(\mathbf{x}, t)}{\partial t} \, dv\, dt = -\int_\Omega \frac{\partial \phi}{\partial t}(\mathbf{x}, t) f(\mathbf{x}, t) \, dv\, dt, \qquad (16)$$

and

$$\int_\Omega \phi(\mathbf{x}, t) \nabla f(\mathbf{x}, t) \, dv\, dt = -\int_\Omega \nabla \phi(\mathbf{x}, t) f(\mathbf{x}, t) \, dv\, dt \qquad (17)$$

for any $\phi \in \Phi$. Here Ω is a compact set in space and time such that the support of ϕ lies in Ω.

9.1.3. Characteristic Function X_k

It is often desirable to isolate each component theoretically, even in the microscale description. To do this, we introduce the *component indicator function*, or *characteristic function*, $X_k(\mathbf{x}, t; \mu)$, in any realization μ. The characteristic function is defined by

$$X_k(\mathbf{x}, t; \mu) = \begin{cases} 1 & \text{if } \mathbf{x} \in k \text{ in realization } \mu, \\ 0 & \text{otherwise.} \end{cases} \qquad (18)$$

This function "picks out" component k, and ignores all other components and interfaces. This function is important in the theoretical description of multicomponent materials.

Interface Delta Function

The quantity ∇X_k plays an important role in the description of multicomponent flow. Let ϕ be a test function and consider

$$\int_\Omega \phi(\mathbf{x}, t) \nabla X_k(\mathbf{x}, t) \, dv \, dt = -\int_\Omega \nabla \phi(\mathbf{x}, t) X_k(\mathbf{x}, t) \, dv \, dt$$
$$= -\int_{\Omega_k} \nabla \phi(\mathbf{x}, t) \, dv \, dt, \quad (19)$$

where Ω_k is the intersection of Ω with component k. Using the divergence theorem gives

$$\int_\Omega \phi(\mathbf{x}, t) \nabla X_k(\mathbf{x}, t) \, dv \, dt = -\int_{\partial\Omega_k} \mathbf{n}_k \phi(\mathbf{x}, t) \, dv \, dt$$
$$= -\int_\Omega \mathbf{n}_k \delta(\mathbf{x} - \mathbf{x}_i, t) \phi(\mathbf{x}, t) \, dv \, dt, \quad (20)$$

where $\partial\Omega_k$ denotes the interface between components, $\mathbf{n}_k$ is the unit normal to it in the direction exterior to component k, and $\delta(\mathbf{x} - \mathbf{x}_i, t)$ denotes the Dirac delta function for the interface. Thus, we have

$$\nabla X_k = \mathbf{n}_k \delta(\mathbf{x} - \mathbf{x}_i, t). \quad (21)$$

The concept of the directional derivative can be generalized to this case. The result is that the delta function for the interface can be written as

$$\frac{\partial X_k}{\partial n} = \mathbf{n}_k \cdot \nabla X_k. \quad (22)$$

The Topological Equation

In the averaging process we will require a result for the term

$$\frac{\partial X_k}{\partial t} + \mathbf{v}_i \cdot \nabla X_k. \quad (23)$$

We shall derive this using the results from generalized functions. Let ϕ be a test function. Then

$$\int_\Omega \phi \frac{\partial X_k}{\partial t} \, dv \, dt = -\int_\Omega \frac{\partial \phi}{\partial t} X_k \, dv \, dt, \quad (24)$$

and

$$\int_\Omega \phi \nabla X_k \, dv \, dt = -\int_\Omega \nabla \phi X_k \, dv \, dt. \quad (25)$$

Thus,

$$\int_\Omega \phi \left[\frac{\partial X_k}{\partial t} + \mathbf{v}_i \cdot \nabla X_k \right] dv \, dt = -\int_\Omega \left[\frac{\partial \phi}{\partial t} + \nabla \cdot (\phi \mathbf{v}_i) \right] X_k \, dv \, dt. \quad (26)$$

Applying the Reynolds Transport Theorem [2] shows that

$$\int_\Omega \left[\frac{\partial \phi}{\partial t} + \nabla \cdot (\phi \mathbf{v}_i)\right] X_k dv\, dt = \int_{t=-\infty}^{t=+\infty} \frac{d}{dt} \int_{V_k(t)} \phi\, dv\, dt$$

$$= \int_{V_k(t)} \phi\, dv \Big|_{t=-\infty}^{t=\infty}$$

$$= 0, \qquad (27)$$

since ϕ vanishes for t sufficiently large. Thus,

$$\frac{\partial X_k}{\partial t} + \mathbf{v}_i \cdot \nabla X_k = 0. \qquad (28)$$

This is the topological equation. Its physical explanation follows. Note that it is the material derivative of X_k following the interface. If we look at a point that is not on the interface, then either $X_k = 1$ or $X_k = 0$. In either case, the partial derivatives both vanish, and hence the left side of the topological equation vanishes identically. Now consider a point on the interface. This point moves with the interface velocity. Then the function X_k is a jump that remains constant. Thus its material derivative following the interface vanishes.

Gauss and Leibniz Rules

In order to average to the exact equations, we need expressions for $\overline{\partial f/\partial t}$ and $\overline{\nabla f}$. If f is adequately well behaved so that the limiting processes of integration and differentiation can be interchanged, then it is clear from the definition of the ensemble average that

$$\overline{\frac{\partial f}{\partial t}} = \frac{\partial \overline{f}}{\partial t}, \qquad (29)$$

and

$$\overline{\nabla f} = \nabla \overline{f}. \qquad (30)$$

Note that if f has discontinuities, then either $\partial f/\partial t$ or ∇f, or both, may have a Dirac delta function character. In this case, care must be taken to interpret $\overline{\partial f/\partial t}$ and $\overline{\nabla f}$.

Consider $\nabla X_k f$, where X_k is the component indicator function for component k

$$\int_\Omega \phi \nabla (X_k f)\, dv\, dt = -\int_\Omega X_k f \nabla \phi\, dv\, dt$$

$$= -\int_{\Omega_k} f \nabla \phi\, dv\, dt. \qquad (31)$$

Assuming that f is well behaved in Ω_k, and applying the divergence theorem, we have

$$-\int_{\Omega_k} f \nabla \phi\, dv\, dt = -\oint_{\partial \Omega_k} \mathbf{n}\phi f_{ki}\, ds\, dt + \int_{\Omega_k} \phi \nabla f\, dv\, dt$$

$$= -\oint_{\partial \Omega_k} \mathbf{n}\phi f_{ki}\,ds\,dt + \int_\Omega \phi X_k \nabla f\,dv\,dt, \tag{32}$$

where f_{ki} is the value of the function f evaluated on the component k side of the interface. Next, note that ∇X_k is defined by

$$\int_\Omega \phi(\mathbf{x},t)\nabla X_k\,dv\,dt = -\int_\Omega X_k \nabla \phi(\mathbf{x},t)\,dv\,dt$$

$$= -\int_{\Omega_k} \nabla \phi(\mathbf{x},t)\,dv\,dt$$

$$= -\oint_{\partial \Omega_k} \mathbf{n}\phi(\mathbf{x},t)\,ds\,dt. \tag{33}$$

Then we see that

$$\int_\Omega \phi \nabla (X_k f)\,dv\,dt = \int_\Omega X_k \nabla f \phi\,dv\,dt + \int_\Omega f_{ki} \nabla X_k \phi\,dv\,dt. \tag{34}$$

Thus, we have the result

$$\nabla(X_k f) = X_k \nabla f + f_{ki} \nabla X_k. \tag{35}$$

Note that this result appears to be similar to the product rule for ordinary derivatives. However, there are some subtleties. Applying the ensemble average to (34) gives

$$\overline{X_k \nabla f} = \overline{\nabla X_k f} - \overline{f \nabla X_k}$$
$$= \nabla \overline{X_k f} - \overline{f_{ki} \nabla X_k}. \tag{36}$$

This is called the Gauss rule.

A similar sequence of calculations shows that

$$\overline{X_k \frac{\partial f}{\partial t}} = \overline{\frac{\partial X_k f}{\partial t}} - \overline{f \frac{\partial X_k}{\partial t}}$$

$$= \frac{\partial \overline{X_k f}}{\partial t} - \overline{f_{ki} \frac{\partial X_k}{\partial t}}. \tag{37}$$

This is called the Leibnitz rule. The second term on the right-hand side in both of these equations is related to the surface average of f, evaluated on the k-component side, over the interface.

We note that $\overline{f} \neq \overline{X_1 f} + \overline{X_2 f}$, since $X_1 + X_2 = 0$ when $\mathbf{x}$ is a point on the interface. Adding (36) for $k = 1, 2$ yields

$$\sum_{i=1}^{2} \overline{X_i \nabla f} = \nabla \sum_{i=1}^{2} \overline{X_i f} + \overline{[f] \nabla X_1}. \tag{38}$$

where the boldface brackets denote the jump from the component 1 side of the interface to the component 2 side of the interface. A similar sequence of operations gives

$$\sum_{i=1}^{2}\overline{X_i\frac{\partial f}{\partial t}} = \frac{\partial}{\partial t}\sum_{i=1}^{2}\overline{X_i f} - \overline{[f]\frac{\partial X_1}{\partial t}}. \tag{39}$$

Correct treatment of generalized functions is essential for multicomponent flow theory.

10
Other Averages

In many multicomponent flows, the variability that justifies the use of an averaging process to make predictions is manifested in a particular realization as rapid variations in space or time. For example, a time trace of some field may show large fluctuations due to the passing of dispersed units in a fluid, or a photograph may show many similar units within a distance which is small compared to the size of the flow domain.

In this chapter, we examine three different averaging processes that are often applied to multicomponent flows, and discuss how they fit into the concept of ensemble averaging. The literature in multicomponent mechanics contains many different types of averages. These are often motivated by the type of application. The common averaging processes used are volume averaging [64], time averaging [45], combinations of time and volume averaging [26], and statistical averaging [16].

Stationary flows are flows which are unsteady, but for which the average variables are steady. Roughly, if we examine a field for a long time interval for given x, we can see no differences in the random behavior of the field at different times. Then, if we use the information from a long time interval T to form statistics, these statistics should be *independent of t*. There is an implicit assumption in the use of the time average. It is that if another realization is studied the same way, its statistics are the same as the first realization. This result is sometimes referred to as an "ergodicity" result, meaning that it relates two averaging procedures. Note, however, that the result is an approximation, since the number of samples must go to ∞, implying that $T \to \infty$. We choose to leave the result as an approximation instead of introducing yet another limiting process in this discussion.

A similar interpretation can be given to the volume average for flows that are statistically spatially uniform, or homogeneous. Consider a spatial scale which is small compared to the size of the flow domain, but which is large compared to the distances between interfaces. If volumes of this size contain the variability due to many dispersed units, we can use the information in such a volume to compute statistical quantities. Again, there is the implicit assumption that the variability through the volume is the same sort of variability that would be seen in other realizations.

A useful conceptualization of the ensemble average assumes that the flow is deterministic, but that randomness arises through the uncertainty in the initial conditions. Another useful conceptualization, equally valid in some applications, is to think of the processes being affected by small random forces throughout the motion. If the dispersed component is made up of distinct units such as particles or bubbles or drops, it may be useful to treat these units as having a distribution of positions, velocities, and sizes. It should be noted, however, that it is not evident how to treat velocity fluctuations in the fluid within the context of this approach.

10.1. Multiparticle Distribution Functions

Consider a system consisting of N discrete units, which may be droplets, particles, or bubbles. For simplicity, we assume that the particles are identical spheres and are therefore indistinguishable. Furthermore, we need not include information about the particle sizes or orientations in this description. In order to describe the randomness in the system, assume that there is a "master" distribution function

$$f^{(N)}(\mathbf{z}_1, \mathbf{v}_1, \mathbf{z}_2, \mathbf{v}_2, \ldots, \mathbf{z}_N, \mathbf{v}_N, t). \tag{1}$$

Roughly speaking, the distribution function $f^{(N)}$ has the interpretation that

$$f^{(N)} d\mathbf{z}_1 d\mathbf{v}_1 \cdots d\mathbf{z}_N d\mathbf{v}_N \tag{2}$$

is the probability of finding a particle within $d\mathbf{z}_1$ of $\mathbf{z}_1$ with velocity within $d\mathbf{v}_1$ of $\mathbf{v}_1$ and finding a particle within $d\mathbf{z}_2$ of $\mathbf{z}_2$ with velocity within $d\mathbf{v}_2$ of $\mathbf{v}_2$, etc., at time t, during a macroscopic process $\mathcal{M}$. Note that this assumes that each particle is identified with a number, and that we can distinguish particle i from particle j, for $i \neq j$. In order to interpret this function in terms of the ensemble, consider a process μ. Suppose that during this process, the position of particle i is given by

$$\mathbf{x}_i = \mathcal{X}_i(\mathbf{X}_i, t; \mu), \tag{3}$$

where $\mathbf{X}_i$ is the initial position of particle i in realization μ. Then $f^{(N)}$ has the more precise interpretation that

$$f^{(N)} d\mathbf{z}_1 d\mathbf{v}_1 \ldots d\mathbf{z}_N d\mathbf{v}_N$$
$$= \Pr_{\mathcal{E}}(|\mathbf{z}_1 - \mathcal{X}_1(\mathbf{X}_1, t; \mu)| < d\mathbf{z}_1, |\mathbf{v}_1 - \dot{\mathcal{X}}_1(\mathbf{X}_1, t; \mu)| < d\mathbf{v}_1, \ldots,$$
$$|\mathbf{z}_N - \mathcal{X}_N(\mathbf{X}_N, t; \mu)| < d\mathbf{z}_N, |\mathbf{v}_N - \dot{\mathcal{X}}_N(\mathbf{X}_N, t; \mu)| < d\mathbf{v}_N), \tag{4}$$

10.1. Multiparticle Distribution Functions 107

where the probability Pr is over the ensemble $\mathcal{E}$. A different way of saying the same thing is to consider the subset of processes in $\mathcal{E}$, denoted by

$$\overline{\mathcal{E}}^{(N)} = \{\mu \in \mathcal{E} | |\mathbf{z}_1 - \mathcal{X}_1(\mathbf{X}_1, t; \mu)| < d\mathbf{z}_1, |\mathbf{v}_1 - \dot{\mathcal{X}}_1(\mathbf{X}_1, t; \mu)| < d\mathbf{v}_1, \ldots,$$
$$|\mathbf{z}_N - \mathcal{X}_N(\mathbf{X}_N, t; \mu)| < d\mathbf{z}_N, |\mathbf{v}_N - \dot{\mathcal{X}}_N(\mathbf{X}_N, t; \mu)| < d\mathbf{v}_N\}. \quad (5)$$

Here $\overline{\mathcal{E}}^{(N)}$ depends on $\mathbf{z}_1, \mathbf{v}_1, d\mathbf{z}_1, d\mathbf{v}_1, \ldots, \mathbf{z}_N, \mathbf{v}_N, d\mathbf{z}_N, d\mathbf{v}_N$. Then

$$m(\overline{\mathcal{E}}^{(N)}) = f^{(N)} d\mathbf{z}_1 d\mathbf{v}_1 \ldots d\mathbf{z}_N d\mathbf{v}_N,$$
$$= \int_{\overline{\mathcal{E}}^{(N)}} dm(\mu). \quad (6)$$

Thus, $m(\overline{\mathcal{E}}^{(N)})$ is also the probability of finding a particle within $d\mathbf{z}_1$ of $\mathbf{z}_1$ with velocity within $d\mathbf{v}_1$ of $\mathbf{v}_1$ and finding a particle within $d\mathbf{z}_2$ of $\mathbf{z}_2$ with velocity within $d\mathbf{v}_2$ of $\mathbf{v}_2$, etc., at time t, during the macroscopic process $\mathcal{M}$.

The other particle distribution functions can be defined similarly. For example, the single particle distribution function $f^{(1)}(\mathbf{z}, \mathbf{v}, t)$ is defined by taking $f^{(1)} d\mathbf{z} d\mathbf{v}$ as the probability of finding a particle within $d\mathbf{z}$ of $\mathbf{z}$ with velocity within $d\mathbf{v}$ of $\mathbf{v}$. The interpretation in terms of subensembles follows. It is interesting to note that we have

$$f^{(1)} = \iint \cdots \int f^{(N)} d\mathbf{z}_2 d\mathbf{v}_2 \cdots d\mathbf{z}_N d\mathbf{v}_N. \quad (7)$$

Nearest-Neighbor Distribution Functions

Suppose that a flow domain $\mathcal{B}$ is occupied by N spherical particles, all with the same radius, and having some distribution

$$f^{(N)}(\mathbf{z}_1, \mathbf{z}_2, \ldots, \mathbf{z}_N)$$

of the positions of the centers. By symmetry, if

$$\{i_1, i_2, \ldots, i_N\},$$

is any rearrangement of the particles, then

$$f^{(N)}(\mathbf{z}_1, \mathbf{z}_2, \ldots, \mathbf{z}_N) = f^{(N)}(\mathbf{z}_{i_1}, \mathbf{z}_{i_2}, \ldots, \mathbf{z}_{i_N}).$$

For each point $\mathbf{x}$ in $\mathcal{B}$, there is a rearrangement of the numbering of the spheres in order of closeness to the point $\mathbf{x}$. This is a "partial ordering" in the sense that sometimes two or more spheres can be the same distance from $\mathbf{x}$. In the case that this sort of equidistance occurs, there is no loss in the context of this particular argument in assigning numbering arbitrarily. That is, sphere number i_1, at location $\mathbf{z}'_1 = \mathbf{z}_{i_1}$, is closest to $\mathbf{x}$, sphere number i_2, at location $\mathbf{z}'_2 = \mathbf{z}_{i_2}$, is next closest, and so on. Note that the order changes as the point $\mathbf{x}$ changes, and will change as t changes. The nearest-neighbor distribution function can be found in terms of the original distribution function by

$$\hat{f}^{(N)}(\mathbf{x}, t; \mathbf{z}'_1, \ldots, \mathbf{z}'_N) = f^{(N)}(\mathbf{z}_1, \ldots, \mathbf{z}_N). \quad (8)$$

For simplicity, we shall suppress the notation for the dependence on t. It should be realized, however, that the nearest-neighbor particle distribution can change in space and time.

Note that ties in order of closeness can occur. Situations involving ties are events of low probability, and in most flows the way in which the tie is broken will not affect the results.

10.1.1. Statistical Averaging

Given a distribution of particles at $z_1, z_2, \ldots, z_N$, the solution to the equations of motion gives the velocity and pressure fields in the fluid as a function of position and time,

$$\mathbf{v}(\mathbf{x}|\mathbf{z}_1, \mathbf{z}_2, \ldots, \mathbf{z}_N)$$

and

$$p(\mathbf{x}|\mathbf{z}_1, \mathbf{z}_2, \ldots, \mathbf{z}_N).$$

We wish to perform the average by multiplying $\mathbf{v}$ or p by $f^{(N)}$ and integrating over all possible particle positions. Thus,

$$\bar{\mathbf{v}} = \int_{\mathbf{z}_1} \int_{\mathbf{z}_2} \cdots \int_{\mathbf{z}_N} f^{(N)}(\mathbf{z}_1, \ldots, \mathbf{z}_N) \mathbf{v}(\mathbf{x}|\mathbf{z}_1, \ldots, \mathbf{z}_N) \, d\mathbf{z}_1 \ldots d\mathbf{z}_N. \quad (9)$$

Conditional Averages

It is also useful to define conditional averages. Batchelor [8] and Hinch [42] write (9) as

$$\bar{\mathbf{v}} = \int_{\mathbf{z}_1} f^{(1)}(\mathbf{z}_1) \bar{\mathbf{v}}^{(1)}(\mathbf{x}|\mathbf{z}_1) \, d\mathbf{z}_1$$

$$= \int_{\mathbf{z}_1} \int_{\mathbf{z}_2} f^{(2)}(\mathbf{z}_1, \mathbf{z}_2) \bar{\mathbf{v}}^{(2)}(\mathbf{x}|\mathbf{z}_1, \mathbf{z}_2) \, d\mathbf{z}_1 \, d\mathbf{z}_2, \quad (10)$$

etc., where

$$\bar{\mathbf{v}}^{(1)}(\mathbf{x}|\mathbf{z}_1) = \frac{1}{f^{(1)}(\mathbf{z}_1)} \int_{\mathbf{z}_2} \int_{\mathbf{z}_3} \cdots \int_{\mathbf{z}_N} f^{(N)}(\mathbf{z}_1, \mathbf{z}_2, \ldots, \mathbf{z}_N)$$
$$\times \mathbf{v}(\mathbf{x}|\mathbf{z}_1, \mathbf{z}_2, \ldots, \mathbf{z}_N) \, d\mathbf{z}_2 \, d\mathbf{z}_3 \ldots d\mathbf{z}_N \quad (11)$$

is the conditionally averaged velocity, averaged on the condition that there is a sphere with its center at $\mathbf{z}_1$. Here, it is assumed that the appropriate definition for $f^{(1)}(\mathbf{z}_1)$ is the unconditional density function for any sphere center being at $\mathbf{z}_1$,

$$f^{(1)}(\mathbf{z}_1) = \int_{\mathbf{z}_2} \int_{\mathbf{z}_3} \cdots \int_{\mathbf{z}_N} f^{(N)}(\mathbf{z}_1, \mathbf{z}_2, \ldots, \mathbf{z}_N) \, d\mathbf{z}_2 \, d\mathbf{z}_3 \ldots d\mathbf{z}_N. \quad (12)$$

Similarly,

$$\bar{\mathbf{v}}^{(2)}(\mathbf{x}|\mathbf{z}_1, \mathbf{z}_2) = \frac{1}{f^{(2)}(\mathbf{z}_1, \mathbf{z}_2)} \int_{\mathbf{z}_3} \cdots \int_{\mathbf{z}_N} f^{(N)}(\mathbf{z}_1, \mathbf{z}_2, \ldots, \mathbf{z}_N) \\ \times \mathbf{v}(\mathbf{x}|\mathbf{z}_1, \mathbf{z}_2, \ldots, \mathbf{z}_N) \, d\mathbf{z}_3 \ldots d\mathbf{z}_N, \quad (13)$$

and

$$f^{(2)}(\mathbf{z}_1, \mathbf{z}_2) = \int_{\mathbf{z}_3} \int_{\mathbf{z}_4} \cdots \int_{\mathbf{z}_N} f^{(N)}(\mathbf{z}_1, \mathbf{z}_2, \ldots, \mathbf{z}_N) \, d\mathbf{z}_3 \, d\mathbf{z}_4 \ldots d\mathbf{z}_N, \quad (14)$$

etc. The distributions $f^{(i)}$ are called the one-particle distribution function, the two-particle distribution function, etc. Note that

$$\bar{\mathbf{v}}^{(1)}(\mathbf{x}|\mathbf{z}_1) = \frac{1}{f^{(1)}(\mathbf{z}_1)} \int_{\mathbf{z}_2} f^{(2)}(\mathbf{z}_1, \mathbf{z}_2) \bar{\mathbf{v}}^{(2)}(\mathbf{x}|\mathbf{z}_1, \mathbf{z}_2) \, d\mathbf{z}_2. \quad (15)$$

In terms of the nearest-neighbor distribution, the average velocity is

$$\bar{\mathbf{v}} = \int_{\mathbf{z}'_1} \cdots \int_{\mathbf{z}'_N} \hat{f}^{(N)}(\mathbf{x}, \mathbf{z}'_1, \ldots, \mathbf{z}'_N) \mathbf{v}(\mathbf{x}|\mathbf{z}'_1, \ldots, \mathbf{z}'_N) \, d\mathbf{z}'_1 \ldots d\mathbf{z}'_N. \quad (16)$$

There are corresponding conditional averages and distributions. Consider

$$\bar{\mathbf{v}}(\mathbf{x}) = \int_{\mathbf{z}'_1} \hat{f}^{(1)}(\mathbf{x}, \mathbf{z}'_1) \hat{\bar{\mathbf{v}}}^{(1)}(\mathbf{x}|\mathbf{z}'_1) \, d\mathbf{z}'_1,$$

$$= \int_{\mathbf{z}'_1} \int_{\mathbf{z}'_2} \hat{f}^{(2)}(\mathbf{x}, \mathbf{z}'_1, \mathbf{z}'_2) \hat{\bar{\mathbf{v}}}^{(2)}(\mathbf{x}|\mathbf{z}'_1, \mathbf{z}'_2) \, d\mathbf{z}'_1 \, d\mathbf{z}'_2,$$

etc., $\quad (17)$

where

$$\hat{\bar{\mathbf{v}}}^{(1)}(\mathbf{x}|\mathbf{z}'_1) = \frac{1}{\hat{f}^{(1)}(\mathbf{x}, \mathbf{z}'_1)} \int_{\mathbf{z}'_2} \int_{\mathbf{z}'_3} \cdots \int_{\mathbf{z}'_N} \hat{f}^{(N)}(\mathbf{x}, \mathbf{z}'_1, \mathbf{z}'_2, \ldots, \mathbf{z}'_N) \\ \times \mathbf{v}(\mathbf{x}|\mathbf{z}'_1, \mathbf{z}'_2, \ldots, \mathbf{z}'_N) \, d\mathbf{z}'_2 \, d\mathbf{z}'_3 \ldots d\mathbf{z}'_N \quad (18)$$

is the conditionally averaged velocity, averaged on the condition that there is a sphere with its center at $\mathbf{z}'_1$.

10.1.2. Computing the Nearest-Neighbor Distribution Functions

We develop a technique for computing the functions $f^{(1)}$, $\hat{f}^{(1)}$, and $\hat{f}^{(2)}$. A *uniform particle concentration* is defined so that $f^{(1)}(\mathbf{z}_1)$ is constant. If the volumetric concentration is given as $\alpha_p(\mathbf{x}, t)$, then

$$f^{(1)}(\mathbf{z}_1) = \frac{\alpha_p(\mathbf{x}, t)}{\frac{4}{3}\pi a^3}.$$

For the nearest-neighbor distributions, note that $\hat{f}^{(1)}(\mathbf{x}, \mathbf{z}_1') \, d\mathbf{z}_1'$ is the probability that the sphere having its center within $d\mathbf{z}_1'$ of $\mathbf{z}_1'$ is the closest to the point $\mathbf{x}$, and

$$\hat{f}^{(2)}(\mathbf{x}, \mathbf{z}_1', \mathbf{z}_2') \, d\mathbf{z}_1' \, d\mathbf{z}_2' \tag{19}$$

is the probability that the sphere having its center within $d\mathbf{z}_1'$ of $\mathbf{z}_1'$ is the closest to the point $\mathbf{x}$ *and* the sphere having its center within $d\mathbf{z}_2'$ of $\mathbf{z}_2'$ is the next closest to the point $\mathbf{x}$, and so on. These distributions are the one-particle and two-particle nearest-neighbor distribution functions.

In order to compute an expression for $\hat{f}^{(1)}(\mathbf{x}, \mathbf{z}_1')$, consider the probability that there is no sphere center within a distance r of $\mathbf{x}$ and that there is a sphere center within the shell between r and $r + dr$. Then it is clear that

$$\hat{f}^{(1)}(\mathbf{x}, \mathbf{z}_1') \, d\mathbf{z}_1' \tag{20}$$

is the probability that there is no sphere within r of $\mathbf{x}$ *and* there is one sphere within $d\mathbf{z}_1'$ of $\mathbf{z}_1'$. Now, the probability that there is no sphere within r of $\mathbf{x}$ and the probability that the nearest sphere is within r of $\mathbf{x}$ sum to unity. Thus,

$$\hat{f}^{(1)}(\mathbf{x}, \mathbf{z}_1') \, d\mathbf{z}_1' = \left(1 - \int_{|\mathbf{x}-\mathbf{z}_1''| \leq |\mathbf{x}-\mathbf{z}_1'|} \hat{f}^{(1)}(\mathbf{x}, \mathbf{z}_1'') \, d\mathbf{z}_1''\right) f^{(1)}(\mathbf{z}_1') \, d\mathbf{z}_1'. \tag{21}$$

If we take $f^{(1)}(\mathbf{z}_1') = \alpha_p(\mathbf{z}_1')/\frac{4}{3}\pi a^3$, and assume that $\alpha_p(\mathbf{z}_1')$ is constant, we see that $\hat{f}^{(1)}$ is a function only of $r_1 = |\mathbf{x} - \mathbf{z}_1'|$. Then, noting that if we replace $d\mathbf{z}_1'$ by $4\pi(r_1')^2 \, dr_1'$, the left side of (21) is the differential of the integral on the right. Then we have

$$\frac{d}{dr_1} \ln\left(1 - 4\pi \int_0^{r_1} \hat{f}^{(1)}(r_1')(r_1')^2 \, dr_1'\right) = -4\pi f^{(1)} r_1^2.$$

Thus,

$$\hat{f}^{(1)}(\mathbf{x}, \mathbf{z}_1') = \exp\left(-\frac{4}{3}\pi f^{(1)}(\mathbf{z}_1')|\mathbf{x} - \mathbf{z}_1'|^3\right) f^{(1)}(\mathbf{z}_1'). \tag{22}$$

In terms of α, we have

$$\hat{f}^{(1)}(\mathbf{x}, \mathbf{z}_1') = \frac{\alpha_p}{\frac{4}{3}\pi a^3} \exp\left[-\alpha_p \left(\frac{|\mathbf{x} - \mathbf{z}_1'|}{a}\right)^3\right]. \tag{23}$$

A similar derivation can be carried out for the two-particle distribution. This distribution is useful for proving convergence results for averages of one-particle distributions.

The distribution function for two nearest particles, $\hat{f}^{(2)}(\mathbf{x}, \mathbf{z}_1', \mathbf{z}_2')$ vanishes for $|\mathbf{x} - \mathbf{z}_2'| \leq |\mathbf{x} - \mathbf{z}_1'|$. For $|\mathbf{x} - \mathbf{z}_2'| > |\mathbf{x} - \mathbf{z}_1'|$, $\hat{f}^{(2)}(\mathbf{x}, \mathbf{z}_1', \mathbf{z}_2') \, d\mathbf{z}_1' \, d\mathbf{z}_2'$ is the probability that the sphere having its center within $d\mathbf{z}_1'$ of $\mathbf{z}_1'$ is the closest to the point $\mathbf{x}$ *and* the sphere within $d\mathbf{z}_2'$ of $\mathbf{z}_2'$ is the next closest to the point $\mathbf{x}$. This can be calculated in terms of the conditional probability that the next nearest sphere to $\mathbf{x}$ is within $d\mathbf{z}_2'$ of $\mathbf{z}_2'$, given that there is a sphere at $\mathbf{z}_1'$. This, in turn, can be calculated in a manner similar to the nearest sphere density above. It is clear that the conditional probability,

$$\hat{f}^{(2)}(\mathbf{x}, \mathbf{z}'_2|\mathbf{z}'_1) \, d\mathbf{z}'_2,$$

satisfies

$$\hat{f}^{(2)}(\mathbf{x}, \mathbf{z}'_2|\mathbf{z}'_1) \, d\mathbf{z}'_2 = 0, \tag{24}$$

for $|\mathbf{x} - \mathbf{z}'_2| < |\mathbf{x} - \mathbf{z}'_1|$. For $|\mathbf{x} - \mathbf{z}'_2| > |\mathbf{x} - \mathbf{z}'_1|$ we have

$$\hat{f}^{(2)}(\mathbf{x}, \mathbf{z}'_2|\mathbf{z}'_1) \, d\mathbf{z}'_2 = \left(1 - \int_{|\mathbf{x}-\mathbf{z}'_1| \leq |\mathbf{x}-\mathbf{z}''_2| \leq |\mathbf{x}-\mathbf{z}'_2|} \hat{f}^{(2)}(\mathbf{x}, \mathbf{z}''_2|\mathbf{z}'_1) \, d\mathbf{z}''_2\right) f^{(1)}(\mathbf{z}'_2) \, d\mathbf{z}'_2. \tag{25}$$

If we then assume that $\alpha_p(\mathbf{z}'_1)$ is constant, we see that $\hat{f}^{(2)}$ is a function only of $r_2 = |\mathbf{x} - \mathbf{z}'_2|$. Then,

$$\hat{f}^{(2)}(r_2|\mathbf{z}'_1) = \left(1 - 4\pi \int_{r_1}^{r_2} \hat{f}^{(2)}(r'_2|\mathbf{z}'_1)(r'_2)^2 \, dr'_2\right) f^{(1)}. \tag{26}$$

Solving for $\hat{f}^{(2)}$ yields

$$\hat{f}^{(2)}(\mathbf{x}, \mathbf{z}'_2|\mathbf{z}'_1) = \frac{\alpha_p}{\frac{4}{3}\pi a^3} \exp\left\{\alpha_p\left[\left(\frac{|\mathbf{x}-\mathbf{z}'_1|}{a}\right)^3 - \left(\frac{|\mathbf{x}-\mathbf{z}'_2|}{a}\right)^3\right]\right\}. \tag{27}$$

The joint probability is then given by

$$\hat{f}^{(2)}(\mathbf{x}, \mathbf{z}'_1|\mathbf{z}'_2) = \begin{cases} 0 & \text{if } |\mathbf{x} - \mathbf{z}'_2| < |\mathbf{x} - \mathbf{z}'_1|, \\ \left(\frac{\alpha_p}{\frac{4}{3}\pi a^3}\right)^2 \exp\left[-\alpha_p\left(\frac{|\mathbf{x}-\mathbf{z}'_2|}{a}\right)^3\right] & \text{otherwise.} \end{cases} \tag{28}$$

Approximate Probability Densities

The concept of volume fraction during conditional averaging is interesting. Consider

$$\alpha_p^{(1)}(\mathbf{x}, t|\mathbf{z})$$
$$= \overline{X_p(\mathbf{x}, t|\mathbf{z}, \mathbf{z}_2, \ldots, \mathbf{z}_N)}^{(1)},$$
$$= \int \cdots \int X_p(\mathbf{x}, t|\mathbf{z}, \mathbf{z}_2, \ldots, \mathbf{z}_N) f^{(N)}(\mathbf{z}, \mathbf{z}_2, \ldots, \mathbf{z}_N) \, d\mathbf{z}_2 \ldots d\mathbf{z}_N. \tag{29}$$

Also, we have

$$\alpha_p^{(2)}(\mathbf{x}, t|\mathbf{z}, \mathbf{z}_2)$$
$$= \overline{X_p(\mathbf{x}, t|\mathbf{z}, \mathbf{z}_2, \ldots, \mathbf{z}_N)}^{(2)},$$
$$= \int \cdots \int X_p(\mathbf{x}, t|\mathbf{z}, \mathbf{z}_2, \ldots, \mathbf{z}_N) f^{(N)}(\mathbf{z}, \mathbf{z}_2, \ldots, \mathbf{z}_N) \, d\mathbf{z}_3 \ldots d\mathbf{z}_N. \tag{30}$$

Note that if $|\mathbf{z} - \mathbf{z}_2| < a$, then $f^{(N)} = 0$ in the first integral. Therefore,

$$\alpha_p^{(2)}(\mathbf{x}, t|\mathbf{z}, \mathbf{z}_2) = 0, \tag{31}$$

there. Also, if we assume that the number density of spheres is given by

$$f^{(2)}(\mathbf{z}, \mathbf{z}_2, t) = \begin{cases} 0 & \text{for } |\mathbf{z} - \mathbf{z}_2| < 2a, \\ \overline{f} = \text{constant} & \text{for } |\mathbf{z} - \mathbf{z}_2| \geq 2a, \end{cases} \tag{32}$$

then we can compute $\alpha_p^{(1)}(\mathbf{x}, t|\mathbf{z})$ by noting that for $|\mathbf{x} - \mathbf{z}| < a$, $\alpha_p^{(1)}(\mathbf{x}, t|\mathbf{z}) = 1$, while for $|\mathbf{x} - \mathbf{z}| > 3a$, $\alpha_p^{(1)}(\mathbf{x}, t, |\mathbf{z}) = 4\pi a^3 \overline{f}/3$, and for $a < |\mathbf{x} - \mathbf{z}| < 3a$, the particle volume fraction can be computed by considering the volume where the center of the sphere could lie, given that $|\mathbf{x} - \mathbf{z}| < 3a$. That volume, divided by $4\pi a^3/3$, multiplied by the number density $\overline{f}$, is the volume fraction. Let $|\mathbf{x} - \mathbf{z}| = x'$. The volume is given by

$$V(x') = \int_{2a-x'}^{a} \int_{0}^{\theta(x')} \int_{0}^{2\pi} r^2 \sin\theta \, dr d\theta \, d\phi$$

$$= 2\pi \int_{2a-x}^{a} r^2 (1 - \cos\theta(x', r)) \, dr, \tag{33}$$

where $\theta(x', r)$ is the angle shown in Figure 10.1. Then $\theta(x', r)$ is given by

$$\cos(\pi - \theta) = -\cos\theta = \frac{(2a)^2 - (x')^2 - r^2}{2rx'}. \tag{34}$$

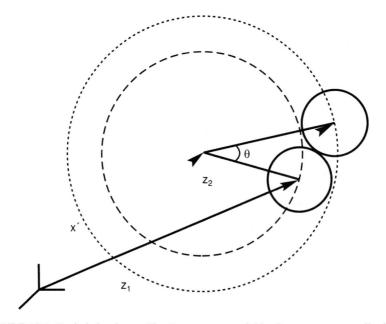

FIGURE 10.1. Excluded volume: The "next nearest-neighbor" cannot occupy all of the sphere at the radius shown.

10.1. Multiparticle Distribution Functions

Thus $V(x')$ is given by

$$V(x') = 2\pi \left(\frac{12a^4 + \cdots}{2x'} \right), \tag{35}$$

and, consequently, the volume fraction is given by

$$\alpha_p^{(1)} = \frac{3(12a^4 + \cdots)}{4a^3 x'} \overline{f}. \tag{36}$$

Note that when $x' = a$, $\alpha_p^{(1)} = 0$, and when $x' = 3a$,

$$\alpha_p^{(1)} = \frac{4\pi a^3 \overline{f}}{3}. \tag{37}$$

Convergence Results

The averages discussed above are used to derive results about constitutive equations from information about exact flows. There are difficulties regarding this procedure, however.

To see this, consider a function

$$g_1(\mathbf{x}, t | \mathbf{z}'_1, \ldots, \mathbf{z}'_N) = g(\mathbf{x}, t | \mathbf{z}'_1) + O\left(\frac{1}{|\mathbf{x} - \mathbf{z}'_2|^\nu} \right),$$

where g_1 is of order 1. We note that the ordinary (i.e., unordered) expression for the probability density leads to averages that have convergence problems in unbounded flows. The average of g is given by

$$\overline{g}(\mathbf{x}, t) = \int_{\mathbf{z}_1} \cdots \int_{\mathbf{z}_N} f^{(N)}(\mathbf{x}, \mathbf{z}_1, \ldots, \mathbf{z}_N) g(\mathbf{x}, t | \mathbf{z}_1, \ldots, \mathbf{z}_N) \, d\mathbf{z}_1 \ldots d\mathbf{z}_N,$$

$$= \int_{\mathbf{z}_1} f^{(1)}(\mathbf{z}_1) g_1(\mathbf{x}, t | \mathbf{z}_1) \, d\mathbf{z}_1 + E, \tag{38}$$

where the error E is due to all the spheres except the one at $\mathbf{z}_1$. If $f^{(1)}$ is constant, the integral will not converge, nor will it converge if the distribution does not decrease sufficiently rapidly.

In terms of the nearest-neighbor distributions, the average of g is given by

$$\hat{\overline{g}}(\mathbf{x}, t) = \int_{\mathbf{z}'_1} \cdots \int_{\mathbf{z}'_N} \hat{f}^{(N)}(\mathbf{x}, \mathbf{z}'_1, \ldots, \mathbf{z}'_N) g(\mathbf{x}, t | \mathbf{z}'_1, \ldots, \mathbf{z}'_N) \, d\mathbf{z}'_1 \ldots d\mathbf{z}'_N,$$

$$= \int_{\mathbf{z}'_1} \hat{f}^{(1)}(\mathbf{x}, \mathbf{z}'_1) g_1(\mathbf{x}, t | \mathbf{z}'_1) \, d\mathbf{z}'_1 + \hat{E}, \tag{39}$$

where $\hat{E}$ is an error, which is of order

$$\int_{\mathbf{z}'_1} \int_{\mathbf{z}'_2} \hat{f}^{(2)}(\mathbf{x}, \mathbf{z}'_1, \mathbf{z}'_2) \left(\frac{1}{|\mathbf{x} - \mathbf{z}'_2|^\nu} \right) d\mathbf{z}'_1 \, d\mathbf{z}'_2.$$

Substituting $\hat{f}^{(2)}$ from (28) shows that the error is of order

$$\int_0^\infty \int_{r_1}^\infty (f^{(1)})^2 \exp(-\frac{4}{3}\pi f^{(1)} r_2^3) 4\pi r_1^3 4\pi r_2^{3-\nu} \, dr_1 \, dr_2 . \tag{40}$$

If we make the change of variables $r_1' = (f^{(1)})^{1/3} r_1$ and $r_2' = (f^{(1)})^{1/3} r_2$, it is apparent that $\hat{E} = O\left((f^{(1)})^{\nu/3}\right)$. Thus, the one-sphere calculation can give results that are accurate for dilute flows, where $f^{(1)}$ is small.

10.2. Time Averaging

Let us now define the time average. Consider the elementary Borel sets

$$\overline{\mathcal{E}}(t, F) = \{\mu \mid f(\mathbf{x}, t; \mu) \le F\} , \tag{41}$$

and

$$\overline{\mathcal{E}}(t - \tau, F) = \{\mu \mid f(\mathbf{x}, t - \tau; \mu) \le F\} . \tag{42}$$

Assume that, if $\mu \in \overline{\mathcal{E}}(t, F)$, then the field defined by shifting the time origin yields another process in the ensemble. Thus, if $f(\mathbf{x}, t; \mu)$ is such a process, then μ_τ defined by $f(\mathbf{x}, t; \mu_\tau) = f(\mathbf{x}, t - \tau; \mu)$ is a process in $\overline{\mathcal{E}}(t - \tau, F)$. Thus, $\overline{\mathcal{E}}(t, F) \subseteq \overline{\mathcal{E}}(t - \tau, F)$. Moreover, it is also clear that $\overline{\mathcal{E}}(t - \tau, F) \subseteq \overline{\mathcal{E}}(t, F)$. Thus, $\overline{\mathcal{E}}(t - \tau, F) = \overline{\mathcal{E}}(t, F)$. In this case, Borel sets, and hence their measures, are independent of time.

The average of f is defined by

$$\overline{f}(\mathbf{x}, t) = \lim_{N \to \infty} \sum_{i=1}^{N-1} F_i m(\overline{\mathcal{E}}(\mathbf{x}, t; F_i, dF_i)) . \tag{43}$$

Since $m(\overline{\mathcal{E}}(\mathbf{x}, t; F_i, dF_i))$ is independent of t, the average of $\overline{f}$ is independent of t.

We shall compute an approximation to $\overline{f}$ using the law of large numbers to approximate $m(\overline{\mathcal{E}}(\mathbf{x}, F, dF))$. That is, we wish to use the ideas of sampling and the law of large numbers to estimate the probability of obtaining a realization that results in F being in the interval $F \le f(\mathbf{x}, t, \mu) \le F + dF$. Consider the time history of one realization of f, as shown in Figure 10.2. Given F and dF, we can find time intervals $t_1^-(F, dF) \le t \le t_1^+(F, dF)$, $t_2^-(F, dF) \le t \le t_2^+(F, dF)$, $t_M^-(F, dF) \le t \le t_M^+(F, dF)$, such that for $t_m^-(F, dF) \le t_M^+(F, dF)$, $F \le f(\mathbf{x}, t; \mu) \le F + dF$. Since the macroscopic process is time-independent, we can sample by choosing random times τ in some interval $t - T \le \tau \le t$ that is sufficiently large that the law of large numbers is valid. Note that we need not sample uniformly; however, there is no reason at this point not to. If the sample distribution is uniform, then the probability of sampling from the set $m(\overline{\mathcal{E}}(\mathbf{x}, F, dF))$ is

$$\frac{1}{T} \Delta_m t(F, dF) = \frac{1}{T} [t_m^+(F, dF) - t_m^-(F, dF)] . \tag{44}$$

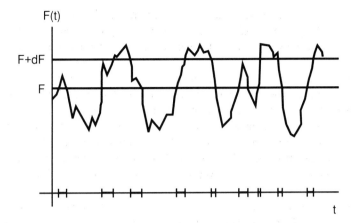

FIGURE 10.2. Time history, showing intervals when $F < f < F + dF$.

Then, the law of large numbers implies that

$$m(\bar{\mathcal{E}}(\mathbf{x}, F, dF)) \approx \sum_{m=1}^{M} \frac{1}{T} \Delta_m t(F, dF). \qquad (45)$$

Then the average of f is given by

$$\bar{f}(\mathbf{x}, t) = \lim_{N \to \infty} \sum_{i=1}^{N-1} F_i m(\bar{\mathcal{E}}(\mathbf{x}, t; F_i, dF_i))$$

$$= \lim_{N \to \infty} \sum_{i=1}^{N-1} F_i \sum_{m=1}^{M} \frac{1}{T} \Delta_m t(F_i, dF_i). \qquad (46)$$

Since F_i is either $f(\mathbf{x}, t_i^{\pm}, \mu)$, the double sum in the last line of (46) converges to the time integral

$$\lim_{N \to \infty} \sum_{i=1}^{N-1} F_i \sum_{m=1}^{M} \frac{1}{T} \Delta_m t(F_i, dF_i)$$

$$= \frac{1}{T} \lim_{N \to \infty} \sum_{i=1}^{N-1} \sum_{m=1}^{M} f(\mathbf{x}, t_m^*; \mu) \Delta_m t(F_i, dF_i)$$

$$= \frac{1}{T} \int_{t-T}^{t} f(\mathbf{x}, \tau; \mu) d\tau . \qquad (47)$$

Thus,

$$\bar{f}^T = \frac{1}{T} \int_{t-T}^{t} f(\mathbf{x}, \tau; \mu) d\tau \qquad (48)$$

is the *time average* of f. It has the following interpretation. From one realization of a stationary macroscopic process, we can obtain an infinite number of other realizations by shifting the time origin by an arbitrary amount τ. Thus, from one realization, we are able to construct an infinite number of realizations. Then, roughly speaking, the probability of observing f is proportional to the time that f spends near each value, so that the weighting in the summation, or integration, is $d\tau/T$.

From a practical point of view, T must be sufficiently large that a large number of independent samples of f can be achieved during the interval. The very special case where $f = \overline{f}$ is constant requires that only the interval from $F = \overline{f} - \epsilon$ to $F + dF = \overline{f} + \epsilon$ be sampled many times; any positive value of T will suffice in that case. If f varies about its mean value, the interval must be long enough to sample all of the different values of f sufficiently often. Thus, for a given value of F, we assume that during the averaging time T, $f(\mathbf{x}, t; \mu) = F$ must occur for many values of t.

The use of the time average has the advantage that we need not sample the ensemble repeatedly but need make only one detailed observation to infer averaged values. This can greatly relieve the burden on an experimenter in this area.

10.3. Volume Averaging

Next, we define volume averaging. Consider the Borel sets $\overline{\mathcal{E}}(\mathbf{x}, t; F, dF)$, which are independent of the spatial variable $\mathbf{x}$. If $\mu \in \overline{\mathcal{E}}(t, F, dF)$, then we can sample other realizations μ_η represented by

$$f(\mathbf{x}, t; \mu_\eta) = f(\mathbf{x} + \eta, t; \mu). \tag{49}$$

Again, a result similar to the previous section shows that we may approximate $\overline{f}$ by integrating over the subensemble represented by μ_η for $\eta \in V$, where V is some sufficiently large volume. With uniform weighting, this gives the *volume average*

$$\overline{f}^V = \frac{1}{V} \int_V f(\mathbf{x} + \eta, t; \mu) \, d\mathbf{x}_\eta. \tag{50}$$

As with the time average, the volume average must sample volumes that contain sufficient variations in f to use the law of large numbers.

In what follows, we outline the derivation of the Gauss and Leibniz rules for time and volume averaging. These results are used in a similar fashion to those for ensemble averaging, but the derivations are more difficult because instead of averaging over a variable that keeps track of the variations in the particle positions, etc., these averaging processes average over an independent variable. The Gauss and Leibniz rules address commutativity of derivatives with respect to those variables with the averaging operator. That is, the variable that is assumed to sample the distribution is either time (for time averaging), or one or more of the spatial variables for space averaging. The variables of interest in mechanics, including the velocity, pressure, etc., are variables for which the equations of motion relate space and time derivatives. Treating either of these variables as one which specifies the random fluctuations in them complicates the mathematical process of averaging.

10.3. Volume Averaging

10.3.1. Gauss and Leibniz Rules for Time and Volume Averaging

The Gauss and Leibniz rules (36) and (37) for ensemble averaging have counterparts for time and volume averaging. We present these results in the following subsections.

Results for Time Averaging

Let $f(t)$ be piecewise continuous with jump discontinuities at $t_1, t_2, \ldots$. Suppose further that for $t_1 < t < t_2, t_3 < t < t_4, \ldots, t_{2n-1} < t < t_{2n}$, the spatial point in question occupies component k. Then

$$\overline{X_k f(t)}^T = \frac{1}{T} \int_{t-T}^{t} X_k f(t)\, dt$$

$$= \frac{1}{T} \left(\int_{t-T}^{t_2} f(t)\, dt + \int_{t_3}^{t_4} f(t)\, dt + \cdots + \int_{t_{2n-1}}^{t} f(t)\, dt \right), \quad (51)$$

where, for convenience, we have taken $t_1 < t - T < t_2$ and $t_{2n-1} < t < t_{2n}$. Then[1]

$$\frac{\partial \overline{X_k f}^T}{\partial t} = f(t) - f(t-T). \tag{53}$$

Next, we note that

$$\overline{X_k \frac{\partial f}{\partial t}}^T = \frac{1}{T} \int_{t-T}^{t} X_k \frac{\partial f}{\partial t}\, dt$$

$$= \frac{1}{T} \left(\int_{t-T}^{t_2} \frac{\partial f}{\partial t}\, dt + \int_{t_3}^{t_4} \frac{\partial f}{\partial t}\, dt + \cdots + \int_{t_{2n-1}}^{t} \frac{\partial f}{\partial t}\, dt \right)$$

$$= \frac{1}{T} \left(f(t) - f(t_{2n-1}^-) + \cdots - f(t_3^-) + f(t_2^+) - f(t-T) \right). \tag{54}$$

Comparing (53) and (54), we see that

$$\frac{\partial \overline{X_k f}^T}{\partial t} = \overline{X_k \frac{\partial f}{\partial t}}^T - \frac{1}{T} \left(-f(t_{2n-1}^-) + \cdots + f(t_4^+) - f(t_3^-) + f(t_2^+) \right). \tag{55}$$

Note that $f(t_j^+)$ and $f(t_j^-)$ are the values of $f(t)$ when the interface is at the averaging point. Introduce $\mathbf{n}_k \cdot \mathbf{v}_i / |\mathbf{n}_k \cdot \mathbf{v}_i|$, where $\mathbf{n}_k$ is the unit normal out of component k at the interface. Then it is evident that $\mathbf{n}_k \cdot \mathbf{v}_i / |\mathbf{n}_k \cdot \mathbf{v}_i|$ is positive when component k arrives at the averaging point, and negative when it leaves there.

---

[1] The other cases can be represented by

$$\frac{\partial \overline{X_k f}^T}{\partial t} = X_k(t) f(t) - X_k(t-T) f(t-T). \tag{52}$$

Thus,

$$\overline{\frac{\partial X_k f}{\partial t}}^T = X_k \overline{\frac{\partial f}{\partial t}}^T - \left(\sum_{t_j \in (t-T,t)} \frac{1}{T|\mathbf{n}_k \cdot \mathbf{v}_i|} \mathbf{n}_k \cdot \mathbf{v}_i f(t_j^k) \right), \tag{56}$$

where $f(t_j^k)$ denotes the value of f on the component k side of the interface. Ishii [45] writes $T|\mathbf{n}_k \cdot \mathbf{v}_i| = l_j$, and states that

$$\sum_{t_j \in (t-T,t)} \frac{1}{l_j} = A \tag{57}$$

is the interfacial area density. Equation (56) is the analog of (37) for time averaging, and is sometimes referred to as the Leibniz rule for time averaging.

The analogous result for spatial gradients is complicated by the fact that t_j depends on $\mathbf{x}$. To see this, note that if the point $\mathbf{x}$ is moved a bit in space, the time at which the interface arrives is also changed.

We shall need a relation for ∇t_j. To get this, consider the form for the interface position as given by $F(\mathbf{x}, t) = 0$. Differentiating this with respect to t gives the standard result $\mathbf{v}_i \cdot \nabla F + \partial F/\partial t = 0$. Taking the gradient gives $\nabla F + \partial F/\partial t \nabla t_j = 0$. Solving for ∇t_j gives

$$\nabla t_j = -\frac{\nabla F}{\partial F/\partial t} = \frac{\nabla F}{\mathbf{v}_i \cdot \nabla F} = \frac{\mathbf{n}_k}{\mathbf{v}_i \cdot \mathbf{n}_k}. \tag{58}$$

Let us now compute the time average of the gradient of f. We have

$$\overline{X_k \nabla f(t)}^T = \frac{1}{T} \left(\int_{t-T}^{t_2} \nabla f(t)\,dt + \int_{t_3}^{t_4} \nabla f(t)\,dt + \cdots + \int_{t_{2n-1}}^{t} \nabla f(t)\,dt \right). \tag{59}$$

The gradient of the time average of $X_k f(t)$ is

$$\nabla \overline{X_k f(t)}^T = \nabla \left[\frac{1}{T} \left(\int_{t-T}^{t_2} f(t)\,dt + \int_{t_3}^{t_4} f(t)\,dt + \cdots + \int_{t_{2n-1}}^{t} f(t)\,dt \right) \right]$$

$$= \frac{1}{T} \left(\int_{t-T}^{t_2} \nabla f(t)\,dt + \int_{t_3}^{t_4} \nabla f(t)\,dt + \cdots + \int_{t_{2n-1}}^{t} \nabla f(t)\,dt \right)$$

$$+ \frac{1}{T} \left(\nabla t_2 f(t_2^-) - \nabla t_3 f(t_3^+) \right.$$

$$\left. + \nabla t_4 f(t_4^-) + \cdots - \nabla t_{2n-1} f(t_{2n-1}^-) \right). \tag{60}$$

Using (59) we have

$$\nabla \overline{X_k f(t)}^T = \overline{\nabla f(t)}^T + \frac{1}{T} \left(\sum_{t_j \in (t-T,t)} \frac{\mathbf{n}_k}{l_j} f(t_j^k) \right). \tag{61}$$

This is the analog of (36) and is called the Gauss rule for time averaging.

10.3. Volume Averaging

Results for Volume Averaging

Now let us consider the volume average of $X_k f$. We have

$$\overline{X_k f}^V(\mathbf{x}, t) = \frac{1}{V} \int_V X_k f(\mathbf{x} + \boldsymbol{\eta}, t) \, d\mathbf{x}_\eta = \frac{1}{V} \int_{V_k(\mathbf{x},t)} f(\mathbf{x}', t) \, d\mathbf{x}'. \tag{62}$$

The volume average of $X_k \nabla f$ is

$$\overline{X_k \nabla f}^V = \frac{1}{V} \int_V X_k \nabla f(\mathbf{x} + \boldsymbol{\eta}, t) \, d\mathbf{x}_\eta$$

$$= \frac{1}{V} \int_{V_k(\mathbf{x},t)} \nabla f(\mathbf{x}', t) \, d\mathbf{x}'$$

$$= \frac{1}{V} \int_{S_k(\mathbf{x},t)} \mathbf{n} f(\mathbf{x}', t) \, dS' + \frac{1}{V} \int_{S_i(\mathbf{x},t)} \mathbf{n}_k f_{ki}(\mathbf{x}', t) \, ds', \tag{63}$$

where $S_k(\mathbf{x}, t)$ is the part of the boundary of V that is inside of component k, and $S_i(\mathbf{x}, t)$ is the interface inside V. If we note that

$$\nabla \overline{X_k f}^V = \frac{1}{V} \nabla \int_V X_k f(\mathbf{x} + \boldsymbol{\eta}, t) \, d\mathbf{x}_\eta$$

$$= \frac{1}{V} \nabla \int_{V_k(\mathbf{x},t)} f(\mathbf{x}', t) \, d\mathbf{x}'$$

$$= \frac{1}{V} \int_{S_k(\mathbf{x},t)} \mathbf{n} f(\mathbf{x}', t) \, ds', \tag{64}$$

then we have

$$\overline{X_k \nabla f}^V = \nabla \overline{X_k f}^V + \frac{1}{V} \int_{S_i(\mathbf{x},t)} \mathbf{n}_k f_{ki}(\mathbf{x}', t) \, ds'. \tag{65}$$

This the Gauss rule for the volume average.

For the time derivative, consider

$$\frac{\partial}{\partial t} \overline{X_k f}^V = \frac{1}{V} \frac{\partial}{\partial t} \int_V X_k f(\mathbf{x} + \boldsymbol{\eta}, t) \, d\mathbf{x}_\eta$$

$$= \frac{1}{V} \frac{\partial}{\partial t} \int_{V_k(\mathbf{x},t)} f(\mathbf{x}', t) \, d\mathbf{x}'$$

$$= \frac{1}{V} \int_{V_k(\mathbf{x},t)} \frac{\partial}{\partial t} f(\mathbf{x}', t) \, d\mathbf{x}' + \frac{1}{V} \int_{S_i(\mathbf{x},t)} \mathbf{n}_k \cdot \mathbf{v}_i f(\mathbf{x}', t) \, ds', \tag{66}$$

where the Reynolds Transport Theorem [2] is used to take the time derivative inside the integral over a moving volume. Noting that

$$\frac{1}{V} \int_{V_k(\mathbf{x},t)} \frac{\partial}{\partial t} f(\mathbf{x}', t) \, d\mathbf{x}' = \overline{X_k \frac{\partial f}{\partial t}}^V, \tag{67}$$

we have

$$\frac{\partial}{\partial t} \overline{X_k f}^V = \overline{X_k \frac{\partial f}{\partial t}}^V + \int_{S_i(\mathbf{x},t)} \mathbf{n}_k \cdot \mathbf{v}_i f(\mathbf{x}', t) \, ds'. \tag{68}$$

This is the Leibniz rule for the volume average.

10.4. Desiderata

The statistical averages discussed above give a way to understand and compute actual averages. We shall use some of these ideas in the chapter on the derivation of constitutive equations. Time and volume averages have the advantage that if we desire to do some experiment to determine some of the averaged variables, the experiment can be done once and, in theory, the detailed measurements can then be used to get the averaged variables.

It has become customary in the literature to use the time and volume averages even when the macroscopic processes are not stationary or statistically uniform. The status of such practices is questionable, and can be justified only when the results are the same as those obtained for ensemble averages. Some researchers [45], [81], [98] derive the averaged equations using time or volume averages. We note that in nonstationary or nonuniform flows, the time and volume averages are at best approximations to the ensemble average. Therefore the ensemble average is fundamental, and specific "averages," such as time and space averages, are appropriate only to approximate the ensemble average in special situations. Although it is convenient that the direct application of space or time averages to the exact equations yields equations of motion which are formally the same as those we shall derive here, it seems inadvisable to attempt a more fundamental interpretation of them.

11
Averaged Equations

11.1. Averaging Balance Equations

Averages of balance equations are obtained by taking the product of the balance equations with X_k, then performing the averaging process. We have:

$$\frac{\partial \overline{X_k \rho \Psi}}{\partial t} + \nabla \cdot \overline{X_k \rho \Psi \mathbf{v}} - \nabla \cdot \overline{X_k \mathbf{J}} - \overline{X_k \rho f} = \overline{\rho \Psi \left(\frac{\partial X_k}{\partial t} + \mathbf{v} \cdot \nabla X_k \right)} - \overline{\mathbf{J} \cdot \nabla X_k}. \tag{1}$$

Subtracting the average of $\rho \Psi$ with the result in (9.1.28) reduces the right-hand side of (1) to

$$\overline{\{\rho \Psi (\mathbf{v} - \mathbf{v}_i) - \mathbf{J}\} \cdot \nabla X_k}. \tag{2}$$

This is the interfacial source of Ψ. It is due to phase change, $(\mathbf{v} - \mathbf{v}_i) \cdot \mathbf{n} \neq 0$, and to the flux $\mathbf{J}$.

Using $\Psi = 1$, $\mathbf{J} = 0$, and $f = 0$ results in the equation of conservation of mass; using $\Psi = \mathbf{v}$, $\mathbf{J} = \mathbf{T}$, and $f = \mathbf{g}$ results in the equation of conservation of momentum; and using $\Psi = e = u + \frac{1}{2}v^2$, $\mathbf{J} = \mathbf{T} \cdot \mathbf{v} - \mathbf{q}$, and $f = \mathbf{g} \cdot \mathbf{v} + r$ gives the equation of conservation of energy. The entropy inequality for component k can be obtained by taking $\Psi = s$, $\mathbf{J} = -\mathbf{q}/\theta$, and $f = r/\theta$, and by changing the equality in (1) to a positive semidefinite (≥ 0) relationship.

The averaged equations are:

Mass

$$\frac{\partial \overline{X_k \rho}}{\partial t} + \nabla \cdot \overline{X_k \rho \mathbf{v}} = \overline{\rho (\mathbf{v} - \mathbf{v}_i) \cdot \nabla X_k}; \tag{3}$$

Momentum

$$\frac{\partial \overline{X_k \rho \mathbf{v}}}{\partial t} + \nabla \cdot \overline{X_k \rho \mathbf{v} \mathbf{v}} = \nabla \cdot \overline{X_k \mathbf{T}} + \overline{X_k \rho \mathbf{g}} + \overline{(\rho \mathbf{v}(\mathbf{v} - \mathbf{v}_i) - \mathbf{T}) \cdot \nabla X_k}; \quad (4)$$

Energy

$$\frac{\partial \overline{X_k \rho (u + \frac{1}{2} v^2)}}{\partial t} + \nabla \cdot \overline{X_k \rho (u + \frac{1}{2} v^2) \mathbf{v}} = \nabla \cdot \overline{X_k (\mathbf{T} \cdot \mathbf{v} - \mathbf{q})}$$

$$+ \overline{X_k \rho (\mathbf{g} \cdot \mathbf{v} + r)} + \overline{[\rho (u + \frac{1}{2} v^2)(\mathbf{v} - \mathbf{v}_i) - (\mathbf{T} \cdot \mathbf{v} - \mathbf{q})] \cdot \nabla X_k}; \quad (5)$$

Entropy

$$\frac{\partial \overline{X_k \rho s}}{\partial t} + \nabla \cdot \overline{X_k \rho s \mathbf{v}} + \nabla \cdot \overline{X_k \frac{\mathbf{q}}{T}} - \overline{X_k \rho \frac{r}{T}} - \overline{\left(\rho s (\mathbf{v} - \mathbf{v}_i) - \frac{\mathbf{q}}{T}\right) \cdot \nabla X_k} \geq 0. \quad (6)$$

Instead of using the equation for the total energy, it is equally valid to use the equation for the internal energy. This gives the equation of balance of

Internal Energy

$$\frac{\partial \overline{X_k \rho u}}{\partial t} + \nabla \cdot \overline{X_k \rho u \mathbf{v}} = -\nabla \cdot \overline{X_k \mathbf{q}} + \overline{X_k \mathbf{T} : \nabla \mathbf{v}}$$

$$+ \overline{X_k \rho r} + \overline{[\rho u (\mathbf{v} - \mathbf{v}_i) + \mathbf{q}] \cdot \nabla X_k}. \quad (7)$$

11.2. Definition of Average Variables

In this section, we define the appropriate average variables describing multicomponent mechanics.

The exact, or microscopic, situation is defined in terms of the component function X_k. The average of X_k is the average fraction of the occurrences of component k at point $\mathbf{x}$ at time t

$$\alpha_k = \overline{X_k}. \quad (8)$$

It has become customary to call this variable the volume fraction, even though the precise volume fraction is the ratio of the volume of component k in a small region to the total volume of that region. We note that this concept is intimately tied to volume averaging. Thus it is exact for spatially homogeneous situations. As discussed in Chapter 1, the correct interpretation in terms of ensembles is that α_k is the expected value of the ratio of the volume of component k to the total volume, in the limit as the volume approaches zero.

The other geometric variable of use here is the interfacial area density of component k. This is defined by

$$A_k = -\overline{\mathbf{n}_k \cdot \nabla X_k}, \quad (9)$$

11.2. Definition of Average Variables

where $\mathbf{n}_k$ is the unit external normal to component k. Again, this variable is the expected value of the ratio of the interfacial area in a small volume to the volume, in the limit as that volume approaches zero.

All the remaining variables are defined in terms of weighted averages. The main, or "component" variables are either component-weighted variables, that is, weighted with the component function X_k or mass-weighted (or Favré) averaged, that is, weighted by $X_k\rho$. Other variables are weighted with the interface variable ∇X_k.

The "conserved" variables are the density

$$\overline{\rho}_k^x = \frac{\overline{X_k\rho}}{\overline{\alpha}_k}, \tag{10}$$

the velocity

$$\overline{\mathbf{v}}_k^{x\rho} = \frac{\overline{X_k\rho\mathbf{v}}}{\overline{\alpha}_k\overline{\rho}_k^x}, \tag{11}$$

the internal energy

$$\overline{u}_k^{x\rho} = \frac{\overline{X_k\rho u}}{\overline{\alpha}_k\overline{\rho}_k^x}, \tag{12}$$

and the entropy

$$\overline{s}_k^{x\rho} = \frac{\overline{X_k\rho s}}{\overline{\alpha}_k\overline{\rho}_k^x}. \tag{13}$$

The definition of the Helmholtz free energy in terms of the internal energy, the temperature, and the entropy is

$$\psi = u - \theta s. \tag{14}$$

For ensemble averaged quantities, we define the Helmholtz free energy in such a way that the relation between the averaged quantities remains the same. Thus, we define the averaged Helmholtz free energy

$$\overline{\psi}_k^{x\rho} = \frac{\overline{X_k\rho\psi}}{\overline{\alpha}_k\overline{\rho}_k^x}. \tag{15}$$

The variables representing the averaged effects of the molecular fluxes are the stress

$$\overline{\mathbf{T}}_k^x = \frac{\overline{X_k\mathbf{T}}}{\overline{\alpha}_k}, \tag{16}$$

the energy flux

$$\overline{\mathbf{q}}_k^x = \frac{\overline{X_k\mathbf{q}}}{\overline{\alpha}_k}, \tag{17}$$

and the entropy flux

$$\overline{\phi}_k^x = \frac{\overline{X_k(\mathbf{q}/\theta)}}{\alpha_k}. \tag{18}$$

The average body sources are the body force

$$\overline{\mathbf{b}}_k^{x\rho} = \frac{\overline{X_k \rho \mathbf{b}}}{\alpha_k \overline{\rho}_k^x}, \tag{19}$$

the energy source

$$\overline{r}_k^{x\rho} = \frac{\overline{X_k \rho r}}{\alpha_k \overline{\rho}_k^x}, \tag{20}$$

and the entropy source

$$\overline{\Sigma}_k^{x\rho} = \frac{\overline{(X_k \rho r/\theta)}}{\alpha_k \overline{\rho}_k^x}. \tag{21}$$

The molecular fluxes to the interface act as sources of mass, momentum, energy, or entropy to the interface. These terms are important in the theory of multi-component flows since they represent the interactions between the materials. The interfacial momentum source is defined by

$$\mathbf{M}_k = -\overline{\mathbf{T} \cdot \nabla X_k}, \tag{22}$$

the interfacial heat source is defined by

$$E_k = \overline{\mathbf{q} \cdot \nabla X_k}, \tag{23}$$

the interfacial work is defined by

$$W_k = -\overline{\mathbf{T} \cdot \mathbf{v} \cdot \nabla X_k}, \tag{24}$$

and the interfacial entropy source is defined by

$$S_k = \overline{(\mathbf{q}/\theta) \cdot \nabla X_k}. \tag{25}$$

The motion of the interfaces gives rise to velocities that are not equal to their average values in general. The velocity fluctuations from the average values may be due to turbulence or to the motion in the components due to the motion of the interfaces. The effect of these velocity fluctuations, whatever their source, on a variable is accounted for by introducing its fluctuating field (denoted by the prime superscript), which is the difference between the complete field and the appropriate mean field. For example,

$$\mathbf{v}_k' = \mathbf{v} - \overline{\mathbf{v}}_k^{x\rho}, \tag{26}$$

$$p_k' = p - \overline{p}_k^x. \tag{27}$$

11.2. Definition of Average Variables

Then, manipulating the momentum flux, we have

$$\begin{aligned}\overline{X_k \rho \mathbf{v}\mathbf{v}} &= \overline{X_k \rho (\overline{\mathbf{v}}_k^{x\rho} + \mathbf{v}'_k)(\overline{\mathbf{v}}_k^{x\rho} + \mathbf{v}'_k)} \\ &= \overline{X_k \rho}\, \overline{\mathbf{v}}_k^{x\rho} \overline{\mathbf{v}}_k^{x\rho} + \overline{X_k \rho \mathbf{v}'_k \mathbf{v}'_k} \\ &= \alpha_k \overline{\rho}_k^x \overline{\mathbf{v}}_k^{x\rho} \overline{\mathbf{v}}_k^{x\rho} - \alpha_k \mathbf{T}_k^{Re}.\end{aligned} \quad (28)$$

The variables defined in this way are the Reynolds stress

$$\mathbf{T}_k^{Re} = -\frac{\overline{X_k \rho \mathbf{v}'_k \mathbf{v}'_k}}{\alpha_k}, \quad (29)$$

and the fluctuation (Reynolds) energy flux,

$$\mathbf{q}_k^{Re} = \hat{\mathbf{q}}_k^{Re} + \mathbf{q}_k^T + \mathbf{q}_k^K. \quad (30)$$

The components of the fluctuation energy flux are the fluctuation (Reynolds) kinetic energy flux

$$\mathbf{q}_k^K = \frac{\overline{X_k \rho \mathbf{v}'_k \tfrac{1}{2}(v'_k)^2}}{\alpha_k}, \quad (31)$$

the fluctuation (Reynolds) shear working

$$\mathbf{q}_k^T = -\frac{\overline{X_k \mathbf{T} \cdot \mathbf{v}'_k}}{\alpha_k}, \quad (32)$$

and the fluctuation (Reynolds) internal energy flux

$$\hat{\mathbf{q}}_k^{Re} = \frac{\overline{X_k \rho \mathbf{v}'_k u'_k}}{\alpha_k}. \quad (33)$$

There is an analogous term in the entropy inequality, the fluctuation (Reynolds) entropy flux

$$\overline{\phi}_k^{Re} = -\frac{\overline{X_k \rho s \mathbf{v}'_k}}{\alpha_k}. \quad (34)$$

One term appears in the equation for the averaged internal energy that has no counterpart in the other equations. It is the dissipation

$$\mathcal{D}_k = \frac{\overline{X_k \mathbf{T} : \nabla \mathbf{v}'_k}}{\alpha_k} \quad (35)$$

Another fluctuation term that appears is the fluctuation (Reynolds) kinetic energy

$$u_k^{Re} = \frac{\overline{X_k \rho v'^2_k}}{2\alpha_k \overline{\rho}_k^x}. \quad (36)$$

Note that

$$\operatorname{tr} \mathbf{T}_k^{Re} = -\frac{\overline{X_k \rho_k \mathbf{v}'_k \cdot \mathbf{v}'_k}}{\alpha_k} = -2 u_k^{Re}. \quad (37)$$

As discussed above, several terms appear representing the actions of the convective and molecular fluxes at the interface. The convective flux terms are the interfacial mass generation source

$$\Gamma_k = \overline{\rho(\mathbf{v} - \mathbf{v}_i) \cdot \nabla X_k}, \tag{38}$$

the interfacial momentum source

$$\mathbf{v}_{ki}^m \Gamma_k = \overline{\rho \mathbf{v}(\mathbf{v} - \mathbf{v}_i) \cdot \nabla X_k}, \tag{39}$$

the interfacial internal energy source

$$\overline{u}_{ki} \Gamma_k = \overline{\rho u (\mathbf{v} - \mathbf{v}_i) \cdot \nabla X_k}, \tag{40}$$

the interfacial kinetic energy source

$$\frac{1}{2} (\overline{v}_{ki}^e)^2 \Gamma_k = \frac{1}{2} \overline{\rho (v_k)^2 (\mathbf{v} - \mathbf{v}_i) \cdot \nabla X_k}, \tag{41}$$

and the interfacial entropy source

$$s_{ki} \Gamma_k = \overline{\rho s (\mathbf{v} - \mathbf{v}_i) \cdot \nabla X_k}. \tag{42}$$

11.3. Averaged Balance Equations

We now present the averaged equations. The averaged equations governing each component are

Mass

$$\frac{\partial \alpha_k \overline{\rho}_k^x}{\partial t} + \nabla \cdot \alpha_k \overline{\rho}_k^x \overline{\mathbf{v}}_k^{x\rho} = \Gamma_k; \tag{43}$$

Momentum

$$\frac{\partial \alpha_k \overline{\rho}_k^x \overline{\mathbf{v}}_k^{x\rho}}{\partial t} + \nabla \cdot \alpha_k \overline{\rho}_k^x \overline{\mathbf{v}}_k^{x\rho} \overline{\mathbf{v}}_k^{x\rho} = \nabla \cdot \alpha_k (\overline{\mathbf{T}}_k^x + \mathbf{T}_k^{Re}) + \alpha_k \overline{\rho}_k^x \overline{\mathbf{b}}_k^{x\rho} + \mathbf{M}_k + \mathbf{v}_{ki}^m \Gamma_k; \tag{44}$$

Energy

$$\frac{\partial}{\partial t} \alpha_k \overline{\rho}_k^x \left(\overline{u}_k^{x\rho} + \frac{1}{2} (\overline{v}_k^{x\rho})^2 + u_k^{Re} \right) + \nabla \cdot \alpha_k \overline{\rho}_k^x \overline{\mathbf{v}}_k^{x\rho} \left(\overline{u}_k^{x\rho} + \frac{1}{2} (\overline{v}_k^{x\rho})^2 + u_k^{Re} \right)$$
$$= \nabla \cdot \alpha_k \left((\overline{\mathbf{T}}_k^x + \mathbf{T}_k^{Re}) \cdot \overline{\mathbf{v}}_k^{x\rho} - \overline{\mathbf{q}}_k^x - \mathbf{q}_k^{Re} \right) + \alpha_k \overline{\rho}_k^x \left(\overline{r}_k^{x\rho} + \overline{\mathbf{b}}_k^{x\rho} \cdot \overline{\mathbf{v}}_k^{x\rho} \right)$$
$$+ E_k + W_k + \left[u_{ki} + \frac{1}{2} (v_{ki}^e)^2 \right] \Gamma_k. \tag{45}$$

The equation for the internal energy is

Internal Energy

$$\frac{\partial \alpha_k \overline{\rho}_k^x \overline{u}_k^{x\rho}}{\partial t} + \nabla \cdot \alpha_k \overline{\rho}_k^x \overline{v}_k^{x\rho} \overline{u}_k^{x\rho} = \alpha_k \overline{\mathbf{T}}_k^x : \nabla \overline{\mathbf{v}}_k^{x\rho} - \nabla \cdot \alpha_k (\overline{\mathbf{q}}_k^x + \hat{\mathbf{q}}_k^{Re})$$
$$+ \alpha_k \overline{\rho}_k^x \overline{r}_k^{x\rho} + E_k + u_{ki} \Gamma_k + \alpha_k \mathcal{D}_k . \quad (46)$$

The entropy inequality becomes

Entropy

$$\frac{\partial \alpha_k \overline{\rho}_k^x \overline{s}_k^{x\rho}}{\partial t} + \nabla \cdot \alpha_k \overline{\rho}_k^x \overline{s}_k^{x\rho} \overline{\mathbf{v}}_k^{x\rho} \geq \nabla \cdot \alpha_k (\overline{\phi}_k^x + \phi_k^{Re}) + \alpha_k \overline{\rho}_k^x \overline{\Sigma}_k^{x\rho} + S_k + s_{ki} \Gamma_k. \quad (47)$$

This entropy inequality corresponds to the microscale entropy inequality of Section 6.4.

11.3.1. Jump Conditions

The jump conditions (8.3.14) hold on sets of volume measure zero. Averaging these quantities weighted with the interfacial delta function yields equations that relate the interfacial sources across the interface in an average sense. The quantities involved, since they are ensemble averaged, hold in a pointwise sense. The ensemble averaged jump conditions are interpreted as describing the ability of the components to interact with one another. In particular, the components can exchange mass, momentum, energy, and entropy. Overall, however, in analogy with the theory of solutions in Chapter 5, mass is conserved, but, in contrast, momentum and energy may not be conserved. The main difference is the presence of an interface separating the components. This interface is able to contribute to momentum and energy balances.

Formally, the jump conditions are obtained by taking the products of the balance equations with $\mathbf{n}_1 \cdot \nabla X_1$, then performing the averaging process. This process yields the following conditions:

Mass

$$\Gamma_1 + \Gamma_2 = 0; \quad (48)$$

Momentum

$$\mathbf{M}_1 + \mathbf{M}_2 + \mathbf{v}_{1i}^m \Gamma_1 + \mathbf{v}_{2i}^m \Gamma_2 = \mathbf{m}; \quad (49)$$

Energy

$$E_1 + W_1 + E_2 + W_2 + \left(u_{1i} + \frac{1}{2}(v_{1i}^e)^2\right) \Gamma_1 + \left(u_{2i} + \frac{1}{2}(v_{2i}^e)^2)\right) \Gamma_2 = \epsilon; \quad (50)$$

Entropy

$$S_1 + S_2 + s_{1i}\Gamma_1 + s_{2i}\Gamma_2 \geq 0; \tag{51}$$

where **m** is the surface tension source, and ϵ is the interfacial energy source.

11.3.2. *Manipulations*

In this section, we give a few miscellaneous relations for the interfacial transfer terms, and derive the Lagrangian forms of the balance equations. We also derive a balance equation for the fluctuation kinetic energy.

Generally, we wish to write the stress as a pressure plus a shear stress,

$$\mathbf{T} = -p\mathbf{I} + \tau. \tag{52}$$

Then, the average stress can be written as

$$\overline{\mathbf{T}}_k^x = -\overline{p}_k^x \mathbf{I} + \overline{\tau}_k^x. \tag{53}$$

The averaged interfacial pressure p_{ki} and shear stress τ_{ki} are introduced to separate mean field effects from local effects in the interfacial force. The interfacial pressure is defined by

$$p_{ki} = \frac{\overline{p\mathbf{n}_k \cdot \nabla X_k}}{a_i}, \tag{54}$$

and the interfacial shear stress is

$$\tau_{ki} = \frac{\overline{\tau_k \mathbf{n}_k \cdot \nabla X_k}}{a_i}. \tag{55}$$

Then the interfacial force density can be written as

$$\begin{aligned}
\mathbf{M}_k &= -\overline{\mathbf{T} \cdot \nabla X_k} \\
&= \overline{p \nabla X_k} - \overline{\tau \cdot \nabla X_k} \\
&= p_{ki} \overline{\nabla X_k} - \tau_{ki} \cdot \overline{\nabla X_k} - \overline{\mathbf{T}'_{ki} \cdot \nabla X_k} \\
&= p_{ki} \nabla \alpha_k - \tau_{ki} \nabla \alpha_k + \mathbf{M}'_k,
\end{aligned} \tag{56}$$

where we define the interfacial extra momentum source

$$\mathbf{M}'_k = \mathbf{M}_k + p_{ki} \nabla \alpha_k - \tau_{ki} \cdot \nabla \alpha_k, \tag{57}$$

and introduce

$$\mathbf{T}'_{ki} = -p'_{ki}\mathbf{I} + \tau'_{ki} = -(p - p_{ki})\mathbf{I} + (\tau - \tau_{ki}). \tag{58}$$

Similarly, the interfacial working is rearranged to read

$$\begin{aligned}
W_k &= -\overline{\mathbf{T} \cdot \mathbf{v} \cdot \nabla X_k} \\
&= -\overline{\mathbf{T} \cdot \overline{\mathbf{v}}_k^{x\rho} \cdot \nabla X_k} - \overline{\mathbf{T} \cdot \mathbf{v}'_k \cdot \nabla X_k} \\
&= W'_k + \mathbf{M}_k \cdot \overline{\mathbf{v}}_k^{x\rho},
\end{aligned} \tag{59}$$

11.3. Averaged Balance Equations

where the interfacial extra working is defined by

$$W'_k = -\mathbf{T} \cdot \mathbf{v}'_k \cdot \nabla X_k. \tag{60}$$

We choose to separate $\mathcal{D}_k$ into two terms, the internal dissipation

$$D_k = \frac{\overline{X_k \tau : \nabla \mathbf{v}'_k}}{\alpha_k}, \tag{61}$$

and the pressure working

$$P_k = \frac{\overline{X_k p \nabla \cdot \mathbf{v}'_k}}{\alpha_k}. \tag{62}$$

Lagrangian Forms

The equation of conservation of mass for component k (43) can be used in the momentum equation (44) to yield the following form of the momentum equation:

$$\alpha_k \overline{\rho}_k^x \frac{D_k \overline{\mathbf{v}}_k^{x\rho}}{Dt} = \alpha_k \overline{\rho}_k^x \left(\frac{\partial \overline{\mathbf{v}}_k^{x\rho}}{\partial t} + \overline{\mathbf{v}}_k^{x\rho} \cdot \nabla \overline{\mathbf{v}}_k^{x\rho} \right)$$
$$= \nabla \cdot \alpha_k (\overline{\mathbf{T}}_k^x + \mathbf{T}_k^{Re}) + \alpha_k \overline{\rho}_k^x \overline{\mathbf{b}}_k^{x\rho} + \mathbf{M}_k + (\mathbf{v}_{ki}^m - \overline{\mathbf{v}}_k^{x\rho}) \Gamma_k, \tag{63}$$

where the operator D_k/Dt is defined by

$$\frac{D_k(\cdot)}{Dt} = \frac{\partial(\cdot)}{\partial t} + \overline{\mathbf{v}}_k^{x\rho} \cdot \nabla(\cdot). \tag{64}$$

Similarly, the equation of conservation of mass applied to the energy equation yields

$$\alpha_k \overline{\rho}_k^x \frac{D_k}{Dt} \left(\overline{u}_k^{x\rho} + \frac{1}{2} (\overline{v}_k^{x\rho})^2 + u_k^{Re} \right)$$
$$= \alpha_k \overline{\rho}_k^x \left[\frac{\partial}{\partial t} \left(\overline{u}_k^{x\rho} + \frac{1}{2} (\overline{v}_k^{x\rho})^2 + u_k^{Re} \right) + \overline{\mathbf{v}}_k^{x\rho} \cdot \nabla \left(\overline{u}_k^{x\rho} + \frac{1}{2} (\overline{v}_k^{x\rho})^2 + u_k^{Re} \right) \right]$$
$$= \nabla \cdot \alpha_k \left((\overline{\mathbf{T}}_k^x + \mathbf{T}_k^{Re}) \cdot \overline{\mathbf{v}}_k^{x\rho} - \overline{\mathbf{q}}_k^x - \mathbf{q}_k^{Re} \right) + \alpha_k \overline{\rho}_k^x (\overline{r}_k^{x\rho} + \overline{\mathbf{b}}_k^{x\rho} \cdot \overline{\mathbf{v}}_k^{x\rho})$$
$$+ E_k + W_k + \left\{ [u_{ki} - \overline{u}_k^{x\rho} - u_k^{Re}] + \frac{1}{2} [(v_{ki}^e)^2 - (\overline{v}_k^{x\rho})^2] \right\} \Gamma_k. \tag{65}$$

This form of the momentum equation is the material, or "Lagrangian" form of the momentum equation. While Lagrangian forms have little advantage in fluid mechanics, some researchers use a Lagrangian form for the motion of the dispersed component. The methods used usually require significant effort to reconcile the material point for the dispersed component with the spatial points for the continuous component. We do not pursue these ideas here.

11.3.3. Balance of Fluctuation Kinetic Energy

With the aid of the momentum equation (44) and the identity

$$\frac{D_k}{Dt}\left(\frac{1}{2}(\overline{v}_k^{x\rho})^2\right) = \overline{\mathbf{v}}_k^{x\rho} \cdot \frac{D_k \overline{\mathbf{v}}_k^{x\rho}}{Dt}, \tag{66}$$

the equation for the conservation of averaged kinetic energy can be derived by taking the inner product of the averaged velocity with the momentum equation. We have

$$\alpha_k \overline{\rho}_k^x \frac{D_k \frac{1}{2}(\overline{v}_k^{x\rho})^2}{Dt} = \overline{\mathbf{v}}_k^{x\rho} \cdot \nabla \alpha_k (\overline{\mathbf{T}}_k^x + \mathbf{T}_k^{Re}) + \alpha_k \overline{\rho}_k^x \overline{\mathbf{v}}_k^{x\rho} \cdot \overline{\mathbf{b}}_k^{x\rho}$$
$$+ \overline{\mathbf{v}}_k^{x\rho} \cdot \mathbf{M}_k + \Gamma_k \overline{\mathbf{v}}_k^{x\rho} \cdot \left(\mathbf{v}_{ki}^m - \overline{v}_k^{x\rho}\right). \tag{67}$$

Using this, the following form of the energy equation can be obtained:

$$\alpha_k \overline{\rho}_k^x \frac{D_k}{Dt}(\overline{u}_k^{x\rho} + u_k^{Re})$$
$$= \alpha_k (\overline{\mathbf{T}}_k^x + \mathbf{T}_k^{Re}) : \nabla \overline{\mathbf{v}}_k^{x\rho} - \nabla \cdot \alpha_k (\overline{\mathbf{q}}_k^x + \mathbf{q}_k^{Re})$$
$$+ \alpha_k \overline{\rho}_k^x \overline{r}_k^{x\rho} + E_k + W_k'$$
$$+ \left[(u_{ki} - \overline{u}_k^{x\rho} - u_k^{Re}) + \frac{1}{2}(v_{ki}^e)^2 + (v_k^{x\rho})^2 - \overline{v}_k^{x\rho} \cdot \mathbf{v}_{ki}^m\right] \Gamma_k. \tag{68}$$

Subtracting the equation for the averaged internal energy from the above gives the equation for the average fluctuation kinetic energy

$$\frac{\partial \alpha_k \overline{\rho}_k^x u_k^{Re}}{\partial t} + \nabla \cdot \alpha_k \overline{\rho}_k^x u_k^{Re} \overline{\mathbf{v}}_k^{x\rho} = \alpha_k \mathbf{T}_k^{Re} : \nabla \overline{\mathbf{v}}_k^{x\rho} - \nabla \cdot \alpha_k (\mathbf{q}_k^K + \mathbf{q}_k^T)$$
$$+ \frac{1}{2}[(v_{ki}^e)^2 + (v_k^{x\rho})^2 - \overline{v}_k^{x\rho} \cdot \mathbf{v}_{ki}^m]\Gamma_k$$
$$+ W_k' + \alpha_k D_k - \alpha_k P_k. \tag{69}$$

This equation is of special interest in multicomponent flows since it is the motivation for the structure of the theory with an entropy, or "disorder" at the scale of the molecular vibrations, and a separate independent entropy, or "disorder" at the scale of the component motion fluctuations. We note that there is an adequate definition of temperature at the molecular level, but that there is no counterpart of "temperature" in the derivation of the averaged equations of motion.

12
Postulational and Averaging Approaches

The approach put forward in Chapter 6, which could be called the "postulational" approach, is appealing in that references to the microstructure do not appear in the formulation. This is analogous to the similar approach to continuum mechanics, where the molecular nature of matter plays no role in the formulation. The existence of the microstructure is often simply ignored, or argued away by saying that the "limit" is stopped at some scale larger than the microstructure, but smaller than any macroscopic scale of interest. Another view of both the continuum approach to both ordinary continua and to multicomponent continua is that they are *models* of reality, and consequently are simplified to the point of omitting some phenomena; namely, they do not govern the evolution of microstructure. These models require taking limits as volumes shrink to zero, so that at some point, the volume is so small that it cannot contain many "corpuscles." This feature does not render the model invalid, but instead suggests that constitutive equations are required to replace the microstructure detail lost in the model.

Nonetheless, many people have some difficulty in the perception of these ideas. The difficulty arises in the limiting operations discussed in Chapter 6. When the size of the volume or flake or tetrahedron is sufficiently small, it becomes comparable with the scale of the microstructure in any particular realization. On this scale, the equations governing the averaged properties of the motion cannot be expected to describe the microscopic dynamics of the situation for any given realization. They do describe the *expected* value of the mass, momentum, energy, and entropy to be found in the small volume, even though in any one realization, that small volume might only have one material or the other in it. In the "flake" and tetrahedron arguments, the resulting stress tensors and heat flux vectors should be interpreted as the expected values, and are valid for the continuum model.

12. Postulational and Averaging Approaches

The equations derived in Chapter 11 and the equations postulated in Chapter 6 are quite similar if the proper interpretations are made of the variables. This requires that each of the variables be interpreted in terms of "expected values." Even so, this is nontrivial. The densities ρ_k in Chapter 6 must be interpreted as the component-weighted density in Chapter 11. This is consistent with the introduction of the effective density $\underline{\rho}_k$ in Chapter 6, which is analogous to $\overline{X_k \rho} = \underline{\rho}_k = \alpha_k \overline{\rho}_k^x$ in Chapter 11. The velocity of component k, $\mathbf{v}_k$, is the barycentric velocity, and corresponds to the mass-weighted (Favré average) velocity $\overline{\mathbf{v}}_k^{x\rho}$. The stress tensor in the postulational approach of Chapter 6, $\mathbf{T}_k$, is seen to correspond to the averaged stress, plus the Reynolds stress, $\overline{\mathbf{T}}_k^x + \mathbf{T}_k^{\text{Re}}$. There is a connection here between Maxwell's definition of stress and these results. Finally, the quantities appearing in the energy equation in Chapter 6 are quite complicated. The internal energy density u_k is analogous to $\overline{u}_k^{x\rho} + u_k^{\text{Re}}$. Other quantities in the energy balance equations in Chapter 6 also have fluctuation quantities associated with them for comparison with the equations in Chapter 11.

Moreover, we note that the microscale and mesoscale stresses, heat fluxes, internal energies, etc. introduced in Section 6.4, can be interpreted as follows:

$$\begin{aligned}
\mathbf{T}_k^m &= \overline{\mathbf{T}}_k^x, \\
\mathbf{T}_k^M &= \mathbf{T}_k^{\text{Re}}, \\
\mathbf{q}_k^m &= \overline{\mathbf{q}}_k^x, \\
\mathbf{q}_k^M &= \mathbf{q}_k^{\text{Re}}, \\
u_k^m &= \overline{u}_k^{x\rho}, \\
u_k^M &= u_k^{\text{Re}}.
\end{aligned} \quad (1)$$

While the structure appears naturally in the averaging approach, it requires some motivational arguments within the postulational approach. Even so, the appearance of an entropy inequality on the mesoscale level is not motivated by averaging arguments.

One substantial difference between the postulational approach and the averaging approach is in the treatment of entropy. The averaged entropy inequalities (11.3.47) correspond to the microscale entropy inequalities (6.4.79). However, there is no corresponding averaged mesoscale entropy inequality. This is very much like the situation in the connection between kinetic theory and continuum mechanics, where the H-Theorem *suggests* the entropy inequality, but there is no derivation thereof.

There is further need of understanding the nature of the averaging and its relation to the macroscopic variables. First, the meaning of velocity in the postulational approach is the time rate of change of the position as given by (6.2.3). This concept essentially defines "particle paths." The meaning of these particle paths is not obvious in terms of the microstructure, but from consideration of the averaging process, it is evident that particle paths have the following meaning. Consider all the realizations μ that allow component k to reside at point $\mathbf{x}$ at time t. Each of these has some (microscopic) particle path of the material of component k passing

Part IV

Modeling Multicomponent Flows

13
Introduction

A goal of many studies of multicomponent fluids is to find field equations that hold over regions containing these fluids. That is, the goal is to represent the fluids as *continua*. The success of classical continuum mechanics in describing single component materials as continua is remarkable. For multicomponent fluids the matter is more complex. For example, there are flow regimes, such as the slug flow regime, where the continuum approximation may not be appropriate. Even if such regimes are not considered, it is clear that the essence of multicomponent fluids is their microstructure. In many flows of multicomponent fluids the microstructure can be described adequately by the equations of continuum mechanics.

In Chapter 6 we have presented the equations of balance of mass, momentum, and energy, along with entropy inequalities for microscale and mesoscale phenomena, for each component. The same equations, along with a microscale entropy inequality for each component can be derived from the exact balance equations. This is done in Chapter 8. The "postulational" approach derives balance equations that apply to large classes of materials. In order to advance the theory to the stage of having a boundary-value problem to solve, the behavior of the material must be specified by constitutive equations. The averaging approach leads to balance equations which are underdetermined. The averaging process "loses information," which must be replaced by supplying "closure conditions." These closure conditions are exactly analogous to constitutive equations; however, we prefer the terminology of the large body of work on constitutive equations. Therefore, whichever approach is taken for obtaining the balance equations, these equations must be supplemented with constitutive equations based on experience with the behavior of the materials and confirmed by rheological measurement.

13. Introduction

The usual approach in studies of the mechanics of single continua is to postulate those constitutive equations, and measure certain parameters in a careful experiment. Thus, for example, in the case of a viscous incompressible fluid, assuming the stress is a linear function of the deformation rate leads to the conclusion that the behavior of the material is controlled by a single constant. It is found that there are exact "universal" solutions to the field equations that allow this constant to be found easily by experiment. For multicomponent materials the number of fields of interest is sufficiently large that this approach leads to systems of equations with coefficient functions so numerous that determining them by experiment is a complex task. The existence of exact solutions of the field equations is quite unusual. No "universal" solutions are known. Thus other techniques are appropriate.

Much of what distinguishes the study of multicomponent fluids from the study of flow of single fluids is that for multicomponent fluids, the fluids can interact. This is expressed by the momentum exchange term in the momentum equation and the energy exchange term in the energy equation. In addition, the presence of a second component alters the stress and the energy flux. Thus, much of understanding the mechanics of multicomponent fluids means understanding the nature of the exchange terms and how the presence of the components modifies the flux.

A significant point must be made about the mechanics of a single fluid. For many fluids of everyday experience, the phenomenology is clear enough so that the general nature of the continuum constitutive equations is known. An example of this is the Navier–Stokes fluid, that is, the incompressible linearly viscous fluid. The whole constitutive behavior of this fluid is controlled by one number, the viscosity. This model is so widely accepted that viscosities are quoted and understood for most fluids in everyday use. For the Navier–Stokes fluid, there exists a class of "universal solutions" to the equations of motion. These are exact solutions in which the kinematics of the flow is independent of the viscosity. Couette flow is a flow of this type. These flows are easily produced experimentally, and a simple set of measurements determines the viscosity. Thus there is a clear path to characterizing the fluid exactly. This path requires continuum concepts only. Moreover, these arguments may be generalized to a much larger class of fluids called "simple fluids." Certain subclasses of these fluids may be characterized entirely by the viscometric experiments quoted above. All simple fluids can be characterized partially by these experiments. Moreover, there are theorems that show that for adequately slow flows, all simple fluids act like the ones that can be characterized exactly.

None of the useful ideas above has been generalized to multicomponent fluids. The same viscometric boundary conditions that are useful for simple fluids do, in general, reduce the equations governing multicomponent fluids to ordinary differential equations. Exact solutions to these equations are rare; universal solutions do not exist. There are no known theorems about asymptotic behavior for slow or other specialized flows. Thus, even in the very simplest physical cases, the determination of the equations governing multicomponent fluids, or the determination of the constants or functions governing putative equations, is a subtle and difficult task.

All correct physical theories must be built on a proper mathematical basis [88]. However, no matter how good the basis of a theory in mathematics, it must agree with physical experience[1] to be a useful engineering tool.

In this Part IV, we discuss the several different approaches to determining the constitutive equations. These include applying general axioms of invariance and using entropy inequalities to obtain restrictions on the terms included in the constitutive equations. We give several calculations that are suggestive of the types of terms that should appear in a theory describing small concentration of the dispersed component. These include the calculation of the flow of an incompressible inviscid fluid around a single sphere; and Stokes, or creeping, flow of an effective medium around a single sphere. In each case, the operation of averaging is discussed and applied. We also investigate a kinetic theory approach. It motivates the inclusion of a mesoscale viscosity and conductivity for the dispersed component. We also give a model for the evolution of the geometry of interfaces that so strongly characterize fluid–fluid mixtures. Finally, in Section 18.1, we suggest forms for constitutive equations for a dispersed flow.

---

[1] Roughly, "physical experience" refers to experiment.

14
Closure Framework

14.1. Completeness of the Formulation

It is physically plausible that the three-dimensional, unsteady two-fluid model given by (6.12), (6.25), and (6.41)[1], must be supplemented with state equations, constitutive equations, and boundary and initial conditions. That is mathematically plausible also, because of the experience that a physical system must be described by the same number of equations as it has unknowns. Such a system is called a *determined system*. *Underdetermined* and *overdetermined* then have the obvious meaning.

Initial conditions specify how the experiment or the flow starts. The boundary conditions specify how the flow interacts with its environment, specifically with the inlet and outlet flow devices and with the walls.

The constitutive equations specify how the individual materials (components) behave and how they interact with each other. State equations specify the thermodynamic state of the material in the usual sense. For example, in a single component fluid, the internal energy may be determined by the density and the entropy. Boundary and initial conditions are important, in the sense that if one is specified incorrectly, the predicted flow will be in error. However, the essence of the difference between multicomponent fluids and single materials is that multicomponent materials require the specification of the component interaction terms ($\mathbf{M}_k$, E_k^m, and E_k^M), the phase change terms (Γ_k, $\mathbf{v}_{ki}^m$, v_{ki}^e, and u_{ki}^m), and the self-

---

[1] Or equivalently, (11.43), (11.44), and (11.45).

interaction terms ($\mathbf{T}_k^m$, $\mathbf{q}_k^m$, $\mathbf{T}_k^M$, and $\mathbf{q}_k^M$) in terms of the state variables such that α_k, $\mathbf{v}_k$, ...

14.2. Constitutive Equations

The variables to be specified involve the component interaction terms, the phase change terms, and the self-interaction terms. These depend on transport properties and on effects caused by velocity fluctuations. In view of the interfacial jump conditions, not all of the component interaction and phase change terms are independent. The first step in completing the formulation, then, requires the choice of a set of dependent variables and a set of independent variables whose values determine them. Once this functional dependence is established, we must find the functions that describe the behavior of the materials. The following ideas are useful in the development of these constitutive relations.

14.2.1. Guiding Principles

The development of constitutive equations is aided by "principles" [23], [87]. These provide rational means for obtaining descriptions of classes of materials, without inadvertently neglecting important dependences and without including irrational dependences. We consider here the principles of well-posedness, equipresence, separation of components, and frame indifference. We shall also require that constitutive equations satisfy appropriate the entropy inequalities. There is quite considerable debate about the validity of the arguments about thermodynamics. They are known to give correct answers for linearly viscous fluids. For more complicated situations, including multicomponent materials, there is more room for debate. In practice, for the physical cases considered in this book, this stipulation is less restrictive than the other four principles mentioned.

The *principle of well-posedness* states that the description of the motion should be such that a solution to the initial-boundary-value problem exists and depends continuously on the initial and boundary conditions.

The *principle of equipresence* is appropriate to the mechanics of single materials and possibly classical mixtures or solutions. It is, all variables should be included in each constitutive equation, unless another axiom rules that a particular dependence cannot occur. Clearly this is not appropriate for multicomponent materials. There, it is replaced by the *principle of separation of components* [26], [68]: a variable expressing a *self*-interaction for component k depends only on variables associated with that component.

The *principle of frame indifference* asserts that the constitutive equations cannot depend on the reference frame. The motivation is that reference frames are inventions of the observer. They have nothing to do with the material being described. Thus, the functional dependence of the constituted variable on the variables describing the mechanical state must be independent of frame, or to be frame indif-

14. Closure Framework

ferent. This principle was discussed in Section 2.4 for single-component motions; however, since it has substantial, and sometimes unexpected, utility in the formulation of the equations of multicomponent flows, we shall discuss it in further detail.

Frame Indifference

Consider two reference frames, so that, for example, $\mathbf{x}$ and $\mathbf{x}'$, are different representations of the same vector. The two representations are related by

$$\mathbf{x}' = \mathbf{Q} \cdot \mathbf{x} + \mathbf{b}, \tag{1}$$

where $\mathbf{Q}(t)$ is an orthonormal transformation, and $\mathbf{b}$ is a translation. A *scalar* s is *objective* if its value is invariant to a change of frame, that is, if $s' = s$. A *vector* $\mathbf{v}$ is objective if it changes according to $\mathbf{v}' = \mathbf{Q} \cdot \mathbf{v}$. Tensors are objective if they transform objective vectors into objective vectors. Thus a second-order *tensor* $\mathbf{T}$ is objective if $\mathbf{T}' = \mathbf{Q} \cdot \mathbf{T} \cdot \mathbf{Q}^\mathsf{T}$.

As discussed in Section 2.4, velocity is not objective. If we differentiate (1) following material k, we see that

$$\mathbf{v}'_k = \mathbf{Q} \cdot \mathbf{v}_k + (\dot{\mathbf{Q}} \cdot \mathbf{x} + \dot{\mathbf{b}}). \tag{2}$$

If (2) is evaluated for two different values of k and the results subtracted, the velocity difference becomes

$$\mathbf{v}'_k - \mathbf{v}'_l = \mathbf{Q} \cdot (\mathbf{v}_k - \mathbf{v}_l). \tag{3}$$

Thus, velocity differences

$$\mathbf{v}_{kl} = \mathbf{v}_k - \mathbf{v}_l \tag{4}$$

are objective. Velocities are not objective because a change of frame changes the velocity; two observers traveling relative to one another observe the same rigid body moving with different velocities. However the relative velocity between two objects remains the same in different frames. In multicomponent flows, any two components form two objects that can have a nontrivial velocity difference.

In multicomponent flows there are reasons to believe that the interactions between two components can depend on the accelerations of each of the components. However, accelerations are not objective. Following component l and differentiating (2) leads to

$$\frac{D_l \mathbf{v}'_k}{Dt} = \mathbf{Q} \cdot \frac{D_l \mathbf{v}_k}{Dt} + \dot{\mathbf{Q}} \cdot (\mathbf{v}_k + \mathbf{v}_l) + (\ddot{\mathbf{Q}} \cdot \mathbf{x} + \ddot{\mathbf{b}}). \tag{5}$$

Again subtracting, we have

$$\frac{D_l \mathbf{v}'_k}{Dt} - \frac{D_k \mathbf{v}'_l}{Dt} = \mathbf{Q} \cdot \left(\frac{D_l \mathbf{v}_k}{Dt} - \frac{D_k \mathbf{v}_l}{Dt} \right). \tag{6}$$

Thus, opposing acceleration differences are objective.

In exactly the same manner as in classical continuum mechanics, since

$$\nabla' \mathbf{v}'_k = \mathbf{Q} \cdot (\nabla \mathbf{v}_k) \cdot \mathbf{Q}^\mathsf{T} + \dot{\mathbf{Q}} \cdot \mathbf{Q}^\mathsf{T}, \tag{7}$$

14.2. Constitutive Equations

we have the result that

$$\mathbf{D}_k = \frac{1}{2}\left(\nabla \mathbf{v}_k + (\nabla \mathbf{v}_k)^T\right) \tag{8}$$

is objective. However, the rotation rate tensor

$$\mathbf{W}_k = \frac{1}{2}\left(\nabla \mathbf{v}_k - (\nabla \mathbf{v}_k)^T\right) \tag{9}$$

is not objective, since

$$\mathbf{W}'_k = \mathbf{Q} \cdot \mathbf{W}_k \cdot \mathbf{Q}^T + \frac{1}{2}\left(\dot{\mathbf{Q}} \cdot \mathbf{Q}^T - (\dot{\mathbf{Q}} \cdot \mathbf{Q}^T)^T\right)$$

$$= \mathbf{Q} \cdot \mathbf{W}_k \cdot \mathbf{Q}^T + \frac{1}{2}\left(\dot{\mathbf{Q}} \cdot \mathbf{Q}^T - \mathbf{Q} \cdot \dot{\mathbf{Q}}^T\right).$$

The observed rate of rotation differs by the relative spin of the two frames. It is interesting that there are objective tensors in multicomponent mechanics that do not occur in the mechanics of single materials. For example, add (7) for different values of k, and we have

$$\nabla' \mathbf{v}'_k + (\nabla' \mathbf{v}'_l)^T = \mathbf{Q} \cdot [\nabla \mathbf{v}_k + (\nabla \mathbf{v}_l)^T] \cdot \mathbf{Q}^T, \tag{10}$$

where we have used the identity $\mathbf{Q} \cdot \mathbf{Q}^T = \mathbf{I}$ to obtain

$$\dot{\mathbf{Q}} \cdot \mathbf{Q}^T + \mathbf{Q} \cdot \dot{\mathbf{Q}}^T = 0.$$

Hence the mixed rate of strain tensor,

$$\mathbf{D}_{kl} = \frac{1}{2}[\nabla \mathbf{v}_k + (\nabla \mathbf{v}_l)^T], \tag{11}$$

is objective for any k and l. It is also clear from (10) that the relative rotation tensor,

$$\mathbf{W}_{kl} = \mathbf{W}_k - \mathbf{W}_l, \tag{12}$$

is objective for any k and l. We also note that

$$\frac{D_k \mathbf{v}'_k}{Dt} - \frac{D_l \mathbf{v}'_l}{Dt} - 2\mathbf{W}'_m \cdot (\mathbf{v}'_k - \mathbf{v}'_l)$$

$$= \mathbf{Q} \cdot \left(\frac{D_k \mathbf{v}_k}{Dt} - \frac{D_l \mathbf{v}_l}{Dt}\right) + 2\dot{\mathbf{Q}} \cdot (\mathbf{v}'_k - \mathbf{v}'_l)$$

$$- 2\mathbf{Q} \cdot \mathbf{W}_m \cdot \mathbf{Q}^T \cdot \mathbf{Q} \cdot (\mathbf{v}_k - \mathbf{v}_l) - 2\dot{\mathbf{Q}} \cdot \mathbf{Q}^T \cdot (\mathbf{v}_k - \mathbf{v}_l)$$

$$= \mathbf{Q} \cdot \left(\frac{D_k \mathbf{v}_k}{Dt} - \frac{D_l \mathbf{v}_l}{Dt} - 2\mathbf{W}_m \cdot (\mathbf{v}_k - \mathbf{v}_l)\right), \tag{13}$$

so that the quantity

$$\frac{D_k \mathbf{v}_k}{Dt} - \frac{D_l \mathbf{v}_l}{Dt} - 2\mathbf{W}_m \cdot (\mathbf{v}_k - \mathbf{v}_l) \tag{14}$$

is objective for any choice of k, l, and m. Since we already know that the combination $\mathbf{W}_{kl}$ is objective, we see that the combination

$$\mathbf{a}_{lk} = \frac{D_k \mathbf{v}_k}{Dt} - \frac{D_l \mathbf{v}_l}{Dt} - 2\mathbf{W}_l \cdot (\mathbf{v}_k - \mathbf{v}_l) \tag{15}$$

is a quantity that has the dimensions of acceleration and is objective.

Representation Theorem

It is sometimes true that the principle of frame indifference alone implies that a frame-indifferent function of an objective variable is an isotropic function. Such is true for vector-valued functions of vectors. More often, if the material described is isotropic and the constitutive functions are frame indifferent, then the function must be isotropic. In the mechanics of single materials, this occurs almost always for fluids and sometimes for isotropic solids. For multicomponent fluids, the matter is more complex. However, the complicated forms assumed for the interfacial interaction terms and the average microscale and Reynolds fluxes usually have this property. Usually, if some material shows nonisotropic behavior, then the variable list in the constitutive equations must include variables which specify the "nonisotropy." It is a theorem of Cauchy (see [95], [96]) that isotropic functions are not arbitrary. Instead, there are exact representation theorems for such functions. For example, if $\mathbf{m}$ is a frame-indifferent vector-valued function of the objective vectors $\mathbf{v}_1, \ldots, \mathbf{v}_n$, and the objective tensors $\mathbf{T}_1, \ldots, \mathbf{T}_m$, then it is a consequence of this representation theorem that it must have the form

$$\mathbf{m} = \sum_{i=1}^{n} A_i \mathbf{v}_i + \sum_{i=1}^{n} \sum_{j=1}^{m} (B_{ij} \mathbf{v}_i \cdot \mathbf{T}_j + C_{ij} \mathbf{T}_j \cdot \mathbf{v}_i), \tag{16}$$

and, if $\mathbf{R}$ is a frame-indifferent tensor-valued function of the same objective quantities, then it must have the form

$$\mathbf{R} = \sum_{i=1}^{m} D_i \mathbf{T}_i + \sum_{i=1}^{n} \sum_{j=1}^{m} E_{ij} \mathbf{v}_i \mathbf{v}_j$$

$$+ \sum_{i=1}^{n} \sum_{j=1}^{n} \sum_{k=1}^{m} \sum_{l=1}^{m} \left(F_{ijkl} \mathbf{v}_i \cdot \mathbf{T}_k \mathbf{v}_j \cdot \mathbf{T}_l + G_{ijkl} \mathbf{T}_k \cdot \mathbf{v}_i \mathbf{v}_j \cdot \mathbf{T}_l \right.$$

$$\left. + H_{ijkl} \mathbf{v}_i \cdot \mathbf{T}_k \mathbf{T}_l \cdot \mathbf{v}_j + I_{ijkl} \mathbf{T}_k \cdot \mathbf{v}_i \mathbf{T}_l \cdot \mathbf{v}_j \right). \tag{17}$$

In each case, the scalar-valued coefficient functions must depend on the independent joint invariants of all of the vectors and tensors.

14.3. Forms for Constitutive Equations

In this section, we give the forms of the general two-component flow model with no energetic effects, either microscale or mesoscale. The stress in a multicomponent mixture represents the force per unit area on component k across an element of area having normal $\mathbf{n}$. Here both materials "flow," so fluid models are appropriate, but the particle stress is due to a different mechanism than the fluid stress. There is no loss in taking the microscale stress to be a pressure plus an extra stress

$$\mathbf{T}_k^m = -p_k^m \mathbf{I} + \tau_k^m. \tag{18}$$

Similarly, the mesoscale stress is written as

$$\mathbf{T}_k^M = -p_k^M \mathbf{I} + \boldsymbol{\tau}_k^M. \tag{19}$$

The interfacial force $\mathbf{M}_k$ arises from stresses acting on the interface. It is the crucial term in modeling multicomponent flows. Indeed, it is the presence of the interface that makes multicomponent mechanics different from ordinary continuum mechanics. In order to describe multicomponent materials, we introduce the interfacial velocity $\mathbf{v}_i$ as a variable for the description of component k.

Let the interfacial force density depend on the objective vectors $\mathbf{v}_{ki} = \mathbf{v}_k - \mathbf{v}_i$, $\mathbf{a}_{ki}$, which is given by

$$\mathbf{a}_{ki} = \left(\frac{\partial \mathbf{v}_k}{\partial t} + \mathbf{v}_i \cdot \nabla \mathbf{v}_k\right) - \left(\frac{\partial \mathbf{v}_i}{\partial t} + \mathbf{v}_k \cdot \nabla \mathbf{v}_i\right)$$

and $\nabla \alpha_k$, and the tensors $\nabla \mathbf{v}_k$, $\nabla \mathbf{v}_i$, and $\nabla \nabla \alpha_k$. Then we can represent the interfacial force density in the form

$$\begin{aligned}
\mathbf{M}_k &= A_{k1}\mathbf{v}_{ki} + A_{k2}\nabla\alpha_k + A_{k3}\mathbf{a}_{ki} \\
&+ B_{k11}\mathbf{v}_{ki} \cdot \mathbf{D}_k + B_{k12}\mathbf{v}_{ki} \cdot \mathbf{D}_i + B_{k13}\mathbf{v}_{ki} \cdot \mathbf{W}_{ki} + B_{k14}\mathbf{v}_{ki} \cdot \nabla\nabla\alpha_k \\
&+ B_{k12}\nabla\alpha_k \cdot \mathbf{D}_k + B_{k22}\nabla\alpha_2 \cdot \mathbf{D}_i + B_{k23}\nabla\alpha_k \cdot \mathbf{W}_{ki} + B_{k24}\nabla\alpha_k \cdot \nabla\nabla\alpha_k \\
&+ B_{k31}\mathbf{a}_{ki} \cdot \mathbf{D}_k + B_{k32}\mathbf{a}_{ki} \cdot \mathbf{D}_i + B_{k33}\mathbf{a}_{ki} \cdot \mathbf{W}_{ki} + B_{k34}\mathbf{a}_{ki} \cdot \nabla\nabla\alpha_k.
\end{aligned} \tag{20}$$

The pressures are determined through an equation of state. There is no need to separate the microscale and mesoscale extra stresses at this point. Consequently, we let

$$\tau_k = \tau_k^m + \tau_k^M. \tag{21}$$

Thus the extra stress can be represented by

$$\begin{aligned}
\tau_k &= \mu_{kk}\mathbf{D}_k + \mu_{ki}\mathbf{D}_i + \mu_{kg}\nabla\nabla\alpha_k \\
&+ E_{k11}\mathbf{v}_{ki}\mathbf{v}_{ki} + E_{k22}\nabla\alpha_k\nabla\alpha_k + E_{k33}\mathbf{a}_{ki}\mathbf{a}_{ki} \\
&+ E_{k12}(\mathbf{v}_{ki}\nabla\alpha_k + \nabla\alpha_k\mathbf{v}_{ki}) + E_{k13}(\mathbf{v}_{ki}\mathbf{a}_{ki} + \mathbf{a}_{ki}\mathbf{v}_{ki}) \\
&+ E_{k23}(\nabla\alpha_k\mathbf{a}_{ki} + \mathbf{a}_{ki}\nabla\alpha_k) + \cdots,
\end{aligned} \tag{22}$$

where $+\cdots$ represents 81 terms of the form

$$\mathbf{v}_i \cdot \mathbf{T}_k \mathbf{v}_l \cdot \mathbf{T}_m. \tag{23}$$

The scalar coefficients in these representations can be functions of any objective scalar variable, including scalar invariants of the vector and tensor quantities.

Let us now restrict the theory to terms having no higher than one derivative with respect to either space or time. This "first grade" theory requires that $B_{14} = 0$, and $B_{2j} = B_{3j} = 0$. Also, $\mu_{k3} = 0$, $E_{k22} = E_{k33} = 0$, and the terms represented by $+\cdots$ should be zero. Thus, the approximated theory reduces to

$$\begin{aligned}
\mathbf{M}_k &= A_{k1}\mathbf{v}_{ki} + A_{k2}\nabla\alpha_k + A_{k3}\mathbf{a}_{ki} \\
&+ B_{k11}\mathbf{v}_{ki} \cdot \mathbf{D}_k + B_{k12}\mathbf{v}_{ki} \cdot \mathbf{D}_i + B_{k13}\mathbf{v}_{ki} \cdot \mathbf{W}_{ki}.
\end{aligned} \tag{24}$$

The stresses can be represented by

$$\tau_k = \mu_{kk}\mathbf{D}_k + \mu_{ki}\mathbf{D}_i + E_{k11}\mathbf{v}_{ki}\mathbf{v}_{ki}$$
$$+ E_{k12}(\mathbf{v}_{ki}\nabla\alpha_k + \nabla\alpha_k\mathbf{v}_{ki}) + E_{k13}(\mathbf{v}_{ki}\mathbf{a}_{ki} + \mathbf{a}_{ki}\mathbf{v}_{ki}). \quad (25)$$

The coefficients in (24) and (25) can be functions of α_k and $|\mathbf{v}_{ki}|$. In addition, the coefficients A_{k1} and E_{k11} can be linear functions of $\operatorname{tr}\mathbf{D}_l = \nabla \cdot \mathbf{v}_l$, $\mathbf{v}_{ki} \cdot \nabla\alpha_k$, $\partial\alpha_k/\partial t$, $\mathbf{v}_{ki} \cdot \mathbf{a}_{ki}$, $\mathbf{v}_{ki} \cdot \mathbf{D}_l \cdot \mathbf{v}_{ki}$ and $|\mathbf{a}_{ki}|$, where

$$\mathbf{v}_{ki} \cdot \mathbf{a}_{ki} = D_k(|\mathbf{v}_{ki}|^2/2)/Dt + \mathbf{v}_{ki} \cdot \mathbf{D}_i \cdot \mathbf{v}_{ki}.$$

We write

$$A_{k1} = A_{k10} + A_{k11}\nabla \cdot \mathbf{v}_k + A_{k12}\nabla \cdot \mathbf{v}_i + A_{k13}\mathbf{v}_{ki} \cdot \nabla\alpha_k$$
$$+ A_{k14}\left(\frac{D_k\alpha_k}{Dt}\right) + A_{k15}\left(\frac{D_k|\mathbf{v}_{ki}|^2}{Dt}\right) + A_{k16}\mathbf{v}_{ki} \cdot \mathbf{D}_k \cdot \mathbf{v}_{ki}$$
$$+ A_{k17}\mathbf{v}_{ki} \cdot \mathbf{D}_i \cdot \mathbf{v}_{ki} + A_{k18}|\mathbf{a}_{ki}|, \quad (26)$$

$$E_{k11} = E_{k110} + E_{k111}\nabla \cdot \mathbf{v}_k + E_{k112}\nabla \cdot \mathbf{v}_i + E_{k113}\mathbf{v}_{12} \cdot \nabla\alpha_k$$
$$+ E_{k114}\left(\frac{D_k\alpha_k}{Dt}\right) + E_{k115}\left(\frac{D_k|\mathbf{v}_{ki}|^2}{Dt}\right) + E_{k116}\mathbf{v}_{ki} \cdot \mathbf{D}_k \cdot \mathbf{v}_{ki}$$
$$+ E_{k117}\mathbf{v}_{ki} \cdot \mathbf{D}_i \cdot \mathbf{v}_{ki} + E_{k118}|\mathbf{a}_{ki}|. \quad (27)$$

Thus, the approximated theory can be specified by determining 37 scalar functions of α_k and $|\mathbf{v}_{ki}|$.

Many flow situations are sufficiently slow that inertia plays no important role. We simplify this theory for the case of no inertia by retaining only the terms that are linear in the velocities. This gives

$$\mathbf{M}_k = A_{k1}\mathbf{v}_{ki} + A_{k2}\nabla\alpha_k, \quad (28)$$

and

$$\tau_k = \mu_{k1}\mathbf{D}_k + \mu_{k2}\mathbf{D}_i + E_{k12}(\mathbf{v}_{ki}\nabla\alpha_k + \nabla\alpha_k\mathbf{v}_{ki}). \quad (29)$$

14.3.1. Dispersed Flow Theory

The formulation thus far is general, in that the description for component k depends on state variables from component k and the interface abutting component k. In this section, we shall specialize this theory to the flow of a single dispersed component in a continuous component which is a fluid. In this situation, the motion of the interface is intimately related to the motion of the dispersed component.

The stress in the continuous component can be represented by

$$\mathbf{M}_c = A_{c1}\mathbf{v}_{cd} - A_{c2}\nabla\alpha_d + A_{c3}\mathbf{a}_{cd}$$
$$+ B_{c11}\mathbf{v}_{cd} \cdot \mathbf{D}_c + B_{c12}\mathbf{v}_{cd} \cdot \mathbf{D}_d + B_{c13}\mathbf{v}_{cd} \cdot \mathbf{W}_{cd}. \quad (30)$$

The continuous component stress can be represented by

$$\boldsymbol{\tau}_c = \mu_{cc}\mathbf{D}_k + \mu_{cd}\mathbf{D}_d + E_{cd1}\mathbf{v}_{cd}\mathbf{v}_{cd}$$
$$+ E_{cd2}(\mathbf{v}_{cd}\nabla\alpha_d + \nabla\alpha_d\mathbf{v}_{cd}) + E_{cd3}(\mathbf{v}_{cd}\mathbf{a}_{cd} + \mathbf{a}_{cd}\mathbf{v}_{cd}). \quad (31)$$

14.4. Entropy Restrictions

In this section, we use the entropy inequalities discussed in Chapter 6 to obtain restrictions on the constitutive equations. We use some assumptions about the dependence of microscale variables, such as the internal energy, on the fields. This yields a structure that is rich in physical restrictions on the constitutive equations, while retaining enough generality to allow interesting models. The advantage of separating the scales in dealing with the internal energy is that it makes it easier to delineate the types of assumptions that are physically appropriate.

14.4.1. Microscale Considerations

The nature of multicomponent materials affects not only the models for their kinematics but also their constitutive models. Such materials are not classical mixtures, where constituents are intermixed at the molecular level. Rather, they consist of distinct bodies, each of which is made of a single material whose properties are modeled by classical theories, and of interfaces between those bodies. The interpretation of this in terms of equations is related to the principle of separation of components. In the terms used here it is that the microscale internal energy u_k^m depends only on the "state" of component k. Thus, it depends on ρ_k and s_k^m. The quantity α_k indicates the portion of constituent k *in the total mixture*, so u_k^m does not depend on it. Moreover, u_k^m does not depend on the density or entropy of the other components.

With that motivation, we assume

$$u_k^m = u_k^m(\rho_k, s_k^m). \quad (32)$$

We write $\mathbf{T}_k^m = -p_k^m \mathbf{I} + \boldsymbol{\tau}_k^m$, and use the equation of balance of mass. Then the microscale entropy inequality (6.4.87) becomes

$$\alpha_k \rho_k \left\{ \left[\theta_k^m - \frac{\partial u_k^m}{\partial s_k^m}\right] \frac{D_k s_k^m}{Dt} + \left[\frac{p_k^m}{(\rho_k)^2} - \frac{\partial u_k^m}{\partial \rho_k}\right] \frac{D_k \rho_k}{Dt} \right\}$$
$$+ p_k^m \frac{D_k \alpha_k}{Dt} + \frac{1}{\theta_k^m} \alpha_k (\mathbf{q}_k^m + \tilde{\mathbf{Q}}_k^m) \cdot \nabla\theta_k^m - \nabla \cdot \alpha_k \tilde{\mathbf{Q}}_k^m$$
$$+ \alpha_k \boldsymbol{\tau}_k^m : \nabla\mathbf{v}_k + \alpha_k \mathcal{D}_k + \mathcal{E}_k^{mi}\left(1 - \frac{\theta_k^m}{\theta_i^m}\right) - \tilde{E}_k^m$$
$$- \left[\theta_k^m(s_{ki}^m - s_k^m) - u_k^m\right]\Gamma_k \geq 0. \quad (33)$$

We use standard thermodynamic arguments on this inequality. The derivative of the microscale temperature does not appear as the argument in any other variable in inequality (33). We assert that, given any numerical value, there is a process where $D_k s_k^m / Dt$ can attain that value, and all other quantities in (33) are zero. Then it follows that the coefficient of the time derivative of temperature must vanish. It follows that

$$\theta_k^m = \frac{\partial u_k^m}{\partial s_k^m}. \tag{34}$$

Similarly, the coefficient of the derivative of the density must vanish. Consequently,

$$p_k^m = \rho_k^2 \frac{\partial u_k^m}{\partial \rho_k}. \tag{35}$$

The remaining microscale entropy inequality is

$$p_k^m \frac{D_k \alpha_k}{Dt} + \frac{1}{\theta_k^m} \alpha_k (\mathbf{q}_k^m + \tilde{\mathbf{Q}}_k^m) \cdot \nabla \theta_k^m - \nabla \cdot \alpha_k \tilde{\mathbf{Q}}_k^m$$
$$+ \alpha_k \mathcal{T}_k^m : \nabla \mathbf{v}_k + \alpha_k \mathcal{D}_k + \mathcal{E}_k^{mi} \left(1 - \frac{\theta_k^m}{\theta_i^m}\right) - \tilde{E}_k^m$$
$$- \left[\theta_k^m (s_{ki}^m - s_k^m) - u_k^m\right] \Gamma_k \geq 0. \tag{36}$$

If we supply constitutive equations for $u_k^m, \mathcal{D}_k, \mathbf{q}_k^m, \mathcal{T}_k^m, \mathcal{E}_k^{mi}$, and Γ_k, this inequality will give information about the signs of the viscous terms, the heat flux coefficient, and the relation between certain coefficients in the constitutive equations. In particular, if we use (25), we have

$$p_k^m \frac{D_k \alpha_k}{Dt} + \frac{1}{\theta_k^m} \alpha_k (\mathbf{q}_k^m + \tilde{\mathbf{Q}}_k^m) \cdot \nabla \theta_k^m - \nabla \cdot \alpha_k \tilde{\mathbf{Q}}_k^m$$
$$+ \alpha_k \mu_{kk}^m |\mathbf{D}_k|^2 + \alpha_k \mu_{ki}^m \mathbf{D}_i : \mathbf{D}_k + \alpha_k E_{k11}^m \mathbf{v}_{ki} \cdot \mathbf{D}_k \cdot \mathbf{v}_{ki}$$
$$+ \alpha_k E_{k12}^m \mathbf{v}_{ki} \cdot \mathbf{D}_k \cdot \nabla \alpha_k + \alpha_k E_{k13}^m \mathbf{v}_{ki} \cdot \mathbf{D}_k \cdot \mathbf{a}_{ki}$$
$$+ \alpha_k \mathcal{D}_k + \mathcal{E}_k^{mi} \left(1 - \frac{\theta_k^m}{\theta_i^m}\right) - \tilde{E}_k^m - \left[\theta_k^m (s_{ki}^m - s_k^m) - u_k^m\right] \Gamma_k \geq 0. \tag{37}$$

If we make the further assumption that the terms $\tilde{E}_k^m$ and $\mathcal{D}_k$ are mechanical, in that they do not involve θ_k^m, θ_i^m, or their derivatives, we conclude that

$$\mathcal{E}_k^{mi} = H_k^{mi} \left(\theta_i^m - \theta_k^m\right), \tag{38}$$

$$\mathbf{q}_k^m + \tilde{\mathbf{Q}}_k^m = \lambda_k^m \nabla \theta_k^m, \tag{39}$$

$$\Gamma_k = h_k \left(\theta_k^m (s_{ki}^m - s_k^m) - u_k^m\right), \tag{40}$$

where $H_k^{mi} \geq 0$, $\lambda_k^m \geq 0$, and $h_k \geq 0$. Moreover, we also see that

$$p_k^m \frac{D_k \alpha_k}{Dt} + \alpha_k \mu_{kk}^m |\mathbf{D}_k|^2 + \alpha_k \mu_{ki}^m \mathbf{D}_i : \mathbf{D}_k + \alpha_k E_{k11}^m \mathbf{v}_{ki} \cdot \mathbf{D}_k \cdot \mathbf{v}_{ki}$$
$$+ \alpha_k E_{k12}^m \mathbf{v}_{ki} \cdot \mathbf{D}_k \cdot \nabla \alpha_k + \alpha_k E_{k13}^m \mathbf{v}_{ki} \cdot \mathbf{D}_k \cdot \mathbf{a}_{ki} + \alpha_k \mathcal{D}_k$$
$$- \tilde{E}_k^m \geq 0. \tag{41}$$

Finally, if we also assume that $\tilde{E}_k^m$ and $\mathcal{D}_k$ have no higher than first derivatives, we see that

$$E_{k13}^m = 0, \tag{42}$$

$$p_k^m \frac{D_k \alpha_k}{Dt} + \alpha_k E_{k11}^m \mathbf{v}_{ki} \cdot \mathbf{D}_k \cdot \mathbf{v}_{ki} + \alpha_k \mathcal{D}_k - \tilde{E}_k^m = 0, \tag{43}$$

and

$$\alpha_k \mu_{kk}^m |\mathbf{D}_k|^2 + \alpha_k \mu_{ki}^m \mathbf{D}_i : \mathbf{D}_k + \alpha_k E_{k12}^m \mathbf{v}_{ki} \cdot \mathbf{D}_k \cdot \nabla \alpha_k \geq 0. \tag{44}$$

Note that the last inequality is satisfied by models that do not account for the interaction between the viscous forces in the components, that is, when $\mu_{ki}^m = 0$ and $E_{k12}^m = 0$, if $\mu_{kk}^m \geq 0$. It can also be satisfied by models which assume a relation between $\mathbf{D}_k$, $\mathbf{D}_i$, and $\mathbf{v}_{ki} \nabla \alpha_k + \nabla \alpha_k \mathbf{v}_{ki}$. For example, if

$$\mathbf{D}_i = a_{ki} \mathbf{D}_k + b_{ki} (\mathbf{v}_{ki} \nabla \alpha_k + \nabla \alpha_k \mathbf{v}_{ki}),$$

then the inequality is satisfied by taking

$$E_{k12}^m + \mu_{ki}^m b_{ki} = 0$$

and

$$\mu_{kk}^m + \mu_{ki}^m a_{ki} \geq 0.$$

14.4.2. Mesoscale Considerations

The analog of density for the mesoscale internal energy u_k^M is $\alpha_k \rho_k$. This contains a crude description of the geometry, but other geometric effects should not be overlooked. In addition, there is experience to motivate assuming that the mesoscale internal energy depends on the velocity difference, $\mathbf{v}_{ki} = \mathbf{v}_k - \mathbf{v}_i$, where $\mathbf{v}_i$ is the velocity of the interface. We require that the mesoscale internal energy be frame indifferent. Then it may be shown that the dependence reduces to dependence only on the variable $v_{ki} = |\mathbf{v}_{ki}|$. Thus, we assume

$$u_k^M(\alpha_k \rho_k, s_k^M, v_{ki}). \tag{45}$$

Note that

$$\frac{\partial u_k^M}{\partial \mathbf{v}_{ki}} = \frac{1}{v_{ki}} \frac{\partial u_k^M}{\partial v_{ki}} \mathbf{v}_{ki}. \tag{46}$$

We write $\mathbf{T}_k^M = -p_k^M \mathbf{I} + \tau_k^M$, and use the equation of balance of mass. Then the mesoscale entropy inequality (6.4.88) becomes

$$\alpha_k \rho_k \left\{ \left[\theta_k^M - \frac{\partial u_k^M}{\partial s_k^M} \right] \frac{D_k s_k^M}{Dt} + \left[\frac{p_k^M}{(\alpha_k \rho_k)^2} - \frac{\partial u_k^M}{\partial (\alpha_k \rho_k)} \right] \frac{D_k(\alpha_k \rho_k)}{Dt} - \frac{\partial u_k^M}{\partial \mathbf{v}_{ki}} \cdot \frac{D_k \mathbf{v}_{ki}}{Dt} \right\}$$

$$+ \alpha_k \frac{1}{\theta_k^M} (\mathbf{q}_k^M + \tilde{\mathbf{Q}}_k^M) \cdot \nabla \theta_k^M - \nabla \cdot \alpha_k \tilde{\mathbf{Q}}_k^M$$

$$+ \mathbf{M}_k \cdot (\mathbf{v}_k - \mathbf{v}_i) - \tilde{E}_k^M + \alpha_k \tau_k^M : \nabla \mathbf{v}_k + \left(-\theta_k^M s_k^\bullet + u_k^M\right) \Gamma_k \geq 0. \tag{47}$$

14. Closure Framework

We again use standard thermodynamic arguments. The derivative of the mesoscale temperature does not appear as the argument in any other variable in inequality (47). We assert that, given any numerical value, there is a process where $D_k s_k^M / Dt$ can attain that value, and all other quantities in (47) are zero. Then it follows that the coefficient of the time derivative of temperature must vanish, and therefore

$$\theta_k^M = \frac{\partial u_k^M}{\partial s_k^M}. \tag{48}$$

Next we substitute the forms (24) and (25) for the interfacial force density and the mesoscale stress. This yields

$$\alpha_k \rho_k \left\{ \left[\frac{p_k^M}{(\alpha_k \rho_k)^2} - \frac{\partial u_k^M}{\partial(\alpha_k \rho_k)} \right] \frac{D_k(\alpha_k \rho_k)}{Dt} + \frac{\partial u_k^M}{\partial \mathbf{v}_{ki}} \cdot \frac{D_k \mathbf{v}_{ki}}{Dt} \right\}$$
$$+ \alpha_k \frac{1}{\theta_k^M} (\mathbf{q}_k^M + \tilde{\mathbf{Q}}_k^M) \cdot \nabla \theta_k^M - \nabla \cdot \alpha_k \tilde{\mathbf{Q}}_k^M$$
$$+ A_{k1} |\mathbf{v}_{ki}|^2 + A_{k2} \nabla \alpha_k \cdot \mathbf{v}_{ki} + A_{k3} \mathbf{a}_{ki} \cdot \mathbf{v}_{ki}$$
$$+ B_{k11} \mathbf{v}_{ki} \cdot \mathbf{D}_k \cdot \mathbf{v}_{ki} + B_{k12} \mathbf{v}_{ki} \cdot \mathbf{D}_i \cdot \mathbf{v}_{ki}$$
$$- \tilde{E}_k^M + \alpha_k \mu_{kk}^M |\mathbf{D}_k|^2 + \alpha_k \mu_{ki}^M \mathbf{D}_i : \mathbf{D}_k + \alpha_k E_{k11}^M \mathbf{v}_{ki} \cdot \mathbf{D}_k \cdot \mathbf{v}_{ki}$$
$$+ \alpha_k E_{k12}^M \mathbf{v}_{ki} \cdot \mathbf{D}_k \cdot \nabla \alpha_k + \alpha_k E_{k13}^M \mathbf{v}_{ki} \cdot \mathbf{D}_k \cdot \mathbf{a}_{ki}$$
$$+ \left(-\theta_k^M s_k^M + u_k^M \right) \Gamma_k \geq 0. \tag{49}$$

We note that

$$\mathbf{a}_{ki} \cdot \mathbf{v}_{ki} = \frac{D_k v_{ki}^2 / 2}{Dt} + \mathbf{v}_{ki} \cdot \mathbf{D}_i \cdot \mathbf{v}_{ki}.$$

If we further assume that the quantities in inequality (49) are of no higher than first order, we see that

$$E_{k13}^M = 0 \tag{50}$$

and

$$\mathbf{q}_k^M + \tilde{\mathbf{Q}}_k^M = \lambda_k^M \nabla \theta_k^M, \tag{51}$$

where $\lambda_k^M \geq 0$.

Consistent with the modeling assumptions just made is

$$\tilde{\mathbf{Q}}_k^M = C_k^M(\alpha_k, v_{ki}) \mathbf{v}_{ki}. \tag{52}$$

Then

$$\nabla \cdot \alpha_k \tilde{\mathbf{Q}}_k^M = \frac{\partial \alpha_k C_k^M}{\partial \alpha_k} \mathbf{v}_{ki} \cdot \nabla \alpha_k + \alpha_k C_k^M \nabla \cdot \mathbf{v}_{ki} + \alpha_k \frac{\partial C_k^M}{\partial v_{ki}} \mathbf{v}_{ki} \cdot \nabla v_{ki} \cdot \alpha_k. \tag{53}$$

But

$$\nabla v_{ki} = \nabla (\mathbf{v}_{ki} \cdot \mathbf{v}_{ki})^{1/2} = \frac{1}{v_{ki}} \mathbf{v}_{ki} \cdot (\mathbf{D}_k - \mathbf{D}_i).$$

14.4. Entropy Restrictions

We next write the rate of deformation tensor as the sum of a deviatoric part and a spherical part, so that

$$\mathbf{D}_k = (\mathbf{D}_k)^D + \frac{1}{3}\text{tr}\,\mathbf{D}_k \mathbf{I} = (\mathbf{D}_k)^D + \frac{1}{3}\nabla \cdot \mathbf{v}_k \mathbf{I}, \tag{54}$$

where $\text{tr}\,\mathbf{D}_k = 0$. Using the equation of balance of mass, we have

$$\nabla \cdot \mathbf{v}_k = -\frac{1}{\alpha_k \rho_k}\left(\frac{D_k \alpha_k \rho_k}{Dt} - \Gamma_k\right).$$

This gives

$$\alpha_k \rho_k \left\{\left[\frac{A_{k3}}{\alpha_k \rho_k}\mathbf{v}_{ki} - \frac{\partial u_k^M}{\partial \mathbf{v}_{ki}}\right] \cdot \frac{D_k \mathbf{v}_{ki}}{Dt}\right.$$

$$+ \frac{\left[p_k^M - (\alpha_k \rho_k/3)\left[(B_{k11} + \alpha_k E_{k11}^M)v_{ki}^2 - \alpha_k C_k^M - \alpha_k (\partial C_k^M/\partial v_{ki})v_{ki}\right]\right]}{(\alpha_k \rho_k)^2}$$

$$\left. - \frac{\partial u_k^M}{\partial (\alpha_k \rho_k)}\right] \times \frac{D_k(\alpha_k \rho_k)}{Dt}\right\}$$

$$+ \alpha_k \frac{1}{\theta_k^M}(\mathbf{q}_k^M + \tilde{\mathbf{Q}}_k^M) \cdot \nabla \theta_k^M - \tilde{E}_k^M + (A_{k1} + \alpha_k C_k^M \nabla \cdot \mathbf{v}_i)|\mathbf{v}_{ki}|^2$$

$$+ (A_{k2} + \alpha_k E_{k12}^M \nabla \cdot \mathbf{v}_k)\nabla \alpha_k \cdot \mathbf{v}_{ki}$$

$$+ \left(B_{k11} + \alpha_k E_{k11}^M - \alpha_k \frac{\partial C_k^M}{\partial v_{ki}}\right)\mathbf{v}_{ki} \cdot \mathbf{D}_k^D \cdot \mathbf{v}_{ki}$$

$$+ \left(B_{k12} + \alpha_k \frac{\partial C_k^M}{\partial v_{ki}}\mathbf{v}_{ki}\right)\mathbf{v}_{ki} \cdot \mathbf{D}_i \cdot \mathbf{v}_{ki} + \alpha_k E_{k12}^M \mathbf{v}_{ki} \cdot \mathbf{D}_k^D \cdot \nabla \alpha_k$$

$$+ \alpha_k \mu_{kk}^M |\mathbf{D}_k|^2 + \alpha_k \mu_{ki}^M \mathbf{D}_i : \mathbf{D}_k$$

$$+ \left[(B_{k11} + \alpha_k E_{k11})v_{ki}^2 - \theta_k^M s_k^M + u_k^M\right]\Gamma_k \geq 0. \tag{55}$$

Since $D_k(\alpha_k \rho_k)/Dt$ and $D_k \mathbf{v}_{ki}/Dt$ can take on arbitrary values, we must have

$$p_k^M = \frac{\alpha_k \rho_k}{3}(B_{k11} + \alpha_k E_{k11}^M)v_{ki}^2 - \alpha_k C_k^M - \alpha_k \frac{\partial C_k^M}{\partial v_{ki}}v_{ki} + (\alpha_k \rho_k)^2 \frac{\partial u_k^M}{\partial (\alpha_k \rho_k)} \tag{56}$$

and

$$\frac{\partial u_k^M}{\partial \mathbf{v}_{ki}} = \frac{A_{k3}}{\alpha_k \rho_k}\mathbf{v}_{ki}. \tag{57}$$

The remaining entropy inequality becomes

$$\alpha_k \frac{1}{\theta_k^M}\lambda_k^M |\nabla \theta_k^M|^2 - \tilde{E}_k^M + (A_{k1} + \alpha_k C_k^M \nabla \cdot \mathbf{v}_i)|\mathbf{v}_{ki}|^2$$

$$+ (A_{k2} + \alpha_k E_{k12}^M \nabla \cdot \mathbf{v}_k)\nabla \alpha_k \cdot \mathbf{v}_{ki}$$

$$+ \left(B_{k11} + \alpha_k E_{k11}^M - \alpha_k \frac{\partial C_k^M}{\partial v_{ki}} \right) \mathbf{v}_{ki} \cdot \mathbf{D}_k^D \cdot \mathbf{v}_{ki}$$

$$+ \left(B_{k12} + \alpha_k \frac{\partial C_k^M}{\partial v_{ki}} \mathbf{v}_{ki} \right) \mathbf{v}_{ki} \cdot \mathbf{D}_i \cdot \mathbf{v}_{ki} + \alpha_k E_{k12}^M \mathbf{v}_{ki} \cdot \mathbf{D}_k^D \cdot \nabla \alpha_k$$

$$+ \alpha_k \mu_{kk}^M |\mathbf{D}_k|^2 + \alpha_k \mu_{ki}^M \mathbf{D}_i : \mathbf{D}_k$$

$$+ \left[(B_{k11} + \alpha_k E_{k11}) v_{ki}^2 - \theta_k^M s_k^M + u_k^M \right] \Gamma_k \geq 0. \tag{58}$$

If we assume

$$A_{k1} + \alpha_k C_k^M \nabla \cdot \mathbf{v}_i = S_k \tag{59}$$

and group the positive definite terms

$$\alpha_k \frac{1}{\theta_k^M} \lambda_k^M |\nabla \theta_k^M|^2 + S_k |\mathbf{v}_{ki}|^2 + \alpha_k \mu_{kk}^M |\mathbf{D}_k|^2 + \alpha_k \mu_{ki}^M \mathbf{D}_i : \mathbf{D}_k \geq 0. \tag{60}$$

We have

$$\tilde{E}_k^M = (A_{k2} + \alpha_k E_{k12}^M \nabla \cdot \mathbf{v}_k) \nabla \alpha_k \cdot \mathbf{v}_{ki}$$

$$+ \left(B_{k11} + \alpha_k E_{k11}^M - \alpha_k \frac{\partial C_k^M}{\partial v_{ki}} \right) \mathbf{v}_{ki} \cdot \mathbf{D}_k^D \cdot \mathbf{v}_{ki}$$

$$+ \left(B_{k12} + \alpha_k \frac{\partial C_k^M}{\partial v_{ki}} \mathbf{v}_{ki} \right) \mathbf{v}_{ki} \cdot \mathbf{D}_i \cdot \mathbf{v}_{ki} + \alpha_k E_{k12}^M \mathbf{v}_{ki} \cdot \mathbf{D}_k^D \cdot \nabla \alpha_k$$

$$+ \left[(B_{k11} + \alpha_k E_{k11}) v_{ki}^2 - \theta_k^M s_k^M + u_k^M \right] \Gamma_k. \tag{61}$$

14.4.3. Conclusion

Determining constitutive equations for multicomponent fluids is facilitated by using representations for the dependences, and subjecting these representations to requirements of invariance and restrictions from entropy considerations. The resulting constitutive equations are still complicated, but such concerns as completeness and correctness are eased by such an approach, which allows the researcher to include or neglect terms based on scaling or measurement, rather than ignorance or faulty intuition.

15
Relation of Microstructure to Constitutive Equations

This chapter is devoted to deriving and using microstructural information to find constitutive equations, giving relationships between certain average quantities and the average fields. The physical situation is one in which corpuscular models can be highly useful in motivating the forms of the constitutive equations. In effect, the study of the mechanics of multicomponent fluids necessitates the study of the form of the momentum exchange term and the stress in the momentum equation. The study of these terms is aided by a physically appropriate corpuscular model, consisting of solutions for flows in this model, which are sometimes exact but usually approximate. Once these solutions are found, they can be averaged to find the force of interaction in the continuum equations.

The general situation with respect to finding such solutions is not satisfactory. The reason is that most solutions are either approximate, or else involve series or integrals which must be treated approximately. Such solutions are not nearly as useful as exact solutions. Even for modestly realistic flows, the physical situation is complex. For example, when the fluid is a Navier–Stokes fluid and the particles are rigid, the ability to produce exact solutions to the equations of hydrodynamics is limited. Controlling factors include the nature of the flow far from the particles, the shape of the particles, the relative motion and interaction of the particles, and the difficulty of dealing with inertial terms that are inherently nonlinear.

Most studies of the force of interaction are limited to the study of the flow of a fluid around a single spherical particle. The richness of the results comes from the complexity of the flow at infinity, and the possibility of approximating particle field effects, such as gradients in the particle concentration and velocity.

Another matter arises naturally. In most of the work in this area, it is assumed that the particles interact through hydrodynamic forces only. That is, they do not collide. This is physically appropriate some of the time, but not all of the time. There are models for collisions of particles in a vacuum, and it is appropriate to consider whether this kind of physics can be incorporated into the study of multicomponent materials.

A solution for a flow in the context of a corpuscular model must be averaged in order to obtain results pertinent to a multicomponent flow model. The appropriate averaging is the ensemble average; however, it is often difficult to describe the ensemble sufficiently accurately to allow the computation of averaged momentum exchange and stress. Even for the simplest situation involving the flow around a single spherical particle, the probability density function for its position is needed. In some situations it may be appropriate to ascribe a probability density function for the velocity of the center of the sphere.

In Chapters 15 and 16, we undertake the program outlined above. In Section 15.1, we describe ways in which the concepts of ensemble averaging can be applied to solutions of corpuscular flows. Use of the cell model for computing ensemble averaged values for the momentum interaction and the stress is described. In Section 15.3, the solution for the motion of an incompressible, inviscid fluid around a single sphere is used to obtain expressions for momentum exchange and stress. In Section 15.4, the solution for Stokes flow of an incompressible linearly viscous fluid is given, and effective medium theory is used to obtain averaged viscous stress and drag. Finally, in Chapter 16, a model for the apparent viscosity of a particle–fluid mixture is derived using a kinetic theory model.

15.1. Averaging Techniques

The flow of a multicomponent fluid usually involves scales that can be observed without complicated instrumentation. This is both a blessing and a curse. It is an advantage to see the fine structure, and to have the sorts of corpuscular models for the solutions on that scale. On the other hand, it is a disadvantage to see the complexity of one realization without observing the effects inherent in the averaging that is necessary to describe the average flow features that allow for practical predictions.

The variations inherent in the corpuscular flows must be averaged to obtain momentum exchange and stress models for use in the balance equations. The averaging method applied to derive expressions for the momentum exchange and the stress must sample the variability in the properties that determine the appropriate solution for the corpuscular flow, and yield results for these expressions at a point in space, as they depend on flow properties.

To do this, we examine a point in space, and describe the probability that there is a particle or a number of particles near it. We shall describe the framework for the description of the probability, and some situations for the practical application of it to specific flows.

Consider the following flow situation. Let an incompressible Navier–Stokes fluid containing N corpuscular units be flowing near the point $\mathbf{x}$ at instant t. Assume that the corpuscular units occupy regions $P_1, \ldots, P_N$. The Navier–Stokes equations of motion are

$$\nabla \cdot \mathbf{v} = 0, \tag{1}$$

$$\rho \dot{\mathbf{v}} = \nabla \cdot \mathbf{T} + \rho_f \mathbf{b}_f, \tag{2}$$

where the stress is given by

$$\mathbf{T} = -p\mathbf{I} + \mu_f \operatorname{sym} \nabla \mathbf{v}. \tag{3}$$

The corpuscular units also have some constitutive equations governing their motion. These may describe, for example, a fluid of a different viscosity and density, or elastic solid particles, or rigid particles. Different models apply to different situations.

One such model is for rigid solid particles. In this case, the motion inside each particle satisfies

$$\operatorname{sym} \nabla \mathbf{v} = 0. \tag{4}$$

It follows that inside each sphere

$$\nabla \cdot \mathbf{v} = 0. \tag{5}$$

The equation of motion in the solid is

$$\nabla \cdot \mathbf{T} + \rho_p \mathbf{b}_p = 0. \tag{6}$$

Since the solid material is assumed rigid, there is no constitutive equation for the stress. We have allowed the body force term to be different for the two components. This will let us consider motions where the particles are held fixed, or move with the fluid.

In the most complicated situations that we shall consider here, $N = 1$ or 2, and the particles are identical spheres. We next develop averaging strategies for these situations.

15.1.1. Particle Distributions

This is the situation in which the description of the statistics in terms of the particle distribution function $f^{(N)}$ is most useful. We recall some of the notation from Section 10.1.

Suppose $g(\mathbf{x}, t | \mathbf{z}_1, \ldots, \mathbf{z}_N)$ is the value of some field g at point $\mathbf{x}$, at time t, when the spheres are at $\mathbf{z}_1, \ldots, \mathbf{z}_N$. Then, the average of g is

$$\bar{g}(\mathbf{x}, t) = \int \cdots \int f^{(N)}(\mathbf{z}_1, \ldots, \mathbf{z}_N, t) g(\mathbf{x}, t | \mathbf{z}_1, \ldots, \mathbf{z}_N) \, d\mathbf{z}_1 \ldots d\mathbf{z}_N. \tag{7}$$

This also can be written as a hierarchy of conditional averages

$$\overline{g}(\mathbf{x},t) = \int f^{(1)}(\mathbf{z},t)\overline{g}^{(1)}(\mathbf{x},t|\mathbf{z}_1)\,d\mathbf{z}_1$$
$$= \iint f^{(2)}(\mathbf{z},\mathbf{z}_2,t)\overline{g}^{(2)}(\mathbf{x},t|\mathbf{z}_1,\mathbf{z}_2)\,d\mathbf{z}_1\,d\mathbf{z}_2, \quad (8)$$

etc., where $\overline{g}^{(1)}$ is the average of g conditioned on there being a sphere at $\mathbf{z}_1$, and $\overline{g}^{(2)}$ is the average of g conditioned on there being spheres at $\mathbf{z}_1$ and $\mathbf{z}_2$. This works because the conditional averages are integrations holding a certain subset of spheres fixed. Upon integration over the fixed spheres, the total, or unconditional average is obtained. Often, we abbreviate $\mathbf{z}_1$ by $\mathbf{z}$.

The average velocity for each component is defined by

$$\overline{\mathbf{v}}_k^{x\rho} = \frac{\overline{X_k \rho \mathbf{v}}}{\alpha_k \rho_k} = \frac{\overline{X_k \mathbf{v}}}{\alpha_k}, \quad (9)$$

since the density of each component is constant. We also define conditionally averaged velocities by

$$\overline{\mathbf{v}}_k^{(1)} = \frac{\overline{X_k \mathbf{v}}^{(1)}}{\alpha_k^{(1)}}, \quad (10)$$

and

$$\overline{\mathbf{v}}_k^{(2)} = \frac{\overline{X_k \mathbf{v}}^{(2)}}{\alpha_k^{(2)}}, \quad (11)$$

etc. In this chapter, we shall assume that there are many particles in the flow, and that

$$\alpha_p^{(1)} = \alpha_p^{(2)} = \alpha_p = \text{constant}$$

except inside the particles being held fixed.

Conditional Averages

Suppose $\mathbf{T}(\mathbf{x},t|\mathbf{z}_1,\ldots,\mathbf{z}_N)$ is the stress $\mathbf{T}$ at point $\mathbf{x}$, at time t, when the spheres are at $\mathbf{z}_1,\ldots,\mathbf{z}_N$. Then, the average stress is

$$\overline{\mathbf{T}}(\mathbf{x},t) = \int\cdots\int f^{(N)}(\mathbf{z}_1,\ldots,\mathbf{z}_N,t)\mathbf{T}(\mathbf{x},t|\mathbf{z}_1,\ldots,\mathbf{z}_N)\,d\mathbf{z}_1\ldots d\mathbf{z}_N.$$

This also can be written as a hierarchy of conditional averages

$$\overline{\mathbf{T}}(\mathbf{x},t) = \int f^{(1)}(\mathbf{z},t)\overline{\mathbf{T}}^{(1)}(\mathbf{x},t|\mathbf{z}_1)\,d\mathbf{z}_1$$
$$= \iint f^{(2)}(\mathbf{z},\mathbf{z}_2,t)\overline{\mathbf{T}}^{(2)}(\mathbf{x},t|\mathbf{z}_1,\mathbf{z}_2)\,d\mathbf{z}_1\,d\mathbf{z}_2,$$

etc., where $\overline{\mathbf{T}}^{(1)}$ is the average stress conditioned on there being a sphere at $\mathbf{z}_1$, and $\overline{\mathbf{T}}^{(2)}$ is the average stress conditioned on there being spheres at $\mathbf{z}_1$ and $\mathbf{z}_2$, where the superscript $^{(1)}$ denotes the conditional average with the sphere centered at $\mathbf{z}$.

The unconditionally averaged interfacial force density can be written as

$$\mathbf{M}_k(\mathbf{x}, t) = -\overline{\nabla X_k \cdot \mathbf{T}}$$
$$= \int_{|\mathbf{z}-\mathbf{x}|=a} \int \cdots \int f^{(N)}(\mathbf{z}, \ldots, \mathbf{z}_2, t) \mathbf{T}(\mathbf{x}, t|\mathbf{z}, \ldots, \mathbf{z}_N) \cdot \mathbf{n}_k$$
$$\times d\Omega_z \, d\mathbf{z}_2 \ldots d\mathbf{x}_N, \tag{12}$$

where $\mathbf{n}_k$ is the unit normal interior to component k and $d\Omega_z$ denotes the area element over the sphere of radius a in the variable $\mathbf{z}$. The conditionally averaged interfacial force density can be written as

$$\mathbf{M}_k^{(1)}(\mathbf{x}, t|\mathbf{z}) = -\overline{\nabla X_k \cdot \mathbf{T}}^{(1)}(\mathbf{x}, t|\mathbf{z})$$
$$= \int_{|\mathbf{z}_2-\mathbf{x}|=a} f^{(2)}(\mathbf{z}, \mathbf{z}_2, t) \overline{\mathbf{T}}^{(2)}(\mathbf{x}, t|\mathbf{z}, \mathbf{z}_2) \cdot \mathbf{n}_k \, d\Omega_{z_2}. \tag{13}$$

Note that $\overline{\mathbf{T}}_k^{x(1)}$ and $\overline{\mathbf{T}}^{(1)}$ are not the same thing.

Again, the interfacial part can be written in terms of stresses computed with more spheres fixed:

$$\mathbf{M}_k^{(2)}(\mathbf{x}, t|\mathbf{z}, \mathbf{z}_2) = -\overline{\nabla X_k \cdot \mathbf{T}}^{(2)}(\mathbf{x}, t|\mathbf{z}, \mathbf{z}_2)$$
$$= \int_{|\mathbf{z}_3-\mathbf{x}|=a} f^{(3)}(\mathbf{z}, \mathbf{z}_2, \mathbf{z}_3, t) \overline{\mathbf{T}}^{(3)}(\mathbf{x}, t|\mathbf{z}, \mathbf{z}_2, \mathbf{z}_3) \cdot \mathbf{n}_k \, d\Omega_{z_3}. \tag{14}$$

Clearly this is a procedure that will terminate only when conditional averages keeping each subset of the spheres fixed can be obtained. It would be desirable to terminate the procedure using arguments about the solutions for the motion about one or a few particles. Such solutions will give insight into the forms to assume for constitutive equations for more general situations.

Batchelor's Approximation

The average stress inside the particles can be found by using a stratagem introduced by Batchelor [8] which is based on the approximation

$$\overline{\mathbf{T}}^{(1)}(\mathbf{x}, t|\mathbf{z}) = \overline{\mathbf{T}}^{(1)}(\mathbf{z} + \mathbf{x}', t|\mathbf{x} - \mathbf{x}')$$
$$= \overline{\mathbf{T}}^{(1)}(\mathbf{z}, t | \mathbf{x}) + O\left(\frac{a}{L}\right), \tag{15}$$

where L is a macroscopic length scale. Then

$$\alpha_p(\mathbf{x}, t)\overline{\mathbf{T}}_p^x(\mathbf{x}, t) = \iiint_{|\mathbf{x}-\mathbf{z}|<a} f^{(1)}(\mathbf{z}, t)\overline{\mathbf{T}}^{(1)}(\mathbf{z}, t|\mathbf{z}) \, d\mathbf{z}$$
$$\approx \iiint_{|\mathbf{x}-\mathbf{z}|<a} f^{(1)}(\mathbf{z}, t)\overline{\mathbf{T}}^{(1)}(\mathbf{z}, t|\mathbf{x}) \, d\mathbf{z},$$

to order a/L. Now, since $\nabla \mathbf{z} = \mathbf{I}$, and $\nabla \cdot \overline{\mathbf{T}}^{(1)}(\mathbf{x}, t|\mathbf{z}) = 0$ inside the particle, we have

$$\alpha_p \overline{\mathbf{T}}_p^x \approx \int_{|\mathbf{x}-\mathbf{z}|<a} f^{(1)}(\mathbf{z}, t)\nabla \mathbf{z} \cdot \overline{\mathbf{T}}^{(1)}(\mathbf{z}, t|\mathbf{x}) \, d\mathbf{z}$$

$$\approx \int_{|\mathbf{x}-\mathbf{z}|<a} f^{(1)}(\mathbf{z},t)\nabla\cdot[\overline{\mathbf{T}}^{(1)}(\mathbf{z},t|\mathbf{x})\mathbf{z}]\,d\mathbf{z}$$

$$\approx \int_{|\mathbf{x}-\mathbf{z}|=a} f^{(1)}(\mathbf{z},t)\mathbf{n}\cdot[\overline{\mathbf{T}}^{(1)}(\mathbf{z},t|\mathbf{x})\mathbf{z}]\,ds_{\mathbf{z}}. \tag{16}$$

This relation allows us to compute the average stress inside the particles using only information on the particle surfaces.

The key to computing the particle stress and the interaction force density is to consider $\overline{\mathbf{T}}^{(1)}(\mathbf{x},t|\mathbf{z})$ as a conditional average in the following sense:

$$f^{(1)}(\mathbf{z},t)\overline{\mathbf{T}}^{(1)}(\mathbf{x},t|\mathbf{z})$$
$$= \int\!\!\int\ldots\int f^{(N)}(\mathbf{z},\mathbf{z}_2,\ldots,\mathbf{z}_N)\overline{\mathbf{T}}(\mathbf{x},t|\mathbf{z},\mathbf{z}_2,\ldots,\mathbf{z}_N)\,d\mathbf{z}_2\ldots d\mathbf{z}_N,$$
$$= \int f^{(2)}(\mathbf{z}_2|\mathbf{z})\overline{\mathbf{T}}^{(2)}(\mathbf{x},t|\mathbf{z},\mathbf{z}_2)\,d\mathbf{z}_2, \tag{17}$$

where $\mathbf{T}(\mathbf{x},t|\mathbf{z},\mathbf{z}_2,\ldots,\mathbf{z}_N)$ is the exact stress in a realization with spheres centered at $\mathbf{z},\mathbf{z}_2,\ldots,\mathbf{z}_N$, and $\overline{\mathbf{T}}^{(2)}(\mathbf{x},t|\mathbf{z},\mathbf{z}_2)$ is the stress averaged over all the spheres except the ones at $\mathbf{z}$ and $\mathbf{z}_2$.

There are essentially three different approaches to obtaining constitutive equations for $\overline{\mathbf{T}}_k^r$ and $\mathbf{M}_p$ using ensemble averaging. All have different treatments for the higher conditional averages needed in the formulas from the previous section. The first two are based on the idea that the presence of the other particles in the flow can be treated by adding their effects one at a time. This leads to a procedure that gives results correct to $O(\alpha_p)$ first, then an $O(\alpha_p^2)$ correction, etc. This is the method developed by Batchelor [8] and Hinch [42], and we shall refer to it as the *hierarchical* method. At some stage, it necessitates the treatment of integrals with respect to $\mathbf{z}$ over all positions that a particle can occupy. If the flow domain is infinite, this can lead to a divergent integral. If the nearest-neighbor distribution is used, the integral converges because of the exponential decay of the distribution function with distance from $\mathbf{x}$. If we assume that the distribution function for the sphere is uniform, then the integrals in (8) are singular, and some other procedure is necessary. One such method is called "renormalization." In this method, the singular integral is replaced by a function that contains the physical effect of the divergent integral. Hierarchical methods give acceptable results for low concentrations of particles, where the approximation of the effect of the particles farthest away is best.

The other approach involves assuming that the integral in (7) has a certain form in terms of the fluid and particle velocities. The resulting equations of motion for the velocity, with certain unknown coefficients, are solved for the motion around a single particle, and the resulting velocity field is used to calculate the integrals defining average variables from (7). These integrals so computed involve the form of the coefficient assumed to get it. The resulting equation is solved for the unknown coefficient. This method is originally attributed to Brinkman [15] and is exploited by Lundgren [60]. It leads to forms of the equations that have terms appearing as an extra viscosity, and are called "effective media" equations. The assumption that the integrated effect of the integral term and its specific value are related is called the assumption (sometimes "approximation") of "self-consistency." This method seems suited to higher concentrations, but has the problem that it approximates the flow near the particles by the (unrealistic) flow of the "effective medium."

It can be argued [76] that the process of renormalization for such problems is moot because the spatial domain Ω is given and finite. We can attempt to derive equations of motion based on increasingly accurate solutions (accounting for more "reflections," for example) of the equations of motion. On the other hand, it is unsettling to use a model whose validity depends on the size of the domain. In using the "nearest-neighbor" definitions of conditional averages, we do not have the convergence problems for which renormalization methods were invented.

15.2. Cell Model

The one- and two-sphere distributions of nearest neighbors given by (10.1.22) and (10.1.28) are complicated to apply in the context of deriving information about constitutive equations. Rather, we use an approximation to the one-sphere distribution, called the "cell model" that is easier to apply. Note that the one-sphere distribution given by (10.1.22) decays exponentially in $|\mathbf{x} - \mathbf{z}'_1|^3$. Thus it is a good approximation to replace the exponential probability distribution for the sphere center location by the characteristic function of the sphere of radius R centered at $\mathbf{x}$, that is, a function that is constant in a sphere of radius R centered at $\mathbf{x}$, and zero outside this sphere. This is depicted in Figure 15.1. We shall call this sphere a "cell," and assume that exactly one spherical particle of the dispersed phase material lies in each cell (see Figure 15.2). The probability of finding the center of the sphere closest to $\mathbf{x}$ within dV of $\mathbf{z}$ vanishes outside the cell, and inside the cell will be taken to be $dV/(4\pi R^3/3)$. Inside that cell, the velocity and pressure fields are given by the solution for the flow around a single sphere. We choose R so that

$$\alpha_p = \frac{\frac{4}{3}\pi a^3}{\frac{4}{3}\pi R^3}. \tag{18}$$

The solution for the flow around a single sphere will be a good approximation to the actual flow in a situation with many particles, if the spheres are sufficiently far apart that the flow disturbances due to the individual spheres do not interact. That is, the flow must be sufficiently dilute. Thus, each sphere is isolated in the sense that it only interacts with its neighbors through the averaged fields. We assume that each sphere has radius a and velocity $\mathbf{v}_p$, and responds to the fluid velocity $\mathbf{v}_f$ at $|\mathbf{x}| \to \infty$.

We now introduce the averaging processes. For the cell approximation to the ensemble average, the ensemble is the set of flows that can occur at $\mathbf{x}$ with the

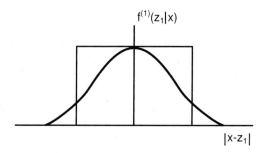

FIGURE 15.1. Comparison of nearest-neighbor and "top hat" distributions.

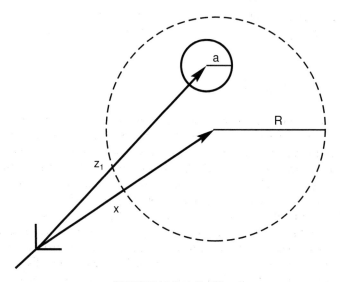

FIGURE 15.2. A "cell"

sphere center occupying different positions **z** in the cell. If f is some microscopic process, we denote the dependence on the ensemble for this case by

$$f = f(\mathbf{x}, t|\mathbf{z}). \tag{19}$$

This is the value of f at the point **x**, at time t, given that the sphere is centered at **z**. For this case, the average is performed by integrating over all possible positions of the sphere center. That is, the center of the sphere at **z** can lie anywhere in the sphere of radius R. Since the density is constant, the mass-weighted average and the volume, or X_k, weighted average are the same. The volume-weighted average of f over the solid component is given by

$$\overline{f}_s^x(\mathbf{x}, t) = \frac{1}{\frac{4}{3}\pi a^3} \int_0^a \int_{\Omega(r)} f(\mathbf{x}, t|\mathbf{z}) \, d\Omega \, dr, \tag{20}$$

where the integration is over the **z** variable, and $\Omega(r)$ is the spherical shell of radius r centered at the point **x**. Thus, the averaging is done over the positions of the center of the sphere in the cell so that the sphere contains the point **x**. Therefore, the integration in the radial variable for **z** runs from 0 to a. For $r > a$, the sphere does not contain the point **x**.

The volume-weighted average of f over the fluid component is given by

$$\overline{f}_f^x(\mathbf{x}, t) = \frac{1}{\frac{4}{3}\pi(R^3 - a^3)} \int_a^R \int_{\Omega(r)} f(\mathbf{x}, t|\mathbf{z}) \, d\Omega \, dr . \tag{21}$$

If $g(\mathbf{x}, t|\mathbf{z})$ is a field defined on the interface, then the interfacial average of g is

$$\overline{g}_i(\mathbf{x}, t) = \overline{\frac{\partial X_k}{\partial n} g} \bigg/ \overline{\frac{\partial X_k}{\partial n}}$$

$$= \frac{1}{4\pi a^2} \int_{\Omega(a)} g(\mathbf{x}, t|\mathbf{z}) \, d\Omega . \tag{22}$$

Let $\mathbf{x}' = \mathbf{x} - \mathbf{z}$. We see that if $|\mathbf{x}'| < a$, the material making up the sphere occupies the field point **x**, if $|\mathbf{x}'| = a$, the interface occupies the material point **x**, and if $|\mathbf{x}'| > a$, the fluid occupies the field point **x**.

15.2.1. Some Mathematical Results

It is convenient to have expressions for the integrals of powers of $\mathbf{x}'$ over a spherical shell $\Omega(r)$ given by $|\mathbf{x}'| = r$. For these integrals, we note that

$$\int_{\Omega(r)} \mathbf{x}' \ldots \mathbf{x}' \, d\Omega = 0 \tag{23}$$

if the factor $\mathbf{x}'$ appears an odd number of times, and

$$\int_{\Omega(r)} d\Omega = 4\pi r^2 , \tag{24}$$

$$\int_{\Omega(r)} \mathbf{x}'\mathbf{x}' \, d\Omega = \frac{4}{3}\pi r^4 \mathbf{I} , \tag{25}$$

$$\int_{\Omega(r)} \mathbf{x}'\mathbf{x}'\mathbf{x}'\mathbf{x}' \, d\Omega = \frac{4}{15}\pi r^6 \mathbf{\Sigma} , \tag{26}$$

where $\mathbf{\Sigma}$ is a fourth-order isotropic tensor which is symmetric in all of its components. It is given in Cartesian coordinates by

$$\Sigma_{ijkl} = \delta_{ij}\delta_{kl} + \delta_{ik}\delta_{jl} + \delta_{il}\delta_{jk}. \tag{27}$$

We further note that if **v** is a vector, and **e** is a symmetric second-order tensor with $e_{ii} = 0$, then

$$\Sigma_{ijkl} v_j e_{kl} = 2v_j e_{ji} , \tag{28}$$

or, in the invariant notation,

$$\mathbf{\Sigma} : \mathbf{ve} = 2\mathbf{v} \cdot \mathbf{e} . \tag{29}$$

15.3. Inviscid Fluid Flowing Around a Sphere

In this section, we use the "cell model" to calculate expressions for the interfacial force and stresses for a dilute suspension of spheres. The cell approximation to the ensemble average gives interesting results for variables related to the fluid, including the interfacial force and the Reynolds stress. It can also can be used with some considerations of elasticity inside the sphere to give results on the dispersed component stress for solid spheres and for bubbles with strong surface tension effects.

The solution for the motion of an incompressible, inviscid fluid around a single sphere is used to obtain expressions for averaged quantities for which constitutive equations are needed. In Section 15.3.2, we give expressions for the interfacial force density, the Reynolds stress in the fluid, and the averaged stress in an elastic sphere. In Section 15.3.3, we calculate the lift force by considering the motion of an inviscid, incompressible fluid around a sphere, with the fluid far from the sphere in rigid rotation. In Section 15.3.4, we allow the probability distribution for the location of the sphere center to be nonuniform, representing the effect of a concentration gradient. Finally, in Section 15.3.5, we include the effect of the particle velocity fluctuations on the computed quantities.

15.3.1. Microscale Solution

We consider the motion of an inviscid fluid outside a single sphere of radius a. The sphere is assumed to have velocity $\mathbf{v}_p$. We assume irrotational flow, so that $\nabla \wedge \mathbf{v} = 0$, outside the sphere. For a point $\mathbf{x}$ outside the sphere, the fluid velocity at $\mathbf{x}$ can be expressed in terms of the velocity potential by

$$\mathbf{v}(\mathbf{x}) = \nabla \phi(\mathbf{x}). \tag{30}$$

The continuity equation becomes

$$\nabla \cdot \mathbf{v} = \nabla \cdot \nabla \phi = \nabla^2 \phi = 0. \tag{31}$$

Substituting (30) into the momentum equation (14) gives

$$\nabla \left(p + \rho_f \left[\frac{\partial \phi}{\partial t} + \frac{1}{2} v^2 \right] \right) = 0, \tag{32}$$

so that

$$p + \rho_f \left(\frac{\partial \phi}{\partial t} + \frac{1}{2} |\nabla \phi|^2 \right) = p_0, \tag{33}$$

where p_0 can depend on time. This is a Bernoulli theorem relating the pressure and the velocity potential.

The boundary condition at the surface of the sphere is

$$\mathbf{n} \cdot \mathbf{v}_p = \mathbf{n} \cdot \mathbf{v} = \mathbf{n} \cdot \nabla \phi \quad \text{at} \quad |\mathbf{x} - \mathbf{z}| = a, \tag{34}$$

15.3. Inviscid Fluid Flowing Around a Sphere

where **n** is the normal to the surface of the sphere. The boundary condition far from the sphere is

$$\phi \to \phi_\infty \quad \text{as} \quad |\mathbf{x} - \mathbf{z}| \to \infty, \tag{35}$$

where

$$\phi_\infty = \mathbf{v}_0(t) \cdot \mathbf{x} + \frac{1}{2} \mathbf{x} \cdot \mathbf{L}_f \cdot \mathbf{x} \tag{36}$$

is the velocity potential that would exist in the fluid if the sphere were not present. Here $\mathbf{v}_0(t)$ is the (unsteady) velocity of the fluid at the origin, and $\mathbf{L}_f$ is the velocity gradient for the fluid. By (10) the trace of $\mathbf{L}_f$ vanishes. For simplicity, we shall assume that $\mathbf{L}_f$ is symmetric. The effects of vorticity are considered in Section 15.3.3. A convenient form for the solution of this problem is [56] and [93],

$$\phi = \mathbf{v}_0(t) \cdot \mathbf{x} + \frac{1}{2} \mathbf{x} \cdot \mathbf{L}_f \cdot \mathbf{x} + \frac{1}{2} \left(\frac{a}{r}\right)^3 (\mathbf{v}_p(\mathbf{z}, t) - \mathbf{v}_0(t) - \mathbf{z} \cdot \mathbf{L}_f) \cdot (\mathbf{x} - \mathbf{z})$$
$$+ \frac{1}{3} \left(\frac{a}{r}\right)^5 (\mathbf{x} - \mathbf{z}) \cdot \mathbf{L}_f \cdot (\mathbf{x} - \mathbf{z}). \tag{37}$$

The fluid velocity at **x** given a sphere at **z** is then

$$\mathbf{v}(\mathbf{x}, t|\mathbf{z}) = \nabla \phi(\mathbf{x}, t|\mathbf{z}),$$
$$= \mathbf{v}_f(\mathbf{x}) + \frac{1}{2} \left(\frac{a}{r}\right)^3 (\mathbf{v}_f(\mathbf{z}) - \mathbf{v}_p(\mathbf{z})) \cdot \left[\mathbf{I} - 3\frac{(\mathbf{x} - \mathbf{z})(\mathbf{x} - \mathbf{z})}{r^2}\right]$$
$$+ \frac{2}{3} \left(\frac{a}{r}\right)^5 (\mathbf{x} - \mathbf{z}) \cdot \mathbf{L}_f \cdot \left[\mathbf{I} - \frac{5}{2}\frac{(\mathbf{x} - \mathbf{z})(\mathbf{x} - \mathbf{z})}{r^2}\right]. \tag{38}$$

Note that

$$\mathbf{v}_f(\mathbf{x}) = \mathbf{v}_0(t) + \mathbf{L}_f \cdot \mathbf{x} \tag{39}$$

is the fluid velocity that would exist at **x** if the sphere were not present, and $\mathbf{v}_f(\mathbf{z}) - \mathbf{v}_p(\mathbf{z})$ is the relative velocity between the sphere and the fluid evaluated at the sphere center.

15.3.2. Averages for Inviscid Fluids

In this section we perform the ensemble average on several parameters of interest using the cell approximation. Consider a sphere located at a point **z** in a flow field, moving with velocity $\mathbf{v}_p(\mathbf{z}, t)$.

First, it is necessary to relate the parameters specifying the microscale flow to the average variables.

Averaged Velocities

The average velocity of the fluid phase is given by

$$\overline{\mathbf{v}}_f^{x\rho} = \frac{1}{\frac{4}{3}\pi(R^3 - a^3)} \int_a^R \int_{\Omega(r)} \mathbf{v}(\mathbf{x}, t|\mathbf{z}) \, d\Omega \, dr, \tag{40}$$

where the integration is over the **z** variable.

In order to evaluate the integrals appearing in (40), we must express the **z** dependence of the velocities in terms of **x** and $\mathbf{x}' = \mathbf{x} - \mathbf{z}$. We have

$$\mathbf{v}_f(\mathbf{z}) = \mathbf{v}_f(\mathbf{x}) - \mathbf{x}' \cdot \mathbf{L}_f, \tag{41}$$

and

$$\mathbf{v}_p(\mathbf{z}) = \mathbf{v}_p(\mathbf{x}) - \mathbf{x}' \cdot \mathbf{L}_p, \tag{42}$$

where $\mathbf{L}_p$ is the velocity gradient tensor for the particle motion. Recall that it is constant, but not necessarily symmetric. This allows the dispersed component to rotate and to dilate on average, by undergoing rigid motions that have a rotation $(\nabla \mathbf{v}_p - (\nabla \mathbf{v}_p)^T \neq 0)$ and dilatation $(\nabla \cdot \mathbf{v}_p \neq 0)$. Substituting this into (38) gives

$$\mathbf{v}(\mathbf{x}, t | \mathbf{z}) = \mathbf{v}_f(\mathbf{x}) + \frac{1}{2}\left(\frac{a}{r}\right)^3 [\mathbf{v}_f(\mathbf{x}) - \mathbf{v}_p(\mathbf{x}) - \mathbf{x}' \cdot (\mathbf{L}_f - \mathbf{L}_p)] \cdot \left[\mathbf{I} - 3\frac{\mathbf{x}'\mathbf{x}'}{r^2}\right]$$
$$+ \frac{2}{3}\left(\frac{a}{r}\right)^5 \mathbf{x}' \cdot \mathbf{L}_f \cdot \left[\mathbf{I} - \frac{5}{2}\frac{\mathbf{x}'\mathbf{x}'}{r^2}\right]. \tag{43}$$

Using these results in (40) gives

$$\bar{\mathbf{v}}_c^{xp}(\mathbf{x}, t) = \mathbf{v}_f(\mathbf{x}, t). \tag{44}$$

The interfacial averaged velocity of the fluid is given by

$$\bar{\mathbf{v}}_{ci}(\mathbf{x}, t) = \frac{1}{4\pi a^2} \int_{\Omega(a)} \mathbf{v}(\mathbf{x}, t | \mathbf{z}) \, d\Omega. \tag{45}$$

Substituting and performing the integrations leads to the result

$$\bar{\mathbf{v}}_{ci}(\mathbf{x}, t) = \mathbf{v}_f(\mathbf{x}, t). \tag{46}$$

This result is a little surprising at first. The fluid at the surface of the sphere satisfies the condition $\mathbf{n} \cdot \mathbf{v} = \mathbf{n} \cdot \mathbf{v}_p$, but is allowed to slip in the tangential direction. After the passage of the sphere, the fluid that was momentarily in contact with the surface of the sphere is again moving with the fluid. The result says that even during the time that it is in contact with the surface of the sphere, its average velocity is still equal to the average velocity of the fluid, and not of the sphere.

Averaged Pressures

Let us compute averaged pressures using this formalism. The exact pressure can be computed by the Bernoulli theorem (33)

$$p = p_0 + \rho_f \left(\frac{1}{2}|\nabla \phi|^2 + \frac{\partial \phi}{\partial t}\right). \tag{47}$$

In order to evaluate the derivatives in (47), we note that **x** is constant during t derivatives, but $\partial \mathbf{z}/\partial t = \mathbf{v}_p(\mathbf{z})$. Also, when evaluating $\nabla \phi$, both t and **z** are held constant.

The expression for the pressure is unwieldy, and we do not give it here. If the expression is averaged over the fluid, and terms of order L_f^2, L_p^2, and $\mathbf{L}_f \cdot \mathbf{L}_p$

15.3. Inviscid Fluid Flowing Around a Sphere

are neglected, we have

$$\overline{p}_c^x = p_0 - \rho_f \frac{\partial \mathbf{v}_f}{\partial t} \cdot \mathbf{x} - \frac{1}{2}\mathbf{v}_f(\mathbf{x}) \cdot \mathbf{v}_f(\mathbf{x}). \tag{48}$$

The interfacial averaged pressure is given by

$$\overline{p}_{ci} = \overline{p}_c^x - \frac{1}{4}\alpha_c \rho_f |\mathbf{v}_f - \mathbf{v}_p|^2. \tag{49}$$

Interfacial Force

The interfacial force density that is calculated is $-\mathbf{M}_c$, and is given by $-\mathbf{M}_c = -\overline{p\nabla X_c}$. Thus,

$$-\mathbf{M}_c = -\frac{1}{\frac{4}{3}\pi R^3} \int_{\Omega(a)} \mathbf{n} p(\mathbf{x}, \mathbf{z}, t) \, d\Omega. \tag{50}$$

Note that the integration is again over the z variable. This corresponds to considering the subensemble where the sphere is tangent to the point **x**, so that **z** is integrated over the spherical shell of radius a centered at **x**. The interfacial force density, then, is the "sampling" over all cases in the ensemble where the interface occupies the point **x**, and therefore, the integration allows the **z** variable to run over the spherical shell of radius a. This is depicted in Figure 15.3. If we substitute (33) for the pressure, and recognize that $\mathbf{n} = \mathbf{x}'/a$, the result is that

$$-\mathbf{M}_c = \alpha_d \rho_c \left(\frac{1}{2} \left[\frac{\partial \mathbf{v}_f}{\partial t} - \frac{\partial \mathbf{v}_p}{\partial t} + \mathbf{v}_f \cdot \mathbf{L}_f - \mathbf{v}_p \cdot \mathbf{L}_p \right] \right.$$
$$\left. + (\mathbf{v}_p - \mathbf{v}_f) \cdot \left[\frac{4}{5}\mathbf{L}_p - \frac{9}{20}\mathbf{L}_p^T + \frac{7}{20}\mathbf{L}_f + \frac{1}{20}(\mathrm{tr}\,\mathbf{L}_p)\mathbf{I} \right] \right). \tag{51}$$

This represents a combination of known forces on a sphere, and some generalizations. The first term is the virtual mass. The rest of the expression can include terms which are calculated by Arnold et al. [5]. They include objective combinations of the "slip" velocity, and the mean shear rate of the two components.

If the fluid and particle velocities are steady, and there is no shearing in the average particle and fluid motions, there is no force on the particles: $-\mathbf{M}_d = 0$ in (51). This result is commonly known as d'Alembert's paradox, that is, there is no net force on a body moving at a constant velocity through an inviscid fluid at rest.

Reynolds Stresses

We next turn to computations of the Reynolds stress using the velocity fluctuations due to the flow of an inviscid fluid around a sphere. Using the expression for the velocity (43), we see that

$$\mathbf{v}_c'(\mathbf{x}, t|\mathbf{z}) = \frac{1}{2}\left(\frac{a}{r}\right)^3 [\mathbf{v}_f(\mathbf{x}) - \mathbf{v}_p(\mathbf{x}) - \mathbf{x}' \cdot (\mathbf{L}_f - \mathbf{L}_p)] \cdot \left[\mathbf{I} - 3\frac{\mathbf{x}'\mathbf{x}'}{r^2}\right]$$
$$+ \frac{2}{3}\left(\frac{a}{r}\right)^5 \mathbf{x}' \cdot \mathbf{L}_f \cdot \left[\mathbf{I} - \frac{5}{2}\frac{\mathbf{x}'\mathbf{x}'}{r^2}\right]. \tag{52}$$

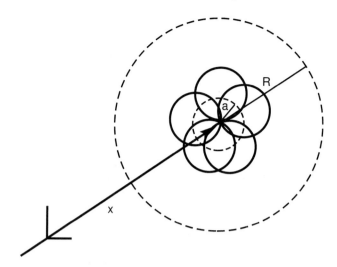

FIGURE 15.3. Cell averaging of the interfacial force.

Therefore $\mathbf{T}_c^{Re} = -\overline{X_f \rho \mathbf{v}' \mathbf{v}'}$ can be computed, and the result is

$$\mathbf{T}_c^{Re} = -\frac{1}{20}\alpha_p \rho_f \left[(\mathbf{v}_f - \mathbf{v}_p)(\mathbf{v}_f - \mathbf{v}_p) + 3(\mathbf{v}_f - \mathbf{v}_p) \cdot (\mathbf{v}_f - \mathbf{v}_p)\mathbf{I} \right]. \quad (53)$$

The fluid fluctuation kinetic energy is $u_c^{Re} = \frac{1}{2}\overline{\mathbf{v}' \cdot \mathbf{v}'}^x$, and can be computed by taking the trace of (53) for $\mathbf{T}_c^{Re}$. The result is

$$u_c^{Re} = \frac{1}{4}\alpha_p (\mathbf{v}_f - \mathbf{v}_p) \cdot (\mathbf{v}_f - \mathbf{v}_p). \quad (54)$$

Interfacial Work

The interfacial work $\mathbf{W}_c = \overline{p\mathbf{v}' \cdot \nabla X_f}$ can be computed in a similar way; the result is

$$\mathbf{W}_c = \frac{1}{2}\alpha_p \rho_f (\mathbf{v}_f - \mathbf{v}_p) \cdot \left(\frac{\partial \mathbf{v}_f}{\partial t} + \mathbf{v}_f \cdot \nabla \mathbf{v}_f - \frac{\partial \mathbf{v}_p}{\partial t} \right). \quad (55)$$

Dispersed Component Elastic Stresses—A

If a distribution of stresses is applied to the surface of an elastic body, there results a distribution of stresses inside the body. These induced stresses can be important in computing constitutive equations for solid–fluid mixtures. The average stress inside the sphere is given by

$$\overline{\mathbf{T}}_d^x(\mathbf{x}, t) = \frac{1}{\frac{4}{3}\pi a^3} \int_0^a \int_{\Omega(r)} \mathbf{T}(\mathbf{x}, t|\mathbf{z}) \, d\Omega, \quad (56)$$

where $\mathbf{T}(\mathbf{x}, t|\mathbf{z})$ is the stress at point $\mathbf{x}$ inside a sphere having its center at $\mathbf{z}$ at time t. We shall assume that the spheres are linearly elastic solids, but we shall assume that the deformation is sufficiently small that the fluid motion is unaffected by the

15.3. Inviscid Fluid Flowing Around a Sphere

deformation of the spheres. Let **u** be the displacement of a material point of the sphere from the unstressed reference configuration. The stress is given by

$$\mathbf{T} = \boldsymbol{\sigma} + \Theta \mathbf{I}. \tag{57}$$

The spherical part of the stress satisfies [59]

$$\nabla^2 \Theta(\mathbf{x}, t|\mathbf{z}) = 0, \tag{58}$$

with boundary conditions

$$\Theta(\mathbf{x}, t|\mathbf{z}) = -p(\mathbf{x}, t|\mathbf{z}) \quad \text{on} \quad |\mathbf{x} - \mathbf{z}| = a. \tag{59}$$

Solving and performing the integration gives

$$\overline{\Theta}_d^r(\mathbf{x}, t) = -\overline{p}_{ci}(\mathbf{x}, t). \tag{60}$$

The solution for σ is also given by Love [59] and can be averaged to give

$$\overline{\sigma}_d^r(\mathbf{x}, t) = \rho_f \left[-\frac{9}{20}(\mathbf{v}_f - \mathbf{v}_p)(\mathbf{v}_f - \mathbf{v}_p) + \frac{2}{5}|\mathbf{v}_f - \mathbf{v}_p|^2 \mathbf{I} \right]. \tag{61}$$

Thus, the dispersed component elastic stress becomes

$$\overline{\mathbf{T}}_d^r = -\overline{p}_{ci}\mathbf{I} + \rho_f \left[-\frac{9}{20}(\mathbf{v}_f - \mathbf{v}_p)(\mathbf{v}_f - \mathbf{v}_p) + \frac{2}{5}|\mathbf{v}_f - \mathbf{v}_p|^2 \mathbf{I} \right]. \tag{62}$$

Dispersed Component Elastic Stresses—B

The result from the previous subsection can be derived in a different manner if we assume that $\mathbf{L}_f = \mathbf{L}_p = 0$. In this case, we have $p(\mathbf{x}, t|\mathbf{z}) = p(\mathbf{x} - \mathbf{z}, t|0)$. The stress inside the elastic solid satisfies the equation

$$\nabla \cdot \mathbf{T}(\mathbf{x}, t|\mathbf{z}) = 0, \tag{63}$$

with the boundary condition

$$\mathbf{n} \cdot \mathbf{T} = -\mathbf{n}p \quad \text{at} \quad |\mathbf{x} - \mathbf{z}| = a. \tag{64}$$

Since the pressure depends on $\mathbf{x} - \mathbf{z}$, the stress $\mathbf{T}$ will also. Consequently, we have

$$\overline{\mathbf{T}}_d^r(\mathbf{x}, t) = \frac{1}{\frac{4}{3}\pi a^3} \int_0^a \int_{\Omega(r)} \mathbf{T}(\mathbf{x} - \mathbf{z}, t) \, d\Omega, \tag{65}$$

which is independent of **x**. In this case, we can make use of a result of Batchelor [8], which is based on the identity

$$\mathbf{T} = \nabla \cdot [\mathbf{T}(\mathbf{x} - \mathbf{z})], \tag{66}$$

which holds because the equation of motion for the solid material is $\nabla \cdot \mathbf{T} = 0$. Then (65) becomes

$$\overline{\mathbf{T}}_d^r(\mathbf{x}, t) = \frac{1}{\frac{4}{3}\pi a^3} \int_{\Omega(a)} \mathbf{n} \cdot \mathbf{T}(\mathbf{x} - \mathbf{z}, t)(\mathbf{x} - \mathbf{z}) \, d\Omega. \tag{67}$$

168   15. Relation of Microstructure to Constitutive Equations

Using the boundary condition (64) and noting that $\mathbf{n} = (\mathbf{x} - \mathbf{z})/a$, we can evaluate the integral to obtain

$$\overline{\mathbf{T}}_d^r(\mathbf{x}, t) = -\overline{p}_{ci}\mathbf{I} + \rho_c \left[-\frac{9}{20}(\mathbf{v}_f - \mathbf{v}_p)(\mathbf{v}_f - \mathbf{v}_p) + \frac{2}{5}|\mathbf{v}_f - \mathbf{v}_p|^2 \mathbf{I} \right]. \quad (68)$$

It is interesting to note that the two methods for calculating the dispersed component elastic stress (62) and (68) give exactly the same stress, and that stress is not related to the elastic properties of the material making up the dispersed component. This result should also be compared with (76).

Model for Bubbly Flow with Surface Tension

Often, surface tension is able to hold sufficiently small bubbles to a nearly spherical shape. However, modeling the effect of surface tension is difficult. A primitive possibility is to replace the effect of surface tension by rigid spherical shells surrounding the dispersed component. In this section, we make that physical approximation. We treat the set of rigid spherical shells surrounding each gas bubble as a third component. As a first approximation, it is plausible to assume that this component is massless. Then, the momentum equation for a shell is

$$\nabla \cdot \mathbf{T} = 0. \quad (69)$$

Let X_s be the characteristic function for the shell. The average momentum equation can be derived by multiplying the exact momentum equation by X_s and integrating. This yields

$$\begin{aligned}\overline{X_s \nabla \cdot \mathbf{T}} &= 0 \\ &= \nabla \cdot \overline{X_s \mathbf{T}} - \overline{\mathbf{T} \cdot \nabla X_s} \\ &= \nabla \cdot \overline{\mathbf{T}} + \mathbf{M}_d + \mathbf{M}_c \end{aligned} \quad (70)$$

where $\overline{\mathbf{T}} = \overline{X_s \mathbf{T}}$. Note the similarity between this equation and (11.3.49), with the source due to surface tension having the form $\mathbf{m} = \nabla \cdot \mathbf{T}$. Indeed, the jump condition represents a momentum equation for the material making up the interface.

Insofar as the fluid comprising the continuous component is concerned, it does not matter what causes the sphere to remain spherical. Thus, average quantities computed for the continuous component are the same as in Section 15.3.2. Thus the interfacial force, $-\mathbf{M}_c$, and the Reynolds stress, $\mathbf{T}_c^{Re}$, are the same as computed for solid particles.

To complete this model, we must compute the stresses in the gas and the shell. First, for the gas, we assume that the averaged pressure is the same as the interfacial averaged pressure. For the stress in the shell, we use Batchelor's computation, as in Section 15.3.2. Again, the stress inside the elastic solid satisfies the equation

$$\nabla \cdot \mathbf{T}(\mathbf{x}, t|\mathbf{z}) = 0, \quad (71)$$

with the boundary conditions

$$\mathbf{n} \cdot \mathbf{T} = -\mathbf{n}p, \quad (72)$$

at $|\mathbf{x} - \mathbf{z}| = a - \delta/2$ and also at $|\mathbf{x} - \mathbf{z}| = a + \delta/2$, where δ is the thickness of the shell. The average of the stress is given by

$$\overline{\mathbf{T}} = \frac{1}{\frac{4}{3}\pi R^3} \int_{a-\delta/2}^{a+\delta/2} \iint_{\Omega(r)} \mathbf{T}(\mathbf{x} - \mathbf{z}, t) \, d\Omega. \tag{73}$$

Again, we use

$$\mathbf{T} = \nabla \cdot [\mathbf{T}(\mathbf{x} - \mathbf{z})], \tag{74}$$

whence (65) becomes

$$\overline{\mathbf{T}} = \frac{1}{\frac{4}{3}\pi R^3} \left(\iint_{\Omega(a-\delta/2)} \mathbf{n} \cdot \mathbf{T}(\mathbf{x} - \mathbf{z}) \, d\Omega + \iint_{\Omega(a+\delta/2)} \mathbf{n} \cdot \mathbf{T}(\mathbf{x} - \mathbf{z}) \, d\Omega \right)$$

$$= \frac{1}{\frac{4}{3}\pi R^3} \left(\iint_{\Omega(a-\delta/2)} \mathbf{n} \, p(\mathbf{x} - \mathbf{z}) \, d\Omega + \iint_{\Omega(a+\delta/2)} \mathbf{n} \, p(\mathbf{x} - \mathbf{z}) \, d\Omega \right). \tag{75}$$

If we assume that the gas pressure is approximately $p = \overline{p}_{di}$, and use (33) for the continuous phase pressure, we have

$$\overline{\mathbf{T}} = \alpha_d (\overline{p}_{di} - \overline{p}_{ci}) \mathbf{I} + \frac{3}{20} \alpha_d \rho_c \left[|\mathbf{v}_d - \mathbf{v}_c|^2 \mathbf{I} - 3(\mathbf{v}_d - \mathbf{v}_c)(\mathbf{v}_d - \mathbf{v}_c) \right]. \tag{76}$$

All of the terms derived in this section are comparable in magnitude to the inertial terms; consequently, a model capable of describing effects that depend on inertia would be incomplete without these terms.

15.3.3. Effects of Rotation

In this section we consider the effects of rotation on the force on a sphere in an incompressible inviscid fluid. We do this by assuming that the fluid has, at the initial instant, $t = 0$, a uniform vorticity. The average fluid vorticity, then, is that constant vorticity. The relative velocity is the velocity of the sphere relative to the fluid. The calculation is valid only at the initial instant. The ensemble average of the interfacial stress is the force on the sphere divided by its volume.

Auton [6] calculates the lift force on a sphere by a different method. He calculates an approximation to the steady-state vorticity by a perturbation method. He assumes that the rotation far from the sphere is small. The smallness that he assumes is sufficient for the present analysis in the case when fluid accelerations are negligible. However, we see from (16) that, in general, a perturbation vorticity that is initially zero will grow to be comparable to the mean fluid vorticity ω in a time of order R/U where U is a velocity scale for the relative motion. We further note that Auton's result breaks down under the same conditions, i.e., when the vorticity is no longer small. Moreover, it is not clear how to average this result.

The method of calculation follows that of Proudman [72], who uses a rotating coordinate system to perform the calculation in a situation where the background vorticity is zero. We use an inertial coordinate system. The results are equivalent.

The Helmholtz representation of the velocity is

$$\mathbf{v} = \nabla\phi + \mathbf{v}' = \nabla\phi + \frac{1}{2}\mathbf{x} \wedge \boldsymbol{\omega}, \tag{77}$$

where ϕ is the velocity potential, and $\boldsymbol{\zeta}$ is the vorticity, which we assume to be a function of t alone. The momentum equation (14) can be written

$$\frac{\partial \mathbf{v}}{\partial t} + \mathbf{v} \cdot \nabla\mathbf{v} = -\nabla\frac{p}{\rho}. \tag{78}$$

Substituting the velocity results in

$$\frac{\partial \mathbf{v}'}{\partial t} + \mathbf{v} \cdot \nabla\mathbf{v}' + \mathbf{v}' \cdot \nabla\nabla\phi = -\nabla P. \tag{79}$$

where

$$P = \frac{p}{\rho} + \frac{1}{2}|\nabla\phi|^2 + \frac{\partial \phi}{\partial t}.$$

Taking the divergence of (79) yields

$$-\frac{1}{2}|\boldsymbol{\zeta}|^2 = -\nabla^2 P. \tag{80}$$

The momentum equation (79) can be written

$$\frac{1}{2}\left(\mathbf{x} \wedge \frac{\partial \boldsymbol{\zeta}}{\partial t}\right) + \frac{1}{2}\mathbf{v} \wedge \boldsymbol{\zeta} + \frac{1}{2}(\mathbf{x} \wedge \boldsymbol{\zeta}) \cdot \nabla\nabla\phi = -\nabla P.$$

Using (16)

$$\frac{1}{2}\mathbf{x} \wedge (\boldsymbol{\zeta} \cdot \nabla\mathbf{v}) + \frac{1}{2}\mathbf{v} \wedge \boldsymbol{\zeta} + \frac{1}{2}(\mathbf{x} \wedge \boldsymbol{\zeta}) \cdot \nabla\nabla\phi = -\nabla P.$$

Simplifying,

$$\frac{1}{2}\mathbf{x} \wedge (\boldsymbol{\zeta} \cdot \nabla\nabla\phi) + \frac{1}{2}\nabla\phi \wedge \boldsymbol{\zeta} + \frac{1}{4}(\mathbf{x} \wedge \boldsymbol{\zeta}) \wedge \boldsymbol{\zeta} + \frac{1}{2}(\mathbf{x} \wedge \boldsymbol{\zeta}) \cdot \nabla\nabla\phi = -\nabla P. \tag{81}$$

If we take $\mathbf{n} \cdot$ (81), we have

$$\mathbf{n} \cdot \nabla P = \frac{1}{2}\mathbf{n} \cdot \nabla\phi \wedge \boldsymbol{\zeta} + \frac{1}{2}\mathbf{n} \cdot \nabla\nabla\phi \cdot (\mathbf{x} \wedge \boldsymbol{\zeta}) + \frac{1}{4}\left(\mathbf{n} \cdot \boldsymbol{\zeta}\mathbf{x} \cdot \boldsymbol{\zeta} - \mathbf{n} \cdot \mathbf{x}|\boldsymbol{\zeta}|^2\right). \tag{82}$$

Note that (80) and (82) define a problem for P where the auxiliary quantities are known. If we take the particle to be spherical, and moving relative to the fluid with velocity $\mathbf{v}_r$, the potential can be written as (37)

$$\phi = -\frac{1}{2}\left(\frac{a}{r}\right)^3 \mathbf{v}_r \cdot \mathbf{x}. \tag{83}$$

Then

$$\nabla\phi|_{r=a} = -\frac{1}{2}\left(\mathbf{I} - 3\frac{\mathbf{xx}}{a^2}\right) \cdot \mathbf{v}_r. \tag{84}$$

15.3. Inviscid Fluid Flowing Around a Sphere

The force on the particle is given by

$$\mathbf{F}_L = \int_{\partial V} \mathbf{n} p \, ds.$$

This can be written in terms of P as

$$\mathbf{F}_L = -\rho \int_{\partial V} \mathbf{n} \left(\frac{\partial \phi}{\partial t} + \frac{1}{2} |\nabla \phi|^2 \right) ds + \rho \int_{\partial V} \mathbf{n} P \, ds. \quad (85)$$

The term

$$-\rho \int_{\partial V} \mathbf{n} \left(\frac{\partial \phi}{\partial t} + \frac{1}{2} |\nabla \phi|^2 \right) ds$$

corresponds to the interfacial force density calculated in the previous section. We shall consider the term corresponding to lift, which is included in the term

$$\rho \int_{\partial V} \mathbf{n} P \, ds.$$

Green's theorem for arbitrary smooth functions can be written as

$$\int_V (P \nabla^2 Q - Q \nabla^2 P) \, dv = \int_{\partial V} (P \mathbf{n} \cdot \nabla Q - Q \mathbf{n} \cdot \nabla P) \, ds. \quad (86)$$

Choosing V to be the spherical particle, and $Q = \mathbf{x}$, we have

$$\int_{r=a} \mathbf{n} P \, ds = - \int_V \mathbf{x} \nabla^2 P \, dv + \int_{r=a} \mathbf{x} \mathbf{n} \cdot \nabla P \, ds. \quad (87)$$

The term

$$\int_V \mathbf{x} \nabla^2 P \, dv$$

is a centrifugal force. The term

$$\int_{r=a} \mathbf{x} \mathbf{n} \cdot \nabla P \, ds$$

also contains a centrifugal term. The lift is

$$\mathbf{F}_L = \int_{r=a} \mathbf{x} \frac{1}{2} \mathbf{n} \cdot \nabla \phi \wedge \zeta \, ds + \int_{r=a} \frac{1}{2} \mathbf{x} \mathbf{n} \cdot \nabla \nabla \phi \cdot (\mathbf{x} \wedge \zeta) \, ds. \quad (88)$$

Evaluating the integral results in

$$\mathbf{F}_L = \frac{1}{4} \left(\frac{4}{3} \pi a^3 \right) \rho \mathbf{v}_r \wedge \zeta. \quad (89)$$

Then $\mathbf{M}_d^L$ is given by

$$\mathbf{M}_d^L = \frac{1}{4} \alpha_d \rho_c (\nabla \wedge \mathbf{v}_f) \wedge (\mathbf{v}_f - \mathbf{v}_p). \quad (90)$$

15.3.4. Effect of Concentration Gradients

In this section, we calculate the effects of the nonuniform distribution of particles on the stresses and interfacial force density. To do this, we assume that the distribution of positions within a cell depends on the gradient of the volume fraction of the particles. We use the approximation that the distribution is linear in the gradient, so that in effect we use the first term of a polynomial expansion. This accounts for nonuniform distributions of particles in a natural way, in determining the probability density function.

Let $\mathbf{x}' = \mathbf{x} - \mathbf{z}$. We assume that the probability of a sphere center being located in a small volume dV around the point $\mathbf{z}$ is now

$$\frac{dV}{\frac{4}{3}\pi R^3}\left[\frac{\alpha_d(\mathbf{z},t)}{\alpha_d(\mathbf{x},t)}\right] \approx \frac{dV}{\frac{4}{3}\pi R^3}\left[1 - \mathbf{x}' \cdot \frac{\nabla \alpha_d(\mathbf{x},t)}{\alpha_d(\mathbf{x},t)}\right]. \tag{91}$$

The result of including the effects of volume fraction gradients is to add terms to the quantities computed in the previous sections. For convenience of notation, we shall denote the average taken without allowing particle velocity fluctuations by $\overline{g}_0^x(\mathbf{x},t;\mathbf{v}_p)$. Thus, for example, the average interfacial force density, as calculated before, evaluated without accounting for a gradient in the volume fraction, will be denoted by

$$\mathbf{M}_{d0}(\mathbf{x},t) = \alpha_d \rho_c \left(\frac{1}{2}\left[\frac{\partial \mathbf{v}_f}{\partial t} - \frac{\partial \mathbf{v}_p}{\partial t} + \mathbf{v}_f \cdot \mathbf{L}_f - \mathbf{v}_p \cdot \mathbf{L}_p\right]\right.$$
$$\left. + (\mathbf{v}_f - \mathbf{v}_p) \cdot \left[-\frac{7}{20}\mathbf{L}_p - \frac{9}{20}\mathbf{L}_p + \frac{5}{4}\mathbf{L}_f^{\mathsf{T}} + \frac{1}{10}\left(\mathrm{tr}\,\mathbf{L}_p\right)\mathbf{I}\right]\right). \tag{92}$$

Using the probability distribution given by (91), the resulting integrals can be computed. Since we use only the linear approximation to the velocity fields, and the volume fraction gradient, we shall neglect the products of $\mathbf{L}_f$, $\mathbf{L}_p$, and $\nabla \alpha_p$. The results are that all averaged quantities are unchanged except the interfacial force density, and it is given by

$$\mathbf{M}_d(\mathbf{x},t) = \mathbf{M}_{d0}(\mathbf{x},t) + \overline{p}_{ci0}\nabla \alpha_d$$
$$+ \rho_f \nabla \alpha_d \cdot \left[-\frac{1}{20}(\mathbf{v}_d - \mathbf{v}_c)(\mathbf{v}_d - \mathbf{v}_c) + -\frac{3}{20}|\mathbf{v}_d - \mathbf{v}_c|^2 \mathbf{I}\right] \tag{93}$$

15.3.5. Effect of Particle Velocity Fluctuations

Generally, models for flows of multicomponent materials assume that the particles are in relatively smooth motion and that they do not collide. Part of the motivation for this assumption is the calculation that infinite force is required to make two particles collide in a slow flow of a viscous fluid. There is copious experimental evidence that particles do collide in many actual flows. This provides a mechanism for momentum transfer. An analogous situation has been studied extensively; that is, the kinetic theory of flowing granular materials. There the physics is dominated by particle collisions and the vibrational energy of particles [50]. The collisions

15.3. Inviscid Fluid Flowing Around a Sphere

give rise to the stress tensor. Here we consider some of the consequences of such physics in the context of multicomponent materials.

Models such as those of Jenkins and Savage [50], that treat assemblages of solid particles have been proposed and studied. In such models it is assumed that the particles act like the atoms in the kinetic theory of gases, with a pressure due to the fluctuations in the velocities that is attributed to collisional motions of the individual particles. The model in Section 15.3.2 for inertial effects does not include this effect. In this section we derive terms representing the effects of the particle velocity fluctuations when the particle motions are assumed to be at a sufficiently high Reynolds number that the viscosity in the bulk fluid can be neglected. In addition, effects of viscosity from the boundary, such as boundary layer separation, are neglected.

Let the sphere located at a point $\mathbf{z}$ move with velocity $\mathbf{v}_p$. We consider an averaging process that includes the particle velocities. The first aspect of the ensemble average when there are particle velocity fluctuations present is to average over the sphere velocities, with a distribution function $f^{(1)}(\mathbf{v}_p, \mathbf{x}, t)$. The second aspect is to allow the sphere to lie anywhere in the "cell" discussed in Section 15.3. Thus, the average of a quantity $g(\mathbf{x}, t; \mathbf{z}, \mathbf{v}_p)$ is performed in two parts. First we perform a conditional average of g for given sphere position $\mathbf{z}$, integrating over the velocity space $V_{\mathbf{v}_p}$, then we average over the spatial positions the sphere can have. Thus, for the average over the fluid of a quantity g, we have

$$\overline{g}(\mathbf{x}, t | \mathbf{z}) = \int g(\mathbf{x}, t; \mathbf{z}, \mathbf{v}_p) f^{(1)}(\mathbf{v}_p, \mathbf{z}, t) \, d\mathbf{v}_p . \tag{94}$$

Here $\overline{g}(\mathbf{x}, t | \mathbf{z})$ is the conditional average assuming the sphere is located at $\mathbf{z}$. The average of g over the fluid phase is then given by

$$\overline{g}_c^x(\mathbf{x}, t) = \frac{1}{\frac{4}{3}\pi(R^3 - a^3)} \int_a^R \int_{\Omega(r)} \overline{g}(\mathbf{x}, t | \mathbf{z}) \, d\Omega \, dr . \tag{95}$$

We introduce the average particle velocity and the fluctuation particle velocity as

$$\overline{\mathbf{v}}_p(\mathbf{x}, t) = \int \mathbf{v}_p f^{(1)}(\mathbf{v}_p, \mathbf{x}, t) \, d\mathbf{v}_p , \tag{96}$$

$$\mathbf{v}'_p(\mathbf{x}, t) = \mathbf{v}_p - \overline{\mathbf{v}}_p(\mathbf{x}, t) . \tag{97}$$

Note that the average of the fluctuating part of the velocity vanishes

$$\overline{\mathbf{v}'_p} = 0 . \tag{98}$$

The particle kinetic energy per unit particle mass is

$$u_d^{Re}(\mathbf{x}, t) = \frac{1}{2} \int |\mathbf{v}'_p|^2 f^{(1)}(\mathbf{v}_p, \mathbf{x}, t) \, d\mathbf{v}_p , \tag{99}$$

and the Reynolds stress for the particles is defined by

$$\mathbf{T}_d^{Re}(\mathbf{x}, t) = -\rho_d \int \mathbf{v}'_p \mathbf{v}'_p f^{(1)}(\mathbf{v}_p, \mathbf{x}, t) \, d\mathbf{v}_p . \tag{100}$$

The result of including the effects of particle velocity fluctuations is to add terms to the quantities computed in the previous section. For convenience of notation, we shall denote the average taken without allowing particle velocity fluctuations by $\overline{g}^x(\mathbf{x}, t; \mathbf{v}_p)$. Thus, for example, the average interfacial pressure, as calculated before, evaluated at the average particle velocity, will be denoted by

$$\overline{p}^x_{ci}(\mathbf{x}, t, \overline{\mathbf{v}}_p) = \overline{p}^x_c(\mathbf{x}, t, \overline{\mathbf{v}}_p) - \frac{1}{4}\rho_c |\overline{\mathbf{v}}_c - \overline{\mathbf{v}}_p|^2. \tag{101}$$

We have

$$\mathbf{v}(\mathbf{x}, t; \mathbf{z}, \mathbf{v}_p) = \nabla \phi(\mathbf{x}, t; \mathbf{z}, \mathbf{v}_p), \tag{102}$$

so that

$$\mathbf{v}(\mathbf{x}, t; \mathbf{z}, \mathbf{v}_p) = \mathbf{v}_f(\mathbf{x})$$
$$+ \frac{1}{2}\left(\frac{a}{r}\right)^3 [\mathbf{v}_f(\mathbf{x}) - \overline{\mathbf{v}}_p(\mathbf{x}) - \mathbf{v}'_p - \mathbf{x}' \cdot (\mathbf{L}_f - \mathbf{L}_p)] \cdot \left(\mathbf{I} - \frac{3}{r^2}\mathbf{x}'\mathbf{x}'\right)$$
$$+ \frac{2}{3}\left(\frac{a}{r}\right)^5 \mathbf{x}' \cdot \mathbf{L}_f \cdot \left(\mathbf{I} - \frac{5}{2r^2}\mathbf{x}'\mathbf{x}'\right). \tag{103}$$

In order to obtain the average velocities from (105), we note that the average over the velocity fluctuations gives no contribution. Therefore,

$$\overline{\mathbf{v}}_c^{x\rho}(\mathbf{x}, t) = \mathbf{v}_f(\mathbf{x}, t), \tag{104}$$

and

$$\overline{\mathbf{v}}_{ci}(\mathbf{x}, t) = \mathbf{v}_f(\mathbf{x}, t). \tag{105}$$

The exact pressure is given by (33). Substituting $\mathbf{v}_p = \overline{\mathbf{v}}_p(\mathbf{z}, t) + \mathbf{v}'_p$ and averaging over the velocity fluctuations gives

$$\overline{p}(\mathbf{x}, t|\mathbf{z}) = p(\mathbf{x}, t|\mathbf{z}, \overline{\mathbf{v}}_p)$$
$$- \rho_c \left\{ u_d^{Re}(\mathbf{z}, t) \left(\frac{a^3}{r^3}\right) - \frac{3}{2} \frac{\mathbf{x}' \cdot \mathbf{T}_d^{Re}(\mathbf{z}, t) \cdot \mathbf{x}'}{\rho_d} \left(\frac{a^3}{r^5}\right) \right.$$
$$\left. + \frac{1}{4} u_d^{Re}(\mathbf{z}, t) \left(\frac{a^6}{r^6}\right) - \frac{9}{8} \frac{\mathbf{x}' \cdot \mathbf{T}_d^{Re}(\mathbf{z}, t) \cdot \mathbf{x}'}{\rho_d} \left(\frac{a^6}{r^{10}}\right) \right\}. \tag{106}$$

The spatial integration is now quite similar to that done in Section 15.3, and results in

$$\overline{p}_c^x = p_0 - \rho_c \frac{\partial \mathbf{v}_f}{\partial t} \cdot \mathbf{x} - \frac{1}{2}\mathbf{v}_f(\mathbf{x}) \cdot \mathbf{v}_f(\mathbf{x}) = \overline{p}_c^x(\mathbf{x}, t, \overline{\mathbf{v}}_p), \tag{107}$$

correct to $O(a/R)$. Thus, there is no contribution to the average fluid pressure due to the particle velocity fluctuations. We also obtain

$$\overline{p}_{ci}(\mathbf{x}, t) = \overline{p}_c^x - \frac{\rho_c}{4}|\overline{\mathbf{v}}_c(\mathbf{x}, t) - \overline{\mathbf{v}}_d(\mathbf{x}, t)|^2 + \frac{1}{2}\rho_c u_d^{Re},$$
$$= \overline{p}_{ci}(\mathbf{x}, t, \overline{\mathbf{v}}_p) + \frac{1}{2}\rho_c u_d^{Re}. \tag{108}$$

15.3. Inviscid Fluid Flowing Around a Sphere

For the interfacial force density $\mathbf{M}_d$, we also must expand the terms

$$u_d^{Re}(\mathbf{z}, t) = u_d^{Re}(\mathbf{x}, t) - \mathbf{x}' \cdot \nabla u_d^{Re}(\mathbf{x}, t)$$

and

$$\mathbf{T}_d^{Re}(\mathbf{z}, t) = \mathbf{T}_d^{Re}(\mathbf{x}, t) - \mathbf{x}' \cdot \nabla \mathbf{T}_d^{Re}(\mathbf{x}, t).$$

The result is that

$$\mathbf{M}_d(\mathbf{x}, t) = \mathbf{M}_d(\mathbf{x}, t, \overline{\mathbf{v}}_p) - \nabla \left(\frac{4}{5} \alpha_d \rho_c u_d^{Re} \right) - \nabla \cdot \left(\frac{9}{20} \alpha_d \frac{\rho_c}{\rho_d} \mathbf{T}_d^{Re} \right). \tag{109}$$

The average stress inside the sphere is given by

$$\overline{\mathbf{T}}_d^x(\mathbf{x}, t) = \overline{\mathbf{T}}_d^x(\mathbf{x}, t, \overline{\mathbf{v}}_p) - \frac{9}{20} \frac{\rho_c}{\rho_d} \mathbf{T}_d^{Re} + \frac{4}{5} \rho_c u_d^{Re} \mathbf{I}. \tag{110}$$

The Reynolds stress can be computed using the velocity fluctuations derived from (104) and (105). We have

$$\mathbf{T}_c^{Re}(\mathbf{x}, t|\mathbf{z}) = -\rho_c \overline{\mathbf{v}'_c(\mathbf{x}, t|\mathbf{z}) \mathbf{v}'_c(\mathbf{x}, t|\mathbf{z})}$$

$$+ \rho_c \left\{ \frac{1}{4} \left(\frac{a^6}{r^6} \right) \frac{\mathbf{T}_d^{Re}}{\rho_d} + \frac{9}{4} \left(\frac{a^6}{r^{10}} \right) \mathbf{x}' \mathbf{x}' \left(\mathbf{x}' \cdot \frac{\mathbf{T}_d^{Re}}{\rho_d} \cdot \mathbf{x}' \right) \right.$$

$$\left. - \frac{3}{4} \left(\frac{a^6}{r^8} \right) [\mathbf{x}' \left(\mathbf{x}' \cdot \frac{\mathbf{T}_d^{Re}}{\rho_d} \right) + \left(\mathbf{x}' \cdot \frac{\mathbf{T}_d^{Re}}{\rho_d} \right) \mathbf{x}'] \right\}, \tag{111}$$

whence

$$\mathbf{T}_c^{Re}(\mathbf{x}, t) = \mathbf{T}_c^{Re}(\mathbf{x}, t, \overline{\mathbf{v}}_p) + \frac{1}{20} \alpha_d \frac{\rho_c}{\rho_d} \mathbf{T}_d^{Re} + \frac{3}{10} \alpha_d \rho_c u_d^{Re}. \tag{112}$$

The fluid fluctuation kinetic energy can be computed by taking the trace of (112). The result is

$$u_c^{Re} = \frac{1}{4} \alpha_d |\overline{\mathbf{v}}_c^{xp} - \overline{\mathbf{v}}_d^{xp}|^2 + \frac{1}{2} \alpha_d u_d^{Re}. \tag{113}$$

15.3.6. Comparison of Averaging Methods

We now show that the cell approximation to the ensemble average is different from the volume average, at least in the way that the volume average is most commonly applied. Volume averaging integrates expressions over large volumes centered at different points $\mathbf{x}$, and divides by the volume of the region. See Figure 15.4. For example, the averaging volume could be a sphere of radius L, with $L \gg a$. We assign averaged values of the flow quantities at different points in the flow field by moving the averaging volume around so that its center is at different points $\mathbf{x}$.

The cell model is often used to approximate the volume average. If the averaging volume is sufficiently large that there are many cells in the averaging volume, then it is customary to average over only one "representative" cell, and assign that average to the field point $\mathbf{x}$. There must be some decision made on where the cell is relative to the field point $\mathbf{x}$. The practice seems to be that one performs the

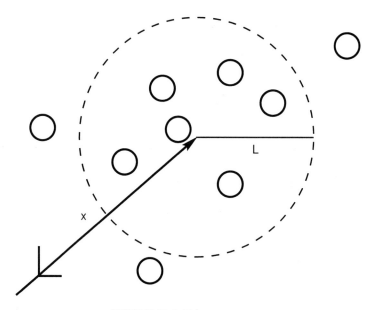

FIGURE 15.4. Volume average.

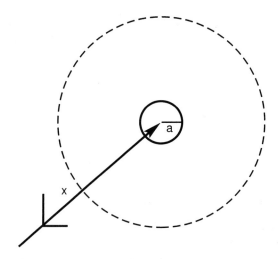

FIGURE 15.5. Cell approximation to the volume average.

cell approximation to the volume average at some point **x** by assuming that the center of the particle coincides with the center of the volume, **x**. See Figure 15.5. This clearly treats the response to the mean fields incorrectly, in that variables are evaluated at **z** that should be evaluated at **x**.

These two averaging processes are different. The essential difference lies in the "background" motion to which the sphere responds. To see this, consider a flow field with a hydrostatic pressure gradient. If we wish to compute the average pressure on an unsymmetric object, the ensemble average puts the object in different places, and the pressure is always equal to the exact pressure. The volume average assumes that the object is fixed, and samples the field over all appropriate points. If the object is unsymmetric then, for example, the sample will contain more points with smaller pressure than with the larger pressure. In this case, the volume average will yield incorrect results. However, we note many calculations presented here have the same results from volume averaging and from ensemble averaging.

In order to illustrate the difference between the ensemble average and the volume average, let us compute the volume average of $\mathbf{M}_d^V = \overline{p \nabla X_d}$. We have

$$\mathbf{M}_d^V = -\frac{1}{\frac{4}{3}\pi R^3} \int_{\Omega(a)} \mathbf{n}\, p(\mathbf{x}, z, t)\, d\Omega, \tag{114}$$

where now the integration is over $\mathbf{x} = \mathbf{z} + \mathbf{x}'$ keeping $\mathbf{z}$ fixed. The result is

$$\mathbf{M}_d^V = \alpha_d \rho_f \left(\frac{\partial \mathbf{v}_0}{\partial t} + \mathbf{v}_f \cdot \mathbf{L}_f \right) + \frac{1}{2} \alpha_d \rho_f \left[\frac{\partial \mathbf{v}_0}{\partial t} - \frac{\partial \mathbf{v}_p}{\partial t} + \mathbf{v}_f \cdot \mathbf{L}_f - \mathbf{v}_p \cdot \mathbf{e}_p \right]. \tag{115}$$

It is interesting to note that this term by itself gives Taylor's result [84] for the force on a sphere in a spatially accelerating flow. This force is given by

$$\frac{3}{2} \left(\frac{4}{3} \pi a^3 \right) \rho_f \mathbf{v}_f \cdot \nabla \mathbf{v}_f. \tag{116}$$

This means that the result (115) will be *incorrect* if used in an equation where a force due to the fluid pressure gradient is also present. The ultimate result will include the fluid acceleration twice. A similar thing happens if we include the potential due to gravity. Using the ensemble average, no gravitational force appears in the interfacial force. However, if we use the volume average, the buoyancy term $\alpha_p \rho_f \mathbf{g}$ appears in the interfacial force. Then the force due to the fluid pressure gradient will also include the same force, and hence it, too, will appear twice. This emphasizes the difficulties inherent in the volume average.

15.4. Viscous Flow Around a Sphere

Several papers (notably Batchelor [8], Brenner [13], Hinch [42], and Lundgren [60]) have addressed the problem of slow viscous flow of a suspension of spheres, and of the flow of an inertialess fluid through a fixed bed of spheres. These papers usually derive sedimentation velocities or effective viscosities for these mixtures. We shall adapt the arguments used in the literature to derive constitutive equations for the stresses and interfacial momentum transfer terms in (11.3.44) that are valid for slow viscous flow. We assume that the dispersed component consists of rigid elastic spheres of uniform composition, and that the continuous component is a fluid of constant density undergoing a motion that is sufficiently slow to be inertia-free.

178   15. Relation of Microstructure to Constitutive Equations

The Stokes equations for the velocity field and the pressure are

$$\nabla \cdot \mathbf{v} = 0, \tag{117}$$

$$\nabla \cdot \mathbf{T} + \rho_f \mathbf{b}_f = 0, \tag{118}$$

where the stress is given by

$$\mathbf{T} = -p\mathbf{I} + \mu_f \operatorname{sym} \nabla \mathbf{v}. \tag{119}$$

The solid spheres are assumed to be rigid. Consequently, the motion inside each sphere satisfies

$$\operatorname{sym} \nabla \mathbf{v} = 0. \tag{120}$$

It follows that inside each sphere

$$\nabla \cdot \mathbf{v} = 0. \tag{121}$$

The equation of motion in the solid is

$$\nabla \cdot \mathbf{T} + \rho_p \mathbf{b}_p = 0. \tag{122}$$

Since the solid material is assumed rigid, there is no constitutive equation for the stress. We have allowed the body force term to be different for the two components. This will let us consider motions where the particles are held fixed, or move with the fluid.

The averaged equations for this system can be obtained by ignoring the inertia terms in the averaged balance equations. However, it is instructive to apply the averaging operator directly to the continuity equation, to derive the equations of balance of mass for each component. If we apply the averaging operator to the product of X_k with (117), we have

$$\overline{X_k \nabla \cdot \mathbf{v}} = \nabla \cdot \overline{X_k \mathbf{v}} - \overline{\mathbf{v} \cdot \nabla X_k} = 0. \tag{123}$$

If we average the topological equation (9.1.28), assuming that $\mathbf{v}_i = \mathbf{v}$ at the interface, we have

$$\overline{\frac{\partial X_k}{\partial t} + \mathbf{v} \cdot \nabla X_k} = \frac{\partial \alpha_k}{\partial t} + \overline{\mathbf{v} \cdot \nabla X_k} = 0, \tag{124}$$

so that

$$-\overline{\mathbf{v} \cdot \nabla X_k} = \frac{\partial \alpha_k}{\partial t}. \tag{125}$$

Thus, the equation of conservation of mass for component k is

$$\frac{\partial \alpha_k}{\partial t} + \nabla \cdot (\alpha_k \overline{\mathbf{v}}_k^{x\rho}) = 0. \tag{126}$$

The equation of balance of momentum for component k is

$$\nabla \cdot \alpha_k \overline{\mathbf{T}}_k^x + \mathbf{M}_k + \alpha_k \rho_k \mathbf{b}_k = 0, \tag{127}$$

where

$$\mathbf{M}_k = -\overline{\mathbf{T} \cdot \nabla X_k}. \tag{128}$$

Note that $\mathbf{M}_p + \mathbf{M}_f = 0$.

Adding (126) for $k = p$ and $k = f$ gives

$$\nabla \cdot (\alpha_f \bar{\mathbf{v}}_f^{x\rho} + \alpha_p \bar{\mathbf{v}}_p^{x\rho}) = 0. \tag{129}$$

Adding the momentum equations (127) for $k = p$ and $k = f$ gives

$$\nabla \cdot (\alpha_f \bar{\mathbf{T}}_f^x + \alpha_p \bar{\mathbf{T}}_p^x) + (\alpha_f \rho_f \mathbf{b}_f + \alpha_p \rho_p \mathbf{b}_p) = 0. \tag{130}$$

Notice that since the equations of motion for the microscale (117, 118, 121) and (122) have no time derivative terms, no initial conditions for either component need be supplied, and the entire randomness in this problem is in the positions of the spheres.

Conditionally Averaged Equations

The exact stress satisfies

$$\nabla \cdot \mathbf{T} + \rho_k \mathbf{b}_k = 0. \tag{131}$$

We derive equations for the conditionally averaged stresses [42]. If we multiply the above equation by X_k and average over all the positions of the spheres *except* the one at $\mathbf{z}$, we obtain

$$\nabla \cdot \overline{X_k \mathbf{T}}^{(1)}(\mathbf{x}, t|\mathbf{z}) - \overline{\nabla X_k \cdot \mathbf{T}}^{(1)}(\mathbf{x}, t|\mathbf{z}) = 0, \tag{132}$$

where the superscript $^{(1)}$ denotes the conditional average with the sphere centered at $\mathbf{z}$.

The conditionally averaged momentum equations can be written as

$$\nabla \cdot \alpha_k^{(1)} \bar{\mathbf{T}}_k^{x(1)} + \mathbf{M}_k^{(1)} + \alpha_k^{(1)} \rho_k \mathbf{b}_k = 0. \tag{133}$$

Multiply (131) by X_k and average over all of the positions of the spheres *except* the one at $\mathbf{z}$ *and* the one at $\mathbf{z}_2$. We obtain

$$\nabla \cdot \alpha_k^{(2)} \bar{\mathbf{T}}_k^{x(2)} + \mathbf{M}_k^{(2)} + \alpha_k^{(2)} \rho_k \mathbf{b}_k = 0, \tag{134}$$

where the superscript $^{(2)}$ denotes the conditional average with spheres centered at $\mathbf{z}$ and $\mathbf{z}_2$.

15.4.1. *Interfacial Force and Stresses*

A major problem in the study of suspensions is to evaluate the viscosity of a suspension in terms of the viscosity of the suspending fluid and of the properties of the particles. Even when the particles are spherical and rigid, attempts at such evaluations lead to extreme theoretical difficulties. In Chapter 21 we argue that it is difficult to compare such evaluations with experimental results measured with viscometers in the same straightforward way as those for conventional fluids. For now we set that difficulty aside, and consider theoretical evaluations.

15.4.2. Effective Viscosity

In an elegant classic paper, Einstein [31] evaluated the viscosity μ_{eff} of a linearly viscous fluid of viscosity μ_f containing a dilute suspension of small particles with volume fraction α. The method consisted of calculating the energy dissipated by the flow around the particles, and associating that with the work done in moving the particles relative to the fluid. Einstein's result is

$$\mu_{\text{eff}} = \mu_f \left(1 + \frac{5}{2}\alpha\right). \tag{135}$$

Comparison of this equation with experimental data is not entirely gratifying. For small values of α, that is, for dilute suspensions, the value of the viscosity and its increase with respect to α are correct. Deviation is soon noted, the trend being a vast underestimation of the apparent viscosity as the solid volume fraction α increases. There is a large and interesting literature devoted to "correction" of the result of Einstein.

For all of the relevant literature, it is assumed that the flow is slow enough so that inertial terms can be ignored for the suspending fluid. This renders the equations of motion linear. Two factors are used to "correct" Einstein's results, the first being that the particles may not be small, the second being that the structure, or arrangement, of the particles may affect the effective viscosity.

Most of the analysis is based on a calculation of Brenner [13]. There, the viscous dissipation rate due to two identical spheres approaching one another is computed. Frankel and Acrivos [34] show that this quantity is large compared to the dissipation rate associated with two spheres "passing" one another. For small distances of approach, the calculation is consistent with the lubrication approximation. They calculate the rate of dissipation of the suspension in an extensional flow, assuming that the suspended particles do not diffuse with respect to the fluid. That dissipation is set equal to the dissipation rate of another pure fluid, the "effective fluid." Since the flow is infinite in extent, a finite control volume is used. The choice of this control volume causes some difficulty in subsequent calculations. The result is an effective viscosity for the suspension. In this calculation, an assumption must be made for the structure, or arrangement, of the particles. Furthermore, an averaging process is required, and the result depends on the details of that process. Emphasis is placed on concentrated suspensions, the result being

$$\frac{\mu_{\text{eff}}}{\mu_f} = C' \left\{ \frac{(\alpha/\alpha_m)^{1/3}}{1 - (\alpha/\alpha_m)^{1/3}} \right\}, \tag{136}$$

where α_m is a maximum packing fraction. It is to be determined experimentally. The functional form appears to be robust, the constant C' being dependent on the details of the assumptions about the volume over which the dissipation is calculated. For example, the value

$$C' = \frac{9}{8} \tag{137}$$

results from assuming that a spherical shell is the "region of influence" around a given sphere, while

$$C' = \frac{3\pi}{16} \tag{138}$$

is the value computed from a cubic "region of influence." Frankel and Acrivos choose

$$C' = \frac{9}{8} \tag{139}$$

as the value that fits the data best. They note a number of other expressions in the literature, some empirical, some theoretical but with empirical parameters. None fit the data for large concentrations as well as theirs does.

Furthermore, there is a generalization of the formula of Frankel and Acrivos [34] that agrees with Einstein's for small α. It is given by Graham [37] as

$$\frac{\mu_{\text{eff}}}{\mu_f} = 1 + \frac{5}{2}\alpha + \frac{9}{2}\left[\frac{1}{\left(\frac{h}{a}\right)\left(2 + \frac{h}{a}\right)\left(1 + \frac{h}{a}\right)^2}\right], \tag{140}$$

where a is the particle radius and h is the interparticle spacing. For a simple cubic packing,

$$\frac{h}{a} = 2\frac{1 - (\alpha/\alpha_m)^{1/3}}{(\alpha/\alpha_m)^{1/3}}, \tag{141}$$

The effective viscosity of a mixture of solid particles in a viscous fluid is dependent on the viscosity of the fluid and the volume fraction of solids in the fluid.

Relation to Constitutive Equations for Stress

The general expression for the stresses for slow flow (29) can be expressed as

$$\mathbf{T}_p^m = \mu_{pp}\mathbf{D}_p + \mu_{pf}\mathbf{D}_f + E_{ppf}(\mathbf{v}_{pf}\nabla\alpha_p + \nabla\alpha_p\mathbf{v}_{pf}), \tag{142}$$

$$\mathbf{T}_f^m = \mu_{fp}\mathbf{D}_p + \mu_{ff}\mathbf{D}_f + E_{fpf}(\mathbf{v}_{pf}\nabla\alpha_p + \nabla\alpha_p\mathbf{v}_{pf}). \tag{143}$$

Assume that the particles move with the fluid, so that $\mathbf{D}_p = \mathbf{D}_f$. Further assume that the mixture is uniform, so that $\nabla\alpha_p = 0$. Then adding (142) and (143) gives

$$\alpha\mathbf{T}_p^m + (1-\alpha)\mathbf{T}_f^m = \left[\mu_{pp}(\alpha) + \mu_{pf}(\alpha) + \mu_{fp}(\alpha) + \mu_{ff}(\alpha)\right]\mathbf{D}_f. \tag{144}$$

Under these assumptions, there is an effective viscosity. It is given by

$$\mu_{\text{eff}} = \mu_{pp}(\alpha) + \mu_{pf}(\alpha) + \mu_{fp}(\alpha) + \mu_{ff}(\alpha). \tag{145}$$

15.4.3. Computing Momentum Exchange and Stress

Lundgren's Stress

In this section, we shall compute the viscous stress in a fluid component by averaging the stress in the fluid, assuming the particles are rigid. This results in a general expression for $\overline{\mathbf{T}}_c^r$, first discovered by Lundgren [60].

The exact stress in the fluid component is given by

$$\mathbf{T} = -p\mathbf{I} + \mu_f \operatorname{sym} \nabla \mathbf{v}. \tag{146}$$

Multiplying this by X_f and averaging gives

$$\alpha_f \overline{\mathbf{T}}_f^x = -\alpha_f \overline{p}_f^x \mathbf{I} + \mu_f \overline{X_f \operatorname{sym} \nabla \mathbf{v}}. \tag{147}$$

Since the spheres are rigid, the deformation rate in them vanishes,

$$\operatorname{sym} \nabla \mathbf{v} = 0. \tag{148}$$

Thus,

$$\overline{X_p \operatorname{sym} \nabla \mathbf{v}} = 0. \tag{149}$$

Now

$$\overline{X_k \nabla \mathbf{v}} = \nabla \alpha_k \overline{\mathbf{v}}_k^{x\rho} - \overline{(\nabla X_k) \mathbf{v}}, \tag{150}$$

and

$$\overline{X_k [(\nabla \mathbf{v})^\mathsf{T}]} = (\nabla \alpha_k \overline{\mathbf{v}}_k^{x\rho})^\mathsf{T} - \overline{\mathbf{v}(\nabla X_k)}. \tag{151}$$

If we add (147) and (149), use the no-slip condition at the interface, which states that $\mathbf{v}$ is continuous across the interface between the particles and the fluid; and (150) and (151) for $k = p$; and the result that $\nabla X_f = -\nabla X_p$, we have

$$\overline{X_f \operatorname{sym} \nabla \mathbf{v}} = \operatorname{sym} \nabla (\alpha_f \overline{\mathbf{v}}_f^{x\rho} + \alpha_p \overline{\mathbf{v}}_p^{x\rho}). \tag{152}$$

Thus,

$$\alpha_f \overline{\mathbf{T}}_f^x = -\alpha_f \overline{p}_f^x \mathbf{I} + \mu_f \operatorname{sym} \nabla (\alpha_f \overline{\mathbf{v}}_f^{x\rho} + \alpha_p \overline{\mathbf{v}}_p^{x\rho}). \tag{153}$$

It is interesting that the assumption of no deformation inside the particles allows a derivation of the average *fluid* stress in terms of the "volumetric velocity" $\mathbf{v}_v = \alpha_f \overline{\mathbf{v}}_f^{x\rho} + \alpha_p \overline{\mathbf{v}}_p^{x\rho}$. We have

$$\nabla \cdot \mathbf{v}_v = 0, \tag{154}$$

$$\mu_f \nabla^2 \mathbf{v}_v - \nabla \alpha_f \overline{p}_f^x + \mathbf{M}_f + \alpha_f \rho_f \mathbf{g} = 0. \tag{155}$$

Equations (154) and (155) do not form a complete model for any situation; at least a model for the interfacial force $\mathbf{M}_f$ is needed. Under most circumstances, the interaction will also involve the particle momentum equation. Thus, we must supply the constitutive equation for $\overline{\mathbf{T}}_p^x$, as well as $\mathbf{M}_f$.

15.4. Viscous Flow Around a Sphere

Note that Lundgren's derivation of the averaged fluid stress applies when we use the conditional averages. If we multiply (146) for the exact stress in the fluid by X_f and average over all realizations that have sphere 1 at $\mathbf{z}$, we have

$$\overline{X_f \mathbf{T}}^{(1)}(\mathbf{x}, t|\mathbf{z}) = -\overline{X_f p}^{(1)}\mathbf{I} + \mu_f \overline{X_f \operatorname{sym} \nabla \mathbf{v}}^{(1)}. \tag{156}$$

Again, the deformation rate vanishes inside the particles. Applying the conditional average to (148) gives

$$\overline{X_p \operatorname{sym} \nabla \mathbf{v}}^{(1)} = 0. \tag{157}$$

A series of manipulations analogous to those for the unconditional average leads to

$$\overline{X_f \operatorname{sym} \nabla \mathbf{v}}^{(1)} = \operatorname{sym} \nabla [\overline{X_f \mathbf{v}}^{(1)} + \overline{X_p \mathbf{v}}^{(1)}]. \tag{158}$$

Consequently, the conditionally averaged stress is

$$\overline{X_f \mathbf{T}}^{(1)}(\mathbf{x}, t|\mathbf{z}) = -\overline{X_f p}^{(1)}\mathbf{I} + \mu_f \operatorname{sym} \nabla (\overline{X_f \mathbf{v}}^{(1)} + \overline{X_p \mathbf{v}}^{(1)}). \tag{159}$$

If we average with two spheres fixed, the operations are the same, and the result is

$$\overline{X_f \mathbf{T}}^{(2)}(\mathbf{x}, t|\mathbf{z}) = -\overline{X_f p}^{(2)}\mathbf{I} + \mu_f \operatorname{sym} \nabla [\overline{X_f \mathbf{v}}^{(2)} + \overline{X_p \mathbf{v}}^{(2)}]. \tag{160}$$

Note that if we assume that $\alpha_p^{(1)} = \alpha_p = $ constant, then

$$\overline{\mathbf{T}}_k^{x(1)} = \overline{\mathbf{T}}_k^{(1)}. \tag{161}$$

This allows us to use the same stress field in the fluid outside the sphere as in the calculation of the interfacial force density.

Stokes Flow—Hierarchical Method

If the suspension is dilute, the stress conditioned on the presence of a sphere at $\mathbf{z}$, $\overline{\mathbf{T}}^{(1)}(\mathbf{x}, t|\mathbf{z})$, is assumed to be the flow of the clear fluid around a single sphere in an unbounded fluid satisfying certain boundary conditions far from the sphere. Thus, we approximate by letting $\alpha_p \to 0$, and $\mathbf{M}_p^{(2)} = 0$. The equations are identical with the single fluid equations, and are

$$\nabla \cdot \overline{\mathbf{v}}^{x(1)} = 0, \tag{162}$$

$$0 = -\nabla p^{(1)} + \mu_f \nabla^2 \overline{\mathbf{v}}^{(1)}. \tag{163}$$

We abbreviate $\overline{\mathbf{v}}^{(1)}$ as $\mathbf{v}$.

Assume that the sphere is translating with respect to the fluid with velocity $\mathbf{v}_p$, and is rotating at angular velocity $\boldsymbol{\omega}$. Then

$$\mathbf{v} = \mathbf{v}_p + \mathbf{x} \wedge \boldsymbol{\omega} \quad \text{at} \quad r = a, \tag{164}$$

where $r = |\mathbf{x}|$. As $r \to \infty$, we require

$$\mathbf{v} \to \mathbf{v}_f + \mathbf{x} \cdot \nabla \mathbf{v}_f, \tag{165}$$

where $\mathbf{v}_f$ is the velocity of the fluid far from the sphere. It is also the velocity that would exist at the sphere if the sphere were not present.

We shall assume that no external torques are exerted on the sphere. Then, the rotation rates of the fluid and the sphere are the same

$$\boldsymbol{\omega} = \nabla \wedge \mathbf{v}_f. \tag{166}$$

The solution to the fluid motion problem is in standard fluid mechanics references [41], [56]. The form that we shall use is

$$p = p^t + p^s, \tag{167}$$
$$\mathbf{v} = \mathbf{v}_f + \mathbf{v}^t + \mathbf{v}^s, \tag{168}$$

where the translation part of the solution is given by

$$p^t = \frac{3\mu fa}{2}(\mathbf{v}_p - \mathbf{v}_f) \cdot \frac{\mathbf{x}'}{|\mathbf{x}'|^3}, \tag{169}$$

$$\mathbf{v}^t = \frac{3a}{4}(\mathbf{v}_p - \mathbf{v}_f) \cdot \left(\frac{\mathbf{x}'\mathbf{x}'}{|\mathbf{x}'|^3} + \frac{\mathbf{I}}{|\mathbf{x}'|}\right) + \frac{a^3}{4}(\mathbf{v}_p - \mathbf{v}_f) \cdot \left(\frac{\mathbf{I}}{|\mathbf{x}'|^3} - \frac{3\mathbf{x}'\mathbf{x}'}{|\mathbf{x}'|^5}\right) \tag{170}$$

and the linear shearing part is given by

$$p^s = \frac{5\mu fa^5}{2}\mathbf{L}_f : \frac{\mathbf{x}'\mathbf{x}'\mathbf{x}'}{|\mathbf{x}'|^7}, \tag{171}$$

$$\mathbf{v}^s = -\frac{5a^3\mathbf{L}_f}{2} : \left(\frac{\mathbf{x}'\mathbf{x}'\mathbf{x}'}{|\mathbf{x}'|^5}\right) + \mathbf{L}_f : \left(\frac{5a^5\mathbf{x}'\mathbf{x}'\mathbf{x}'}{2|\mathbf{x}'|^7} - \frac{a^5(\mathbf{x}'\mathbf{I} + \mathbf{I}\mathbf{x}')}{2|\mathbf{x}'|^5}\right). \tag{172}$$

Note that the most slowly decaying part of the translational solution involves the "Stokeslet"

$$\mathbf{V}_{St} = \frac{\mathbf{x}'\mathbf{x}'}{|\mathbf{x}'|^3} + \frac{\mathbf{I}}{|\mathbf{x}'|}, \tag{173}$$

and the most slowly decaying part of the shearing solution involves the "stresslet"

$$\mathbf{V}_\tau = \frac{\mathbf{x}'\mathbf{x}'\mathbf{x}'}{|\mathbf{x}'|^5}. \tag{174}$$

First-Order Results

First, we note that the solution given by (170) and (172) cannot be correct if some other sphere is closer to $\mathbf{x}$ than the one at $\mathbf{z}$. Thus we replace $f^{(1)}(\mathbf{z}, t)$ by $\hat{f}^{(1)}(\mathbf{z}, t; \mathbf{x})$, which is the distribution function of the position of the nearest neighbor to $\mathbf{x}$. Then the integral converges because of the exponential decay in $\hat{f}^{(1)}$. As in Section 15.3, we shall approximate that distribution by

$$\hat{f}^{(1)}(\mathbf{z}, t; \mathbf{x}) = \begin{cases} \frac{1}{(4\pi/3)(R^3 - a^3)} & \text{if } |\mathbf{x} - \mathbf{z}| < R, \\ 0 & \text{otherwise.} \end{cases} \tag{175}$$

15.4. Viscous Flow Around a Sphere

Here the cell radius R is chosen so that

$$\alpha_p = \frac{\frac{4}{3}\pi a^3}{\frac{4}{3}\pi R^3}. \tag{176}$$

If we perform the average in (12), we have the result that the interfacial force density is given by the Stokes drag

$$\mathbf{M}_p = \frac{1}{\frac{4}{3}\pi R^3} \int_{|\mathbf{x}'|=a} \mathbf{n} \cdot \overline{\mathbf{T}}^{(1)}(\mathbf{x}, t|\mathbf{x} - \mathbf{x}') d\Omega_{\mathbf{x}'}$$

$$= \frac{9}{2}\frac{\alpha_p}{a^2}(\mathbf{v}_f - \mathbf{v}_p). \tag{177}$$

This result is considered classical [83]. It can be motivated in many different ways.

The interfacial averaged stress can be found by

$$\overline{\mathbf{T}}_{ci} = \frac{1}{4\pi a^2} \int_{\Omega(a)} \overline{\mathbf{T}}^{(1)}(\mathbf{x}, t|\mathbf{x} - \mathbf{x}') d\Omega. \tag{178}$$

Carrying out the integration using (24), (25), and (26) gives

$$\overline{\mathbf{T}}_{ci} = \mu_f \mathbf{L}_f. \tag{179}$$

The average stress inside the particles can be found from (16) to be

$$\overline{\mathbf{T}}_p^r = \frac{5}{2}\mu_f \mathbf{L}_f. \tag{180}$$

We note that the calculation of Lundgren [60] of the fluid stress, to within neglecting terms of $O(\alpha_p)$, gives

$$\overline{\mathbf{T}}_f^r = \mu_f \mathbf{L}_f. \tag{181}$$

Higher-Order Approximations

In the previous section, $\overline{\mathbf{T}}^{(1)}(\mathbf{x}, t|\mathbf{z})$ is computed by neglecting the presence of the other spheres. This gives a result for the stresses correct to order 1, and for the interaction force correct to order α_p. Note that the cell model and renormalization methods are both unnecessary, since no integrals with convergence problems are encountered. This is because the fluid stress can be computed without averaging the specific fields.

The solution for the velocity field in an unbounded fluid with two spheres with no shearing far from the spheres is given by Jeffrey and Onishi [49]. Hurwitz [44] gives an expression that does contain the liquid shearing. If we are interested in corrections for small concentrations, we can assume that the second sphere is far from the first sphere. In that case, the velocity field can be found by considering the perturbation of the velocity in the neighborhood of the sphere at $\mathbf{z}$ due to the sphere at $\mathbf{z}_2$. The velocity in the unbounded fluid due to the sphere at $\mathbf{z}$ is given by (170) and (172). This induces a perturbation velocity at $\mathbf{z}_2$ of

$$\mathbf{v}_2' = \frac{3a}{4}(\mathbf{v}_p - \mathbf{v}_f) \cdot \left(\frac{\mathbf{z}_2'\mathbf{z}_2'}{|\mathbf{z}_2'|^3} + \frac{\mathbf{I}}{|\mathbf{z}_2'|}\right) - \frac{5a^3 \mathbf{L}_f}{2} : \left(\frac{\mathbf{z}_2'\mathbf{z}_2'\mathbf{z}_2'}{|\mathbf{z}_2'|^5}\right) \tag{182}$$

to order $1/|\mathbf{z}_2'|^3$ where

$$\mathbf{z}_2' = \mathbf{z}_2 - \mathbf{z}. \tag{183}$$

By a result of Faxén [33], [41], there results a force on the sphere at $\mathbf{z}_2$ given by

$$\mathbf{f}_2'(\mathbf{z}_2') = 6\pi\mu_f a\left(1 + \frac{a^2}{6}\nabla^2\right)\mathbf{v}_2', \tag{184}$$

to order $1/|\mathbf{z}_2'|^3$. This force results in a velocity field

$$\mathbf{v}''(\mathbf{x}) = \frac{\mathbf{f}_2'(\mathbf{z}_2')}{8\pi\mu_f} \cdot \left(\frac{\mathbf{x}''\mathbf{x}''}{|\mathbf{x}''|^3} + \frac{\mathbf{I}}{|\mathbf{x}''|}\right) \tag{185}$$

to order $1/|\mathbf{x}''|^2$, where $\mathbf{x}'' = \mathbf{x} - \mathbf{z}_2$. Note that since $\mathbf{f}_2'$ is $O(1/|\mathbf{z}_2|^2)$ and the Stokeslet is $O(1/|\mathbf{z}_2|)$, the stress due to $\mathbf{v}''$ is $O(1/|\mathbf{z}_2|^3)$, and the integral over $\mathbf{z}_2$ must be treated carefully. The "offending" part comes from the Stokeslet in (185), and can be treated in two different ways, namely, by renormalization or by effective media.

Renormalization

The second way to treat the possibility of nonconvergence is to "renormalize" the problem. This method consists of recognizing that the origin of the singular term is due to a drag force on the sphere at $\mathbf{z}_2$, and should be treated with the other drag terms. Batchelor [8], and others using the hierarchical method, treat this problem by adding and subtracting the singular part of the stress from the stress $\mathbf{T}$ in the interfacial force density, and evaluating the singular part of the stress contribution by a distributed force density. As we see from (185), the singular part is the velocity caused by a point force

$$\mathbf{f}' = \frac{9\pi\mu_f}{2}(\mathbf{v}_p - \mathbf{v}_f)\delta(\mathbf{x} - \mathbf{z}_2), \tag{186}$$

resulting in a Stokeslet behavior:

$$\mathbf{v}_s = \frac{\mathbf{f}'}{8\pi} \cdot \left(\frac{\mathbf{x}''\mathbf{x}''}{|\mathbf{x}''|^3} + \frac{\mathbf{I}}{|\mathbf{x}''|}\right). \tag{187}$$

If this point force is multiplied by $f^{(1)}(\mathbf{z}_2)$ and integrated over all space, the result is

$$\overline{\mathbf{T}_s \cdot \nabla X_f} = \frac{9}{2a^2}\alpha_p(\mathbf{v}_p - \mathbf{v}_f), \tag{188}$$

the Stokes drag on a single sphere per unit mixture volume.

The average stress inside the particle can be found from (16) to be

$$\overline{\mathbf{T}}_p^x = \left(5 + \frac{515}{32}\alpha_p\right)\mu_f \mathbf{L}_f. \tag{189}$$

The remaining integral in (13) can be evaluated to give the total drag on the dispersed phase to order α_p^2. It is

$$\mathbf{M}_p = \frac{9\alpha_p \mu_f}{2a}\left(1 - \frac{845}{128}\alpha_p\right)(\mathbf{v}_f - \mathbf{v}_p). \tag{190}$$

Effective Medium Theory

The forms of the interfacial force and of the effective viscosity derived in the previous section suggest that a single sphere does not experience forces as if it were in the presence of one (or a few) other spheres, but instead as if it were embedded in an *effective medium*, that is, one where the mechanics is given by equations with terms representing effective properties. The original idea for this method is due to Brinkman [15]. It has been used extensively in studies of electromagnetic and acoustic wave propagation in inhomogeneous media. There are difficulties with this method. First, the method is predicated on the replacement of the effects of the dynamics of *two* components with the dynamics of one effective medium. There is no unique way to do this. At least two separate ways are obvious, holding the particles fixed, and letting the particles move with the fluid. The second difficulty is in the treatment of the effect of the field near the fixed particle. This is related to the difficulty with what to use for $\alpha_p^{(1)}$. We note that we know of no exact solutions of the equations of slow viscous flow around a sphere when the viscosity is variable. Therefore, we expect that the effective medium equations are not soluble if we allow nonuniform particle density. Clearly, however, the fixed particle is actually in contact with the clear fluid almost all of the time, over almost all of its surface area. Lundgren [60] uses the effective medium right up to the sphere. This is not entirely accurate. Recent researchers [44], [48] have instead introduced a "blending" function accounting for the presence of the clear fluid near the sphere, and gradually becoming the effective medium far from the sphere. However, they solve the conditionally averaged equations without considering the nonuniform volume fraction. Itoh [48] uses a linear distribution.

If we now wish to calculate the interfacial force density

$$\mathbf{M}_p = \int_{|\mathbf{z}-\mathbf{x}|=a} f^{(1)}(\mathbf{z},t)\overline{\mathbf{T}}^{(1)}(\mathbf{x},t|\mathbf{z})\cdot\mathbf{n}\,d\Omega, \tag{191}$$

we must derive an expression for $\overline{\mathbf{T}}^{(1)}(\mathbf{x},t|\mathbf{z})$, the averaged stress at $\mathbf{x}$ given that there is a sphere centered at $\mathbf{z}$. It is convenient to start with (126) and (132) for the dynamics of the fluid with one sphere held fixed. These can be written as

$$\nabla\cdot\overline{\mathbf{v}}_v^{(1)} = 0, \tag{192}$$

$$\mu_f\nabla^2\overline{\mathbf{v}}_v^{(1)} + \mathbf{M}_p^{(1)} - \nabla\alpha_f^{(1)}\overline{p}_f^{(1)} = 0, \tag{193}$$

where

$$\overline{\mathbf{v}}_v^{(1)} = \alpha_p^{(1)}\overline{\mathbf{v}}^{r(1)} + \alpha_f^{(1)}\overline{\mathbf{v}}_f^{r(1)} \tag{194}$$

is the volumetric velocity. We assume that there is no body force on the fluid component, so that $\mathbf{g}_f = 0$. We shall use the body force on the particles to justify certain special assumptions about the particle velocity. We shall use the assumption that $\alpha_p^{(1)} = \alpha_p =$ constant. Furthermore, we shall abbreviate $\overline{\mathbf{v}}_v^{(1)}$ by $\mathbf{v}_v$. The remainder of the procedure for obtaining expressions for the effective viscosity and the drag is to assume a form for $\mathbf{M}_p$ in terms of α_p, $\mathbf{v}_p$, and $\mathbf{v}_f$, *assume the same form* for $\mathbf{M}_p^{(1)}$, solve the resulting boundary-value problem for the flow around the sphere centered at $\mathbf{z}$, and use the stress to compute $\mathbf{M}_p$ from (191). The resulting equation will contain coefficients from the form for $\mathbf{M}_p$. The strategy is to get one equation for one unknown coefficient at a time. Then it is simple to solve for the coefficient.

For slow viscous flow with constant particle density, the form

$$\mathbf{M}_p = \alpha_p \overline{S}^e(\alpha_p)(\mathbf{v}_f - \mathbf{v}_p) \tag{195}$$

appears to be adequate. Also, it is less obvious at this point that we need an expression for $\overline{\mathbf{T}}_p^r$; however, we give one here. We assume that

$$\overline{\mathbf{T}}_p^x = -\overline{p}_f^x \mathbf{I} + \overline{\mu}_p^{ep}(\alpha_p)\mathbf{D}_p + \overline{\mu}_p^{ef}(\alpha_p)\mathbf{D}_f . \tag{196}$$

Effective Viscosity

For this calculation of the effective viscosity, we assume that the particles move with the fluid, so that $\mathbf{v}_p = \mathbf{v}_f$. Then the continuity equation (129) gives

$$\nabla \cdot \mathbf{v}_v = 0, \tag{197}$$

and the momentum equation (130) for the fluid becomes

$$\nabla^2[\overline{\mu}^e(\alpha_p)\mathbf{v}_v] - \nabla p_v = 0, \tag{198}$$

where

$$\overline{\mu}^e(\alpha_p) = \mu_f + \overline{\mu}^{ep} + \overline{\mu}^{ef}, \tag{199}$$

and

$$p_v = \alpha_f p_f . \tag{200}$$

The boundary conditions that we use are

$$\mathbf{v}_v = 0 \quad \text{on} \quad |\mathbf{x} - \mathbf{z}| = a, \tag{201}$$

and

$$\mathbf{v}_v \to \mathbf{v}_f \quad \text{as} \quad |\mathbf{x} - \mathbf{z}| \to \infty . \tag{202}$$

We again take $\mathbf{v}_f = \mathbf{v}_{f0} + \mathbf{L}_f \cdot (\mathbf{x} - \mathbf{z})$, where $\mathbf{L}_f$ is a constant tensor with tr $\mathbf{L}_f = 0$. We assume that the particle momentum equation will yield $\mathbf{v}_p = \mathbf{v}_f$ when $\mathbf{g}_p = 0$. Then we must have

$$\mu_p^{ep} = \mu^{ep} \quad \text{and} \quad \mu_p^{ef} = \mu^{ef} . \tag{203}$$

15.4. Viscous Flow Around a Sphere

Equation (198) is just the equation for Stokes flow, with an effective viscosity. The solution to this problem is given by (169), (170), (171), and (172). Thus, proceeding as before, we find that the average stress in the fixed particle is given by

$$\overline{\mathbf{T}}_p^x = \frac{5}{2}\mu^e(\alpha_p)\mathbf{L}_f. \qquad (204)$$

Thus, we must have

$$\frac{5}{2}\mu^e = \mu^{ep} + \mu^{ef}. \qquad (205)$$

The equation defining the fluid stress is (153). Thus, the average stress in the fluid is

$$(1-\alpha_p)\overline{\mathbf{T}}_f^x = \mu_f \mathbf{L}_f. \qquad (206)$$

The effective stress is the mixture stress,

$$\begin{aligned}\overline{\mathbf{T}}^x &= (1-\alpha_p)\overline{\mathbf{T}}_f^x + \alpha_p \overline{\mathbf{T}}_p^x \\ &= \mu_f \mathbf{L}_f + \frac{5}{2}\alpha_p \mu^e(\alpha_p)\mathbf{L}_f \\ &= \mu^e(\alpha_p)\mathbf{L}_f. \end{aligned} \qquad (207)$$

Solving for $\mu^e(\alpha_p)$ gives

$$\mu^e(\alpha_p) = \frac{\mu_f}{1 - \frac{5}{2}\alpha_p}, \qquad (208)$$

and, from (204)

$$\mu_p^{ep}(\alpha_p) + \mu_p^{ef}(\alpha_p) = \frac{\frac{5}{2}\mu_f}{1 - \frac{5}{2}\alpha_p}. \qquad (209)$$

This result for $\mu^e(\alpha_p)$ is shown in Figure (15.6), along with some other models and some experimental data. Lundgren notes that this expression becomes infinite at $\alpha_p = 0.4$. He claims that at such concentrations, the assumption that a suspension behaves as a Newtonian fluid is probably not valid. Note that for small α_p, the effective viscosity $\mu^e(\alpha_p)$ can be expanded in a Taylor series in α_p. The result is

$$\mu^e(\alpha_p) \approx \mu_f\left(1 + \frac{5}{2}\alpha_p + \frac{25}{4}\alpha_p^2 + O(\alpha_p^3)\right). \qquad (210)$$

When terms of $O(\alpha_p^2)$ are neglected, this reduces to Einstein's [31] result.

Drag

In order to calculate the drag by effective medium theory, consider the problem given by (197) and (198), with boundary conditions

$$\mathbf{v}_v = \mathbf{v}_p \quad \text{on} \quad |\mathbf{x} - \mathbf{z}| = a \qquad (211)$$

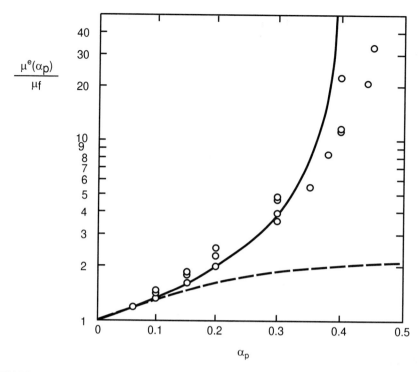

FIGURE 15.6. Effective viscosity *versus* volume fraction (adapted from Lundgren [60]). The dashed curve is Einstein's viscosity, and the solid curve is Lundgren's.

and

$$\mathbf{v}_v \to \mathbf{v}_f \quad \text{as} \quad |\mathbf{x} - \mathbf{z}| \to \infty. \tag{212}$$

We now take

$$\mathbf{v}_f = \mathbf{v}_{f0} = V_0 \mathbf{e}_z, \tag{213}$$

where V_0 is a constant. Without loss of generality, we can take $\mathbf{v}_p = 0$. For the calculation of the effective drag coefficient, we assume that α_p is constant. Thus, we must solve the problem

$$\nabla \cdot \mathbf{v}_v = 0, \tag{214}$$

$$(1 - \alpha_p)\mu_f(\alpha_p)\nabla^2 \mathbf{v}_v - \nabla p_v - \alpha_p \overline{S}^e(\alpha_p)\mathbf{v}_v = 0, \tag{215}$$

with the boundary condition that the velocity must approach

$$\mathbf{v}_{f0} = V_0 \mathbf{e}_z. \tag{216}$$

We see that $\nabla^2 p_v = 0$. Itoh [48] gives the solution for the velocity and pressure as follows. First, assume that the pressure is of the form

$$p_v = (1 - \alpha_p)\mu_f(\alpha_p)V_0\chi(r)\cos\theta, \tag{217}$$

15.4. Viscous Flow Around a Sphere

where

$$r = |\mathbf{x} - \mathbf{z}|,$$

and θ is the angle between $\mathbf{v}_{f0}$ and $\mathbf{x} - \mathbf{z}$. Since the pressure is harmonic, the function χ must satisfy Laplace's equation in spherical coordinates, so that its r dependence is given by

$$\chi_{rr} + \frac{2}{r}\chi_r - \frac{2}{r^2}\chi = 0, \tag{218}$$

with solution

$$\chi = C_0 r + \frac{C_1}{r^2}, \tag{219}$$

where C_0 and C_1 are constants.

Writing

$$\mathbf{v}_{f0} = V_0 \mathbf{e}_z = V_0(\cos\theta\, \mathbf{e}_r - \sin\theta\, \mathbf{e}_\theta)$$

suggests that the solution for the velocity should be of the form

$$\mathbf{v}_f = V_0\left[\phi \cos\theta\, \mathbf{e}_r - \left\{\phi + \frac{r}{2}\phi_r\right\}\sin\theta\, \mathbf{e}_\theta\right]. \tag{220}$$

Substituting (220) into (215) results in the following equation for ϕ:

$$\phi_{rr} + \frac{4}{r}\phi_r - \hat{S}\phi - \chi_r = 0, \tag{221}$$

where $\hat{S} = \alpha_p \overline{S}^e(\alpha_p)/(1-\alpha_p)\mu_f(\alpha_p)$. The boundary conditions for ϕ are

$$\phi(a) = 0, \tag{222}$$

$$\phi_r(a) = 0, \tag{223}$$

$$\phi \to 1 \quad \text{as} \quad r \to \infty. \tag{224}$$

The solution can be written as

$$\phi = -\frac{C_0}{\hat{S}} + \frac{2C_1}{r^3 \hat{S}} + C_2 \frac{1+\sqrt{\hat{S}}r}{r^3} e^{-\sqrt{\hat{S}}r}, \tag{225}$$

where C_2 is a constant of integration. There is a second solution of the homogeneous equation that grows like $\exp(\hat{S}^{1/2}r)$. It has been left out. The boundary conditions give

$$C_0 = -\hat{S}, \tag{226}$$

$$C_1 = -\frac{3a + 3a^2\sqrt{\hat{S}} + a^3\hat{S}}{2}, \tag{227}$$

$$C_2 = \frac{3ae^{a\sqrt{\hat{S}}}}{\hat{S}}. \tag{228}$$

15. Relation of Microstructure to Constitutive Equations

In terms of ϕ and χ, the stress is given by

$$\mathbf{T} = \mu_f(\alpha_p)V_0\Big\{ -\chi\cos\theta\mathbf{I} + (1-\alpha_p)2\phi_r\cos\theta\mathbf{e}_r\mathbf{e}_r$$
$$- \frac{1}{2}\left(r\phi_{rr} + \frac{2}{r}\phi\right)\sin\theta(\mathbf{e}_r\mathbf{e}_\theta + \mathbf{e}_\theta\mathbf{e}_r) + \frac{1}{r}\phi\cos\theta(\mathbf{e}_\theta\mathbf{e}_\theta + \mathbf{e}_\omega\mathbf{e}_\omega)\Big\} \quad (229)$$

where $\mathbf{e}_r$, $\mathbf{e}_\theta$, and $\mathbf{e}_\omega$ are the unit vectors in spherical coordinates. Taking the inner product of the stress with $\mathbf{n}$, and integrating over $r = a$ gives

$$-\overline{\mathbf{T}\cdot\nabla X_p} = \frac{4}{3}\pi a(1-\alpha_p)\mu_f(\alpha_p)V_0[\phi_{rr}|_{r=a} + \chi(a)]\mathbf{e}_z.$$

Substituting for χ and ϕ, and setting this equal to $\alpha_p \overline{S}^e(\alpha_p)V_0\mathbf{e}_z$, and solving for $\overline{S}^e$ gives

$$\overline{S}^e(\alpha_p) = \frac{9}{2a^2}\frac{1 + \frac{3}{4}\left[3\alpha_p + (8\alpha_p + 9\alpha_p^2)^{1/2}\right]}{\left(1 - \frac{3}{2}\alpha_p\right)^2}. \quad (230)$$

16
Maxwell–Boltzmann Dynamics

Kinetic theory modeling can be applied to the dispersed component in a dispersed multicomponent flow. It motivates the use of an entropy associated with the dispersed component, and leads to constitutive equations for the dispersed component Reynolds stress.

Particle velocity fluctuations cause the dispersed phase to behave much like a gas, or, at least, a gas as modeled by a Boltzmann description of identical solid spheres. The kinetic energy contained in the fluctuations of the velocity of the particles around the mean is analogous to the temperature in a kinetic theory model for a gas. The random motions of the particle around the mean motion cause transfers of momentum and energy. The random motions of the particles can be caused by collisions or by the random motions of the fluid, which influence the particles through the interfacial force.

Let $f^{(1)}(\mathbf{z}, \mathbf{v}, \mathbf{u}, t)$ be the number density of identical spheres in phase space for both the particle velocity $\mathbf{v}$ and the fluid velocity $\mathbf{u}$. That is, the number of particles within $d\mathbf{z}$ of $\mathbf{z}$ having velocity within $d\mathbf{v}$ of $\mathbf{v}$, when the fluid velocity is within $d\mathbf{u}$ of $\mathbf{u}$ is

$$f^{(1)}(\mathbf{z}, \mathbf{v}, \mathbf{u}, t)\, d\mathbf{x}\, d\mathbf{v}\, d\mathbf{u}.$$

Following Reeks [74], we assume that $f^{(1)}$ satisfies Boltzmann's equation in the form

$$\frac{\partial f^{(1)}}{\partial t} + \nabla \cdot (\mathbf{v} f^{(1)}) + \frac{\partial}{\partial \mathbf{v}} \cdot \left[f^{(1)}(\mathbf{F}_p/m_p) \right] = \mathcal{C}(f^{(1)}), \qquad (1)$$

where $\mathbf{F}_p$ is the force experienced by the sphere, which depends on the particle and fluid velocities. Also, $\mathcal{C}(f^{(1)})$ is the operator representing the collisions.

16.1. Collision Effects

There is no existing theory to justify superposing the results from collisions and those from fluid velocity fluctuations. Even so, we shall present the results analogous to those in Chapter 4. The stress and heat flux are given by

$$\mathbf{T}_d^{Re} + p_d^{Re}\mathbf{I} = \mu_p^{\text{coll}}\left[\nabla\bar{\mathbf{v}}_d^{x\rho} + (\nabla\bar{\mathbf{v}}_d^{x\rho})^T - \frac{1}{3}(\nabla\cdot\bar{\mathbf{v}}_d^{x\rho})\mathbf{I}\right], \tag{2}$$

$$\mathbf{q}_d^{Re} = -\lambda_p^{\text{coll}}\nabla u_p^{Re}, \tag{3}$$

$$\mu_p^{\text{coll}} = \frac{\left[1 + \frac{8}{5}\alpha_p G(\alpha_p)\right]}{G(\alpha_p)}\frac{5\sqrt{3}m_p}{64\sqrt{2\pi}a^2}\sqrt{u_p^{Re}}, \tag{4}$$

$$\lambda_p^{\text{coll}} = \frac{\left[1 + \frac{12}{5}\alpha_p G(\alpha_p)\right]}{G(\alpha_p)}\frac{225\sqrt{3}m_p}{512\sqrt{2\pi}a^2}\sqrt{u_p^{Re}}, \tag{5}$$

where u_p^{Re} is the particle fluctuation kinetic energy, a is the particle radius, α_p is the volume fraction of the particle component, and $G(\alpha_p)$ is the shielding factor, calculated to be

$$G(\alpha_p) = \frac{1 - \frac{11}{2}\alpha_p}{1 - 8\alpha_p} \approx 1 + \frac{5}{2}\alpha_p. \tag{6}$$

16.2. Fluid Velocity Effects

We wish to use simple yet reasonable assumptions about the interaction of the particles with the fluid to calculate the Reynolds stress, the energy flux, the interfacial force density, and the interfacial work. To do so, we shall derive the equations of balance of mass, momentum, and energy for the particle component from (1) by averaging over the particle velocities and positions, and the fluid turbulence.

16.2.1. Balance Equations

Neglecting the collision operator results in the equation

$$\frac{\partial f^{(1)}}{\partial t} + \nabla\cdot(\mathbf{v}f^{(1)}) + \frac{\partial}{\partial\mathbf{v}}\cdot\left[f^{(1)}(\mathbf{F}_p/m_p)\right] = 0. \tag{7}$$

If we integrate (7) over all particle velocities and fluid velocities, we have

$$\frac{\partial n_p}{\partial t} + \nabla\cdot n_p\bar{\mathbf{v}}_p = 0, \tag{8}$$

16.2. Fluid Velocity Effects

where n_p is the particle number density,

$$n_p = \int f^{(1)} \, d\mathbf{v} \, d\mathbf{u},$$

and $\bar{\mathbf{v}}_p$ is the average particle velocity,

$$\bar{\mathbf{v}}_p = \frac{1}{n_p} \int \mathbf{v} f^{(1)} \, d\mathbf{v} \, d\mathbf{u}.$$

If we multiply (7) by $\mathbf{v}$, integrate over the velocities, and use the relation

$$\int \mathbf{v} \frac{\partial}{\partial \mathbf{v}} \cdot [f^{(1)}(\mathbf{F}_p/m_p)] \, d\mathbf{v} = -\int f^{(1)}(\mathbf{F}_p/m_p) \, d\mathbf{v},$$

we have

$$\frac{\partial n_p \bar{\mathbf{v}}_p}{\partial t} + \nabla \cdot n_p \overline{\mathbf{v}\mathbf{v}} = \overline{f^{(1)}(\mathbf{F}_p/m_p)}. \tag{9}$$

The term $n_p \overline{\mathbf{v}\mathbf{v}}$ can be written as

$$n_p \overline{\mathbf{v}\mathbf{v}} = n_p \bar{\mathbf{v}}_p \bar{\mathbf{v}}_p - n_p \mathbf{T}_p,$$

where $\mathbf{T}_p$ is the particle stress, which corresponds to the Reynolds stress in the dispersed component.

Finally, if we multiply (7) by $v^2/2$, integrate over the velocities, and use the relation

$$\int \frac{v^2}{2} \frac{\partial}{\partial \mathbf{v}} \cdot [f^{(1)}(\mathbf{F}_p/m_p)] \, d\mathbf{v} = -\int f^{(1)}(\mathbf{v} \cdot \mathbf{F}_p/m_p) \, d\mathbf{v},$$

we have

$$\frac{\partial n_p \overline{v^2/2}}{\partial t} + \nabla \cdot n_p \overline{\mathbf{v} \tfrac{1}{2} v^2} = \overline{f^{(1)}(\mathbf{v} \cdot \mathbf{F}_p/m_p)}. \tag{10}$$

Again, we can write

$$n_p \overline{\mathbf{v} \tfrac{1}{2} v^2} = n_p \bar{\mathbf{v}}_p \tfrac{1}{2} \bar{v}_p^2 - n_p \bar{\mathbf{v}}_p \cdot \mathbf{T}_p - n_p \bar{\mathbf{q}}_p,$$

where $\mathbf{q}_p$ is the energy flux.

These equations are analogous to the dispersed component equations of balance of mass, momentum, and energy.

16.2.2. Correlation Assumptions

Consider the conditional probability $f^{(1)}(\mathbf{z}, \mathbf{v}, t|\mathbf{u})$ for the particle velocity given the fluid velocity. Then

$$f^{(1)}(\mathbf{z}, \mathbf{v}, \mathbf{u}, t) = f^{(1)}(\mathbf{z}, \mathbf{v}, t|\mathbf{u}) P(\mathbf{u}),$$

where $P(\mathbf{u})$ is the fluid velocity density function.

16. Maxwell–Boltzmann Dynamics

We assume that the particle momentum equation is dominated by drag, and that the particles spend random amounts of time in different fluid eddies, each having a constant, uniform velocity [1]. The force $\mathbf{F}_p$ is taken to be

$$\mathbf{F}_p = S(\mathbf{u} - \mathbf{v}), \tag{11}$$

where S is the drag coefficient. For Stokes drag,

$$S = 6\pi \mu a.$$

The parameter S is also related to the particle relaxation time,

$$\tau_p = \frac{m_p}{S} = \beta^{-1}.$$

Then the microscale motions of the particles satisfies a Langevin equation

$$\frac{d\mathbf{v}}{dt} = \beta(\mathbf{u} - \mathbf{v}).$$

The particle motion in an eddy with velocity $\mathbf{u}$ is given by

$$\mathbf{v} = \omega e^{-\beta(t-\xi)} + \mathbf{u}(1 - e^{-\beta(t-\xi)}), \tag{12}$$

where ω is the particle velocity at time ξ. We further assume that the velocity ω is a random variable from the *unconditional* particle velocity distribution. Given the particle velocity and the position of the particle at time t, the starting position of the particle is

$$\zeta = \mathbf{z} - \int_\xi^t \mathbf{v}(t')\,dt' = \mathbf{u}(t - \xi) + (\omega - \mathbf{u})\tau_p(1 - e^{\beta(t-\xi)}).$$

The particles that reach $\mathbf{z}$ with velocity $\mathbf{v}$ at time t, come from position ζ with velocity ω, at time ξ. We further assume that the particle starting velocity ω is uncorrelated with the fluid velocity $\mathbf{u}$ and the starting time ξ. Then the conditional probability of having particle velocity $\mathbf{v}$, *given* that the fluid velocity is $\mathbf{u}$, *and* that the particle had velocity ω at time ξ is

$$f^{(1)}(\mathbf{z}, \mathbf{v} | \mathbf{u}, \omega, \xi) = n_p(\zeta)\delta(\mathbf{v} - \omega e^{-\beta(t-\xi)} + \mathbf{u}(1 - e^{-\beta(t-\xi)})), \tag{13}$$

where δ denotes the Dirac delta function. Here, we assume that the distribution for the duration in the eddy is

$$\Pr(t - \xi) = \delta(t - \xi - \tau_e), \tag{14}$$

where τ_e is the eddy time scale. It is consistent with the other modeling assumptions in this chapter to assume that $\overline{\mathbf{u}\omega} = 0$. Finally, we assume that

$$n_p(\zeta) = n_p(\mathbf{z}) + (\zeta - \mathbf{z}) \cdot \nabla n_p(\mathbf{z}). \tag{15}$$

16.2.3. Calculated Quantities

Interfacial Transfers

We assume that
$$\mathbf{F}_p(\mathbf{v}, \mathbf{u}) = \mathbf{F}_p(\bar{\mathbf{v}}_p, \bar{\mathbf{v}}_f) + \mathbf{F}'_p \approx \mathbf{F}_p(\bar{\mathbf{v}}_p, \bar{\mathbf{v}}_f) + S(\mathbf{u}' - \mathbf{v}'), \tag{16}$$
where we use the dominance of drag in the fluctuation part of the interfacial force. Then
$$\overline{f^{(1)} \mathbf{F}_p(\mathbf{v}, \mathbf{u})} = n_p \mathbf{F}_p(\bar{\mathbf{v}}_p, \bar{\mathbf{v}}_f) + \mathbf{M}_p^{TD}, \tag{17}$$
where $\mathbf{M}_p^{TD}$ is the force of turbulent dispersion, given by
$$\mathbf{M}_p^{TD} = \overline{S(\mathbf{u}' - \mathbf{v}') f^{(1)}}.$$
Also,
$$\overline{f^{(1)} \mathbf{v} \cdot \mathbf{F}_p(\mathbf{v}, \mathbf{u})} = n_p \bar{\mathbf{v}}_p \cdot \left(\mathbf{F}_p(\bar{\mathbf{v}}_p, \bar{\mathbf{v}}_f) + \mathbf{M}_p^{TD} \right) + E_p^{TD}, \tag{18}$$
where E_p^{TD} is the interfacial source of energy due to turbulent dispersion, given by
$$E_p^{TD} = \overline{S \mathbf{v}' \cdot (\mathbf{u}' - \mathbf{v}') f^{(1)}}.$$
We note that $\overline{\mathbf{v}' f^{(1)}} = 0$, so that
$$\mathbf{M}_p^{TD} = \overline{S \mathbf{u}' f^{(1)}}.$$
Substituting (13) and (15), we have
$$\mathbf{M}_p^{TD} = \overline{\mathbf{u}'(\zeta - \mathbf{z})} \cdot \nabla n_p = \left[1 - \frac{\tau_e}{\tau_p} - e^{-\tau_e/\tau_p} \right] \overline{\mathbf{u}' \mathbf{u}'} \cdot \nabla n_p. \tag{19}$$
For the energy source, we find
$$E_p^{TD} = -n_p S \overline{(\mathbf{u}')^2} \, e^{-\tau_e/\tau_p}. \tag{20}$$

Stress and Heat Flux

We wish to calculate the particle component Reynolds stress
$$\mathbf{T}_p^{Re} = -m_p \overline{(\mathbf{v} - \bar{\mathbf{v}}_p)(\mathbf{v} - \bar{\mathbf{v}}_p)}, \tag{21}$$
and the heat (fluctuation energy) flux
$$\mathbf{q}_p^{Re} = m_p \frac{1}{2} \overline{(v - \bar{v}_p)^2 (\mathbf{v} - \bar{\mathbf{v}}_p)}. \tag{22}$$
Substituting the assumed and derived relations for the probability density functions gives
$$\mathbf{T}_p^{Re} = -m_p \iiint [(\boldsymbol{\omega} - \mathbf{u}')e^{-\beta(t-\xi)} + \mathbf{u}'][(\boldsymbol{\omega} - \mathbf{u}')e^{-\beta(t-\xi)} + \mathbf{u}']$$
$$\times \Pr(\boldsymbol{\omega}, \xi) \, d\boldsymbol{\omega} \, d\xi \, \Pr(\mathbf{u}) \, d\mathbf{u}$$
$$= -m_p \overline{\boldsymbol{\omega} \boldsymbol{\omega}} e^{-2\tau_e/\tau_p} - m_p \overline{\mathbf{u}' \mathbf{u}'} (1 - e^{-\tau_e/\tau_p})^2. \tag{23}$$

Recognizing that

$$\mathbf{T}_p^{Re} = -m_p \overline{\boldsymbol{\omega}\boldsymbol{\omega}}$$

and

$$\overline{\mathbf{u'u'}} = \mathbf{T}_f^{Re}/\rho_f,$$

and solving for $\mathbf{T}_p^{Re}$ gives

$$\mathbf{T}_p^{Re} = \frac{\rho_p}{\rho_f} \mathbf{T}_f^{Re} \frac{(1 - e^{-\tau_e/\tau_p})^2}{1 - e^{-2\tau_e/\tau_p}}, \tag{24}$$

where $\rho_p = n_p m_p$.

Note that as $\tau_e/\tau_p \to 0$, we have $\overline{\mathbf{T}}_p^{Re} \to 0$. Thus, if the particle relaxation time greatly exceeds the eddy time scale, the particle will not respond to the fluid motions, and the particle Reynolds stress will vanish. On the other extreme, as $\tau_e/\tau_p \to \infty$, the particles are able to follow the fluid velocity fluctuations relatively faithfully, and therefore,

$$\mathbf{T}_p^{Re} \to \frac{\rho_p}{\rho_f} \mathbf{T}_f^{Re}. \tag{25}$$

Proceeding as with the stress calculation yields the result that the heat (fluctuation energy) flux is

$$\mathbf{q}_p^{Re} = \frac{\rho_p}{\rho_f} \mathbf{q}_f^{Re} \frac{(1 - e^{-\tau_e/\tau_p})^3}{1 - e^{-3\tau_e/\tau_p}}. \tag{26}$$

17
Interfacial Area

Multicomponent mechanics is a description of a variety of phenomena, many of which involve the evolution of interfaces separating the components. In some processes the interface may evolve smoothly, remaining approximately planar throughout. In others, the interface may assume complicated shapes such as dimples, dendrites, ripples, and irregular shapes that resemble fractals.

In solidification of molten materials, the interface between the melt and the solidified material is the location where the latent heat of the melt is changed to sensible heat, whence it must be transported to the boundaries of the flow region. As the materials cool, the location, size, and shape of the interface are important in determining the rate at which the process proceeds, as well as the state of the final product. See Marsh et al. [61] for a discussion of the processes.

Gas–liquid flows occur in many energy-production situations. Boilers, heat exchangers, and turbines work on the cycle that uses heat to make steam, and the work of the expanding steam to generate electricity. Again, the processes occurring at the gas–liquid interface determine the rate at which steam is generated or, at the other end of the cycle, how fast water condenses from the steam. One concept for cooling high-density electronic semiconductor circuits involves boiling a coolant. Burning of liquid fuels is facilitated by spraying the liquid fuel into an atmosphere of oxidant. Here, the generation of sufficient surface area governs the rate of burning.

The most crucial factor in heat and mass transfer in gas–liquid flows is the shape taken by the interface. Nucleate boiling involves the generation of small bubbles in crevices at the interface, their subsequent growth and breakaway, and the stirring caused by their rise away from the surface; it is a relatively efficient means of heat transfer. On the other hand, the formation of a vapor blanket on the interface hinders the resupply of liquid to the surface, and the efficiency of the heat transfer process is reduced drastically. Present descriptions of the "flow regime" of a gas–liquid process are highly empirical, and represent the largest conceptual barrier to theoretical description of multicomponent mechanics.

It is clear that the description of the flow regime in a gas–liquid flow, or the evolution of the interface in a solidifying flow depends on having an adequate description of the geometry of the interface. The mechanical models for the dynamics of the different materials, or "components," have terms representing the interactions between the materials across the interfaces. These terms obviously depend on the geometry of the interface, and the internal stresses and heat fluxes. Specifying constitutive equations for these functions in terms of the state variables requires that the set of state variables be sufficient to describe all of the mechanical interactions involved. However, the present status of continuum modeling is such that the treatment of the dependence on the volume fraction of the materials, and a small number of length scales (say the radius of the bubbles or drops) is the best that can be done without introducing constants or functions that are exceedingly difficult to measure.

In this chapter we discuss the modeling of the local geometry of individual elements in the dispersed component of multicomponent fluids. We give a short discussion of an empirical approach to the problem, followed by a derivation of a set of equations governing the evolution of Gaussian and mean curvature of an interface. The derivation is based on the description of the interface as moving normal to itself at a speed that is assumed known. In practice this speed is coupled to the evolution of the interface, and must be determined along with it. In order to understand the implications of the equations so derived, we study their predictions in some examples, including the evolution of two simple objects, namely, spheres and cylinders, and one object that is not so simple, an ellipse. We also study the prediction of the model on general figures in the case when the speed of propagation of the interface normal to itself is independent of position.

17.1. Geometry Models

The dispersed component volume fraction α_d is a geometric variable. That is, it is determined entirely by the geometry of the system independent of its dynamics or thermodynamics. However, two different flows may have the same value of α_d and be drastically different. For example, arrays of body-centered cubic-packed spheres have $\alpha_d = \pi/6$, no matter how big the spheres are. Therefore, values of other geometric quantities are also necessary to specify the state of a given two-component flow. For example, if the dispersed component consists of identical rigid spheres, then the particle radius r_d can be taken as a parameter. In bubbly or droplet flows, the average bubble or drop radius r_d is often employed. If the dispersed component average radius is allowed to evolve, we must use a model governing its rate of change. One way to do this is by relating average radius and number density to volume fraction. We define

$$\alpha_d = \frac{4}{3}\pi n_d r_d^3. \tag{1}$$

If the bubbles are not spherical, the radius so defined is called the *equivalent radius*.

The essence of multicomponent continua is the geometry of the different components. In many cases there is a single dispersed phase, and it consists of particles that are small, approximately spherical, and approximately of the same size. Then their structure may not be important to the solution of certain problems. Many cases are much more interesting; the structure of the individual particles may be complex, it may evolve, and the particles may differ among one another. Then that structure may affect the gross behavior of the material. That the dispersed component volume fraction α_d does not suffice to describe the material is obvious from considering arrays of body-centered cubic-packed spheres. There all such arrays have $\alpha_d = \pi/6$, *no matter how big the spheres are*. Clearly here, assuming the geometry of the array is known, one other quantity such as the sphere diameter suffices to specify the array completely. More complicated geometry of course would require additional parameters.

If no coalescence or breakup of the dispersed component occurs, then the number density of that component is conserved, giving

$$\frac{\partial n}{\partial t} + \nabla \cdot n\,\mathbf{v}_d = 0. \tag{2}$$

If coalescence and breakup do occur, we can model this by considering a size distribution, and how it evolves.

17.1.1. *Bubble Coalescence and Breakup*

Consider a distribution of spherical bubbles, each with the same mass density, with $f(u, \mathbf{x}, t)$ being the number density of spheres of volume u. That is, $f(u, \mathbf{x}, t)\,du\,d\mathbf{x}$ is the probability of finding a bubble having volume between u and $u + du$, within $d\mathbf{x}$ of the point $\mathbf{x}$. We assume an evolution equation of the form

$$\frac{\partial f}{\partial t} + \nabla \cdot f\mathbf{v}_u = \left.\frac{df}{dt}\right|_c + \left.\frac{df}{dt}\right|_b, \tag{3}$$

where $\mathbf{v}_u$ is the velocity associated with the bubbles of volume u, $df/dt|_c$ is the rate of change of the number of bubbles due to coalescence, and $df/dt|_b$ is the rate of change of the number of bubbles due to breakup. The *number density* is the average number of bubbles, per unit volume (in $\mathbf{x}$), over all possible bubble volumes u, and is defined by

$$n(\mathbf{x}, t) = \int_0^\infty f(u, \mathbf{x}, t)\,du. \tag{4}$$

Note that
$$\phi(u; \mathbf{x}, t) = \frac{f(u, \mathbf{x}, t)}{n(\mathbf{x}, t)}, \tag{5}$$
is the probability density function for the volume at each point **x**. This function satisfies the relation
$$\int_0^\infty \phi(u; \mathbf{x}, t) \, du = 1. \tag{6}$$
The volume fraction is taken to be the average
$$\alpha_d(\mathbf{x}, t) = \int_0^\infty u f(u, \mathbf{x}, t) \, du. \tag{7}$$
Note that the average volume at **x** is
$$\bar{u}(\mathbf{x}, t) = \int_0^\infty u \phi(u, \mathbf{x}, t) \, du, \tag{8}$$
so that the volume fraction α_d is given by
$$\alpha_d = \bar{u} n. \tag{9}$$

Assume that if two bubbles coalesce, the volume of the resulting bubble is the same as that of the original two bubbles, with a similar assumption for breakup. Thus, for example, coalescence of a bubble of volume u' with one of volume $u - u'$ results in a bubble of volume u. Thus, we assume

$$\left.\frac{df}{dt}\right|_c = -\int_0^\infty c(u', u) f(u', \mathbf{x}, t) f(u, \mathbf{x}, t) \, du'$$
$$+ \frac{1}{2} \int_0^u c(u', u - u')] f(u', \mathbf{x}, t) f(u - u', \mathbf{x}, t) \, du', \tag{10}$$

where $c(u', u'')$ is the rate of coalescence of bubbles of volumes u' and u''. Assume that a bubble of volume u breaks into a bubble of volume u' and a bubble of volume $u'' = u - u'$ at rate $b(u', u - u')$, then

$$\left.\frac{df}{dt}\right|_b = -\int_0^\infty b(u, u) f(u, \mathbf{x}, t) \, du' + \int_0^u b(u', u - u') f(u', \mathbf{x}, t) \, du', \tag{11}$$

where $b(u', u'')$ is the rate at which bubbles of volume u break into bubbles of volumes u' and $u - u'$.

Constitutive equations for $c(u', u'')$ and $b(u', u'')$ must be supplied. We wish to consider a simple model for coalescence and breakup. For coalescence, assume that the probability of a coalescence of two bubbles in a small volume $d\mathbf{x}$ is proportional to the probability of the bubbles "overlapping" if they were allowed to fall in the volume $d\mathbf{x}$ at random. Thus the probability of a coalescence of a bubble of volume u' with a bubble of volume $u - u'$ in a short time dt is taken to be proportional to the total volume. Thus we assume $c(u', u'') = C(u' + u'')$, where C is a rate constant.

17.1. Geometry Models

Then

$$\left.\frac{df}{dt}\right|_c = -C \int_0^\infty (u' + u) f(u', \mathbf{x}, t) f(u, \mathbf{x}, t) \, du'$$

$$+ \frac{C}{2} \int_0^u [u' + (u - u')] f(u', \mathbf{x}, t) f(u - u', \mathbf{x}, t) \, du'$$

$$= -C f(u, \mathbf{x}, t)(n\, u + \alpha_d) + \frac{C}{2} u \int_0^u f(u', \mathbf{x}, t) f(u - u', \mathbf{x}, t) \, du' \quad (12)$$

For breakup, assume that a bubble of volume u' breaks in half if it is larger than some critical size, u_{cr}. Then the rate is zero, unless $u' + u'' > u_{cr}$, and unless $u' = u''$. Assume that the rate is given by $b(u', u'') = BH(u - u_{cr})\delta(u' - u'')$, where $H(\cdot)$ is the Heaviside function, and $\delta(\cdot)$ is the Dirac delta function. Then

$$\left.\frac{df}{dt}\right|_b = \begin{cases} 4Bf(2u, \mathbf{x}, t), & u < u_{cr}, \\ -Bf(u) + 4Bf(2u, \mathbf{x}, t), & u > u_{cr}. \end{cases} \quad (13)$$

This latter may be written in terms of Heaviside functions as

$$\left.\frac{df}{dt}\right|_b = B[4f(2u, \mathbf{x}, t) - H(u - u_{cr}) f(u, \mathbf{x}, t)]. \quad (14)$$

The evolution equation for f is

$$\frac{\partial f}{\partial t} + \nabla \cdot f \mathbf{v}_u = -C f(u, \mathbf{x}, t)(n(\mathbf{x}, t) u + \alpha(\mathbf{x}, t))$$

$$+ \frac{C}{2} u \int_0^u f(u', \mathbf{x}, t) f(u - u', \mathbf{x}, t) \, du'$$

$$+ B(4f(2u, \mathbf{x}, t) - H(u - u_{cr}) f(u, \mathbf{x}, t)). \quad (15)$$

By integrating (3) with respect to u we obtain

$$\frac{\partial n}{\partial t} + \nabla \cdot n \bar{\mathbf{v}}^{(n)} = \int_0^\infty \left.\frac{df}{dt}\right|_c + \left.\frac{df}{dt}\right|_b \, du,$$

$$= -Cn\alpha_d + B(2n - n^+), \quad (16)$$

where

$$n^+(\mathbf{x}, t) = \int_{u_{cr}}^\infty f(u, \mathbf{x}, t) \, du \quad (17)$$

is the number density of bubbles larger than the critical size for breakup, and $\bar{\mathbf{v}}^{(n)}$ is the number density weighted velocity. If we multiply (3) by u and integrate, we have

$$\frac{\partial \alpha_d}{\partial t} + \nabla \cdot \alpha_d \bar{\mathbf{v}}_d^x = 0, \quad (18)$$

where $\bar{\mathbf{v}}_d^x$ is the x-weighted average velocity, which is equal to the mass-weighted velocity, since the dispersed component density is constant. In addition to the number density and the volume fraction, there are two more physical moments of the size distribution that are of interest. If we assume that

$$u = \frac{4\pi r^3}{3}, \tag{19}$$

then the radius of an individual bubble is

$$r = \left(\frac{3}{4\pi}u\right)^{1/3}, \tag{20}$$

so the average radius is

$$r_d = \frac{1}{n}\int_0^\infty f(u, \mathbf{x}, t)\left(\frac{3}{4\pi}u\right)^{1/3} du, \tag{21}$$

and the average area per unit volume is

$$A = 4\pi \int_0^\infty f(u, \mathbf{x}, t)\left(\frac{3}{4\pi}u\right)^{2/3} du. \tag{22}$$

Steady State

If we assume that the processes of breakup and coalescence are in steady state and the flow is homogeneous, we have

$$-Cf(u)(nu + \alpha_d) + \frac{C}{2}u\int_0^u f(u')f(u - u')du'$$
$$+ B[4f(2u) - H(u - u_{cr})f(u)] = 0. \tag{23}$$

Note that if $u < u_{cr}/2$, then only coalescence can occur, and the only possible steady state is $f = 0$. For $u > u_{cr}$, we can solve (23) for $f(2u)$, and substitute u for $2u$ to get

$$f(u) = H\left(\frac{u}{2} - u_{cr}\right) f\left(\frac{u}{2}\right)$$
$$- \gamma \left[f\left(\frac{u}{2}\right)\left(n\frac{u}{2} + \alpha_d\right) + \int_0^{u/2} f(u') f\left(\frac{u}{2} - u'\right) du' \right], \tag{24}$$

where $\gamma = C/B$ is the ratio of coalescence to breakup rates and is, roughly, the number of coalescences per breakup. This expression can be used to generate $f(u)$ from values of $f(u')$, for $u' < u/2$. Note that the distribution for $u_{cr}/2 < u < u_{cr}$ is not specified by (23). Note further that the distribution must satisfy the boundary condition at ∞ that

$$f(u) \to 0 \quad \text{as} \quad u \to \infty. \tag{25}$$

Kalkach-Navarro [52] obtained numerical approximations to $f(u)$. The distribution was determined by breaking the interval $u_{cr}/2 < u < u_{cr}$ into subintervals and approximating the distribution by a histogram. The histogram values were adjusted until the distribution was sufficiently small for u large. The resulting distribution seemed to be unique in the sense that different initial guesses for the

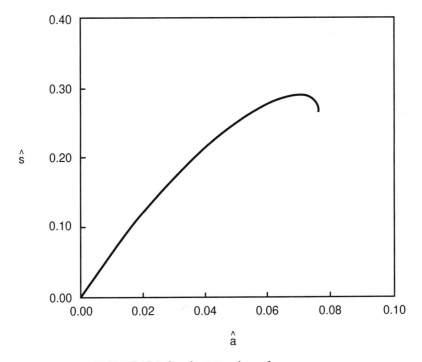

FIGURE 17.1. Steady-state values of s versus α_d.

histogram seemed to converge to the same final distribution. The steady-state distribution was also used to find $s(n, \alpha_d)$ in the steady state. A typical result for this function is shown in Figure 17.1. Note that the area per unit volume increases with increasing volume fraction for small volume fractions, but decreases for larger volume fractions. Indeed, numerical solutions could not be obtained for values of the volume fraction larger than the largest shown on the graph. This could signal the onset of a flow regime transition.

17.2. Evolution of Geometric Statistics

Ishii [45], [47] has proposed an evolution equation for the interfacial area density of the form

$$\frac{\partial A}{\partial t} + \nabla \cdot A\mathbf{v}_i = \beta_i, \qquad (26)$$

where $\mathbf{v}_i$ is the average interfacial velocity, and β_i is a source term due to interfacial deformation.

We shall derive a model for the evolution of interfacial geometry for situations where the interface is complex, and only averaged information is desired. Equations for the evolution of volume fraction, interfacial area density, and the average mean and Gaussian curvatures are derived. Ishii [45] and Kataoka [53] have proposed a model for the evolution of surface area. The present model has a similar equation for the evolution of surface area, but supplements it with the equations for the evolution of Gaussian and mean curvature.

As is usual in averaging processes, after averaging, "residual" terms remain that need to be determined. Thus the model must be supplemented by *closure* assumptions, that is, assumptions for the forms of the residual terms adequate so that a determined system of equations is obtained. Here we study the behavior of this model for specific assumptions about the terms needed for closure. We give particular attention to the models for the loss of surface area due to the coalescence of surfaces, and the changes that occur in average properties due to the coalescence and breakup of a population of spherical bubbles.

This model is a candidate for the description of geometry adequate for multicomponent descriptions of the mechanics of solidification and flow regime transition.

17.2.1. Evolution of Curvature

Consider a region of space divided into two parts by an interface that moves in time. There are two useful descriptions of the interface. The first description is given by

$$F(\mathbf{x}, t) = 0. \tag{27}$$

The second description is given by

$$\mathbf{x}(\alpha, t), \tag{28}$$

where α are surface coordinates. The relation between (27) and (28) is

$$F(\mathbf{x}(\alpha, t), t) = 0. \tag{29}$$

In the first description, the normal vector to the surface is given by

$$\mathbf{n}(\mathbf{x}, t) = \frac{\nabla F}{|\nabla F|}. \tag{30}$$

The kinematic equation for the evolution of the surface is

$$\frac{\partial F}{\partial t} + \mathbf{v}_i \cdot \nabla F = 0, \tag{31}$$

where

$$\mathbf{v}_i = \frac{\partial \mathbf{x}}{\partial t}. \tag{32}$$

There are an infinite number of surface velocity fields that give rise to the same surface motion. To see this, assume that $\mathbf{v}_i^{(1)}$ and $\mathbf{v}_i^{(2)}$ are two surface velocity fields. The corresponding surface evolution equations are

$$\frac{\partial F}{\partial t} + \mathbf{v}_i^{(1)} \cdot \nabla F = 0, \tag{33}$$

and

$$\frac{\partial F}{\partial t} + \mathbf{v}_i^{(2)} \cdot \nabla F = 0. \tag{34}$$

If $\mathbf{v}_i^{(1)} \cdot \mathbf{n} = \mathbf{v}_i^{(2)} \cdot \mathbf{n}$, then $\partial F/\partial t$ is the same in each description. Thus, any two surface velocity fields which have the same normal velocity component give rise to the same surface motion. Therefore, we assume, without loss of generality, that the velocity of every point on the surface is normal to the surface. Note that *material* points on the surface can move tangentially to the surface. However, since surface points are indistinguishable, translation of the surface in the tangential direction does not change the geometry. Thus,

$$\mathbf{v}_i = v\mathbf{n}. \tag{35}$$

Furthermore, we assume that the parametrization is such that any point $\mathbf{x}$ on the surface at time t, parametrized by $\boldsymbol{\alpha}$, moves normal to the surface. That is, we assume that not only does the interface move normal to itself, but the parametrization is carried normally, as well. This implies that the parametrization need not correspond to the motion of any material point. Thus,

$$\frac{\partial \mathbf{x}}{\partial t} = \mathbf{v}_i(\boldsymbol{\alpha}, t) = v(\boldsymbol{\alpha}, t)\mathbf{n}(\boldsymbol{\alpha}, t). \tag{36}$$

If the surface coordinates α_J are chosen to be the principal coordinates in the surface, we have

$$\frac{\partial \mathbf{x}}{\partial \alpha_J} = R_J \frac{\partial \mathbf{n}}{\partial \alpha_J}, \tag{37}$$

where R_J, $J = 1, 2$, are the principal radii of curvature. We assume that α_J are chosen so that $R_1 \leq R_2$. This is depicted in Figure 17.2.

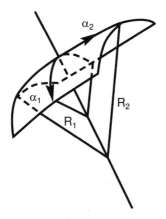

FIGURE 17.2. Principal coordinates and radii of curvature.

Then if we differentiate (37) with respect to t following a point on the surface, we have

$$\frac{\partial}{\partial t}\frac{\partial \mathbf{x}}{\partial \alpha_J} = \frac{\partial}{\partial \alpha_J}\frac{\partial \mathbf{x}}{\partial t}$$
$$= \frac{\partial \mathbf{v}}{\partial \alpha_J} = \frac{\partial v}{\partial \alpha_J}\mathbf{n} + v\frac{\partial \mathbf{n}}{\partial \alpha_J}$$
$$= \frac{\partial R_J}{\partial t}\frac{\partial \mathbf{n}}{\partial \alpha_J} + R_J\frac{\partial}{\partial t}\frac{\partial \mathbf{n}}{\partial \alpha_J}. \tag{38}$$

Thus,

$$\frac{\partial R_J}{\partial t}\frac{\partial \mathbf{n}}{\partial \alpha_J} + R_J\frac{\partial}{\partial t}\frac{\partial \mathbf{n}}{\partial \alpha_J} = \frac{\partial v}{\partial \alpha_J}\mathbf{n} + v\frac{\partial \mathbf{n}}{\partial \alpha_J}. \tag{39}$$

Note that $\mathbf{n} \cdot \mathbf{n} = 1$, so that

$$\frac{\partial \mathbf{n}}{\partial \alpha_J} \cdot \mathbf{n} = 0 \tag{40}$$

and

$$\frac{\partial \mathbf{n}}{\partial t} \cdot \mathbf{n} = 0. \tag{41}$$

Then, if we take the inner product of $\mathbf{n}$ with (39) and use (40), we have

$$\frac{\partial v}{\partial \alpha_J} = R_J \mathbf{n} \cdot \frac{\partial}{\partial t}\frac{\partial \mathbf{n}}{\partial \alpha_J}. \tag{42}$$

Also, if we take the inner product of $\partial \mathbf{n}/\partial \alpha_J$ with (39), we have

$$\frac{\partial R_J}{\partial t} = v - R_J \frac{\dfrac{\partial \mathbf{n}}{\partial \alpha_J} \cdot \dfrac{\partial}{\partial t}\dfrac{\partial \mathbf{n}}{\partial \alpha_J}}{\dfrac{\partial \mathbf{n}}{\partial \alpha_J} \cdot \dfrac{\partial \mathbf{n}}{\partial \alpha_J}}. \tag{43}$$

In order to calculate $\partial \mathbf{n}/\partial t$, we take the time derivative of (30). We have

$$\frac{\partial \mathbf{n}}{\partial t} = \frac{1}{|\nabla F|}\frac{\partial \nabla F}{\partial t} - \nabla F \cdot \frac{\partial \nabla F}{\partial t}\frac{\nabla F}{|\nabla F|^3}$$
$$= \frac{1}{|\nabla F|}\frac{\partial \nabla F}{\partial t} - \mathbf{nn} \cdot \frac{\partial \nabla F}{\partial t}. \tag{44}$$

Since F vanishes identically on the interface, we have

$$\frac{\partial F}{\partial t} + \mathbf{v}_i \cdot \nabla F = 0. \tag{45}$$

Substituting (30) and (35) into (45) gives

$$\frac{\partial F}{\partial t} + v|\nabla F| = 0. \tag{46}$$

Taking the gradient of (46) and using (30), we have

$$\frac{\partial \nabla F}{\partial t} + \nabla v |\nabla F| + v \nabla \nabla F \cdot \mathbf{n} = 0. \tag{47}$$

17.2. Evolution of Geometric Statistics

Substituting (47) into (44) gives

$$\frac{\partial \mathbf{n}}{\partial t} = -\nabla v + \mathbf{n} \cdot \nabla v \mathbf{n} - \frac{1}{|\nabla F|} v \mathbf{n} \cdot \nabla \nabla F + \frac{1}{|\nabla F|} v \mathbf{n} \cdot \nabla \nabla F \cdot \mathbf{nn}. \quad (48)$$

The quantity $\nabla \nabla F$ can be eliminated by noting that

$$\nabla \mathbf{n} = \frac{1}{|\nabla F|} \nabla \nabla F - \frac{1}{|\nabla F|^3} \nabla F \cdot \nabla \nabla F$$

$$= \frac{1}{|\nabla F|} \nabla \nabla F \cdot (\mathbf{I} - \mathbf{nn}). \quad (49)$$

Since $\mathbf{n} \cdot \mathbf{n} = 1$, we have $\nabla \mathbf{n} \cdot \mathbf{n} = 0$, and taking the inner product of (49) with $\mathbf{n}$ gives

$$\mathbf{n} \cdot \nabla \nabla F \cdot (\mathbf{I} - \mathbf{nn}) = 0. \quad (50)$$

Using this in (48) results in

$$\frac{\partial \mathbf{n}}{\partial t} = -\nabla v + \mathbf{nn} \cdot \nabla v. \quad (51)$$

Finally, we note that since $v = v(\alpha, t)$, $\mathbf{n} \cdot \nabla v = 0$, the latter being the directional derivative of v in the direction normal to the surface. Thus, (51) reduces to

$$\frac{\partial \mathbf{n}}{\partial t} = -\nabla v. \quad (52)$$

We differentiate (52) with respect to α_J to obtain

$$\frac{\partial}{\partial \alpha_J} \frac{\partial \mathbf{n}}{\partial t} = -\nabla \nabla v \cdot \frac{\partial \mathbf{x}}{\partial \alpha_J}. \quad (53)$$

Then

$$\frac{\partial}{\partial \alpha_J} \frac{\partial \mathbf{n}}{\partial t} \cdot \frac{\partial \mathbf{n}}{\partial \alpha_J} = -\frac{\partial \mathbf{n}}{\partial \alpha_J} \cdot \nabla \nabla v \cdot \frac{\partial \mathbf{x}}{\partial \alpha_J}. \quad (54)$$

Hence, from (37), we have

$$\frac{\dfrac{\partial \mathbf{n}}{\partial \alpha_J} \cdot \dfrac{\partial}{\partial t} \dfrac{\partial \mathbf{n}}{\partial \alpha_J}}{\dfrac{\partial \mathbf{n}}{\partial \alpha_J} \cdot \dfrac{\partial \mathbf{n}}{\partial \alpha_J}} = R_J \nabla^2 v. \quad (55)$$

Thus,

$$\frac{\partial R_J}{\partial t} = v + R_J^2 \nabla^2 v. \quad (56)$$

The mean curvature H and the Gaussian curvature K are defined by [2]

$$H = \frac{1}{2}\left(\frac{1}{R_1} + \frac{1}{R_2}\right), \quad (57)$$

$$K = \frac{1}{R_1 R_2}. \quad (58)$$

Note that

$$\frac{1}{R_1} = H + \sqrt{H^2 - K}, \tag{59}$$

$$\frac{1}{R_2} = H - \sqrt{H^2 - K}. \tag{60}$$

We differentiate (57) and (58) with respect to t, and eliminate R_1 and R_2 using (59) and (60). We obtain

$$\frac{\partial H}{\partial t} = -(2H^2 - K)v - \frac{1}{2}\nabla^2 v, \tag{61}$$

$$\frac{\partial K}{\partial t} = -2HKv - 2H\nabla^2 v. \tag{62}$$

Equations (61) and (62) give the rate of change of the mean and Gaussian curvature for a surface in terms of the curvatures and the velocity field giving the rate of propagation of the surface normal to itself. In practical examples of interfacial evolution, the velocity field will be determined by some dynamical process occurring in the materials separated by the interface [40]. For example, in a solidifying material, the temperature fields in the solid and the melt determine the rate of heat flux from the surface, and therefore the velocity of propagation of the interface. The problem is further complicated in that the fields in the solid and the melt depend on the interface position and shape. A second example is the evolution of a liquid-gas interface in a two-component flow. The interface moves in response to the pressures in the two components, which, in turn, are coupled to the interface shape through surface tension.

There is a substantial literature on "curve shortening," the two-dimensional version of this problem ([43], for example). The present model reduces to the two-dimensional one by letting $K = 0$. The two-dimensional models assume that $v = v(H)$ and are models for the evolution of a solid–liquid interface during solidification.

17.3. Average Geometrical Properties

Often, in physical situations involving the evolution of interfaces, those interfaces develop complicated shapes. For example, solidification can lead to dendritic growth or multiply connected components. Gas–liquid flows develop wavy interfaces and coalescence and breakup of bubbles. Clearly, quite often it is not necessary to retain the full details of the shape and size of the interfaces in order to describe the gross physics. Thus, a statistical treatment of the interfacial properties is desirable.

We wish to apply the ensemble average to (9.1.28), (61), and (62). The average of the component function is denoted by $\overline{X_1} = \alpha$. Averaging the topological equation (9.1.28) gives

$$\frac{\partial \alpha}{\partial t} + \overline{\mathbf{v} \cdot \nabla X} = 0. \tag{63}$$

17.3. Average Geometrical Properties

Here $\mathbf{v} = v\mathbf{n}$ and the directional derivative[1] $\partial X/\partial n$ is defined by $\partial X/\partial n \equiv \mathbf{n}_1 \cdot \nabla X_1 = \mathbf{n}_2 \cdot \nabla X_2$. Note that $\partial X/\partial n$ is the Dirac delta function on the interface. Thus, $\overline{\mathbf{v} \cdot \nabla X} = \overline{v\, \partial X/\partial n}$ is the surface average of v. We have

$$\frac{\partial \alpha}{\partial t} + Av = 0, \qquad (64)$$

where $A = \overline{\partial X/\partial n}$ is the average of the delta function, and corresponds to the interfacial area per unit volume; and

$$v = \frac{\overline{\mathbf{v} \cdot \nabla X}}{A} \qquad (65)$$

is the average interfacial normal velocity.

Equation (64) expresses the evolution of the volume fraction. It introduces two new variables, the average area per unit volume, and the average interfacial normal velocity.

We next derive an equation for the evolution of interfacial area per unit volume. Consider

$$\frac{\partial \mathbf{n} \cdot \nabla X}{\partial t} + \nabla \cdot (\mathbf{vn} \cdot \nabla X) = \left[\frac{\partial \mathbf{n}}{\partial t} + \nabla \cdot (\mathbf{vn})\right] \cdot \nabla X$$
$$+ \mathbf{n} \cdot \left[\frac{\partial \nabla X}{\partial t} + \mathbf{v} \cdot \nabla \nabla X\right],$$
$$= \left[\frac{\partial \mathbf{n}}{\partial t} + \nabla \cdot (\mathbf{vn})\right] \cdot \nabla X + (\nabla \cdot \mathbf{v})\mathbf{n} \cdot \nabla X$$
$$+ \mathbf{n} \cdot [-\nabla \mathbf{v} \cdot \nabla X], \qquad (66)$$

where we have used the gradient of the topological equation (9.1.28). Now from (40) and (52), we have

$$\frac{\partial}{\partial t}\left(\frac{\partial X}{\partial n}\right) + \nabla \cdot \left(\mathbf{v}\frac{\partial X}{\partial n}\right) = v(\nabla \cdot \mathbf{n})\frac{\partial X}{\partial n},$$
$$= vH\frac{\partial X}{\partial n}. \qquad (67)$$

Applying the averaging process yields

$$\frac{\partial A}{\partial t} + \nabla \cdot \overline{\mathbf{v}}_i A = S_A, \qquad (68)$$

where

$$\overline{\mathbf{v}}_i = \frac{\overline{\mathbf{v}\frac{\partial X}{\partial n}}}{A} \qquad (69)$$

---

[1] The notation for the directional derivative $\partial X/\partial n$ appears as if it were a partial derivative. However, it is not; indeed, directional derivatives and partial derivatives have different properties. For example, directional derivatives in different directions do not commute [78].

is the interfacial averaged velocity (contrast this with the definition of v), and
$$S_A = v\overline{\overline{H}}A, \tag{70}$$
with
$$\overline{\overline{H}} = \frac{\overline{vH\dfrac{\partial X}{\partial n}}}{Av} \tag{71}$$
being the *velocity-weighted* average mean curvature. The proper definition of the average mean curvature is
$$\overline{H} = \frac{\overline{H\dfrac{\partial X}{\partial n}}}{A}. \tag{72}$$

Note that if v is deterministic so that $\overline{v} = v$, then $\overline{\overline{H}} = \overline{H}$. Otherwise, this equality does not hold. We write
$$Av\overline{\overline{H}} = Av\overline{H} + A\Phi_A, \tag{73}$$
where the interfacial source per unit area Φ_A is defined by
$$A\Phi_A = \overline{(H - \overline{H})v\dfrac{\partial X}{\partial n}}. \tag{74}$$
Thus (68) becomes
$$\frac{\partial A}{\partial t} + \nabla \cdot \overline{\mathbf{v}}_i A = v\overline{H}A + A\Phi_A. \tag{75}$$

We now derive evolution equations for the average of the mean and Gaussian curvatures. We start from (61) and (62), written in spatial form,
$$\frac{\partial H}{\partial t} + \mathbf{v} \cdot \nabla H = -(2H^2 - K)v - \frac{1}{2}\nabla^2 v, \tag{76}$$
$$\frac{\partial K}{\partial t} + \mathbf{v} \cdot \nabla K = -2HKv - 2H\nabla^2 v. \tag{77}$$
If we multiply (76) by $\partial X/\partial n$ and add to it the product of (67) with H, we have
$$\frac{\partial}{\partial t}\left(H\frac{\partial X}{\partial n}\right) + \nabla \cdot \left(\mathbf{v}H\frac{\partial X}{\partial n}\right) = -(H^2 - K)v\frac{\partial X}{\partial n} - \frac{1}{2}\nabla^2 v \frac{\partial X}{\partial n}. \tag{78}$$
Performing a similar sequence of operations with (77) yields
$$\frac{\partial}{\partial t}\left(K\frac{\partial X}{\partial n}\right) + \nabla \cdot \left(\mathbf{v}K\frac{\partial X}{\partial n}\right) = -HKv\frac{\partial X}{\partial n} - 2H\nabla^2 v \frac{\partial X}{\partial n}. \tag{79}$$
Applying the averaging process to (78) and (79) gives the results
$$\frac{\partial \overline{H}A}{\partial t} + \nabla \cdot \overline{\mathbf{v}}_i \overline{H}A = -\nabla \cdot A\mathbf{q}_H - (\overline{H}^2 - \overline{K})vA + s\Phi_H, \tag{80}$$
$$\frac{\partial \overline{K}A}{\partial t} + \nabla \cdot \overline{\mathbf{v}}_i \overline{K}A = -\nabla \cdot A\mathbf{q}_K - \overline{H}\,\overline{K}v + A\Phi_K. \tag{81}$$

17.3. Average Geometrical Properties

Here

$$\overline{K} = \frac{\overline{K \frac{\partial X}{\partial n}}}{A}, \tag{82}$$

$$\mathbf{q}_H = \frac{\overline{(\mathbf{v} - \overline{\mathbf{v}}_i) H \frac{\partial X}{\partial n}}}{A}, \tag{83}$$

$$\mathbf{q}_K = \frac{\overline{(\mathbf{v} - \overline{\mathbf{v}}_i) K \frac{\partial X}{\partial n}}}{A}, \tag{84}$$

$$A\Phi_H = -\frac{1}{2}\overline{\nabla^2 v \frac{\partial X}{\partial n}} - \overline{(H^2 - \overline{H}^2) v \frac{\partial X}{\partial n}} - \overline{(K - \overline{K}) v \frac{\partial X}{\partial n}}, \tag{85}$$

$$A\Phi_K = -2H\overline{\nabla^2 v \frac{\partial X}{\partial n}} + \overline{(HK - \overline{H}\,\overline{K}) v \frac{\partial X}{\partial n}}. \tag{86}$$

Equations (64), (75), (80), and (81) are conservation equations for average geometric properties of the multicomponent process. The equations represent evolution equations for volume fraction, interfacial area density, and average mean and Gaussian curvature. As is usual for averaging processes, there is a "closure" problem for these equations. Specifically, there are not closed-form expressions for Φ_A, Φ_H, Φ_K, $\mathbf{q}_H$, or $\mathbf{q}_K$. The quantities v and $\overline{\mathbf{v}}_i$ must come from dynamical equations representing processes occurring in the two materials on either side of the interface.

We start by discussing the physical processes implicit in the quantities that determine the predictive capabilities of the system. First, v is the average velocity of the interface normal to itself. Thus, it is the rate at which component 2 "consumes" component 1, per unit time, per unit area. The velocity $\overline{\mathbf{v}}_i$ is the "convective" velocity of the interface, that is, the rate at which the interfacial area moves around. Note that if $\nabla \cdot \overline{\mathbf{v}}_i > 0$ at some spatial point $\mathbf{x}$ at some instant t, then interface is moving out away from $\mathbf{x}$.

The quantities $\mathbf{q}_H$ and $\mathbf{q}_K$ are fluxes of average mean and Gaussian curvature, respectively, due to differences between $\mathbf{v}$ and $\overline{\mathbf{v}}_i$. This flux results from a correlation of the curvature with the motion. It is not known how to quantify this effect; however, if the curvature is not correlated with the velocity, these terms vanish.

The remaining terms, Φ_A, Φ_H, Φ_K, represent sources of area and of mean and Gaussian curvature, per unit area. In a flow consisting of coalescing and breaking bubbles, the term $v\overline{H}A$ represents the rate of growth of interfacial area due to the growth of bubbles caused by such things as changes in pressure or dissolution or solution of gas. The term $A\Phi_A$ represents the rate of change of interfacial area due to bubble coalescence and breakup.

In a similar manner, the rate of change of mean and Gaussian curvature due to bubble growth due to such things as changes in pressure are contained in the terms $(\overline{H}^2 - \overline{K})vA$ and $\overline{H}\,\overline{K}vA$, respectively. Several effects are further lumped into Φ_H and Φ_K. First, effects resulting from the variation of the normal velocity with position ($v_{,11}$ and $v_{,22}$) are included here. These terms result in interfacial instabilities. This is illustrated in Example 3b. Also included in Φ_H and Φ_K are effects resulting from the correlation of curvature with the interfacial velocity. Specifically, normal growth of wiggly interfaces changes the average curvature.

The equations for the evolution of the geometry, with all notation for the averaging removed, are

$$\frac{\partial \alpha}{\partial t} + Av = 0, \tag{87}$$

$$\frac{\partial A}{\partial t} + \nabla \cdot \mathbf{v}_i A = vHA + A\Phi_A, \tag{88}$$

$$\frac{\partial H}{\partial t} + \mathbf{v}_i \cdot \nabla H = -\frac{1}{s}\nabla \cdot s\mathbf{q}_H - (2H^2 - K)v + \Phi_H - H\Phi_A, \tag{89}$$

$$\frac{\partial K}{\partial t} + \mathbf{v}_i \cdot \nabla K = -\frac{1}{s}\nabla \cdot s\mathbf{q}_K - 2HKv + \Phi_K - K\Phi_A. \tag{90}$$

17.3.1. Examples

In order to render the system determined, we must supply constitutive equations for $\Phi_A, \Phi_H, \Phi_K, \mathbf{q}_H$, and $\mathbf{q}_K$. To be able to test the predictions of the geometric system alone, we shall make reasonable yet simple assumptions about the kinematics, as represented in the quantities v and $\mathbf{v}_i$. Specifically, we shall take $\mathbf{v}_i = 0$, and assume that v is a given function.

1. *No Area Source.* Assume $\mathbf{q}_H = \mathbf{q}_K = 0$, and $\Phi_A = \Phi_H = \Phi_K = 0$. Then we have

$$\frac{dH}{dt} = -(2H^2 - K)v, \tag{91}$$

$$\frac{dK}{dt} = -2HKv. \tag{92}$$

Note that

$$G = \frac{H^2 - K}{K^2} \tag{93}$$

is conserved by (91) and (92). Then

$$H = (K^2 G + K)^{1/2}, \tag{94}$$

so that

$$\frac{dK}{dt} = -2(K^2 G + K)^{1/2} Kv. \tag{95}$$

If we rescale time by defining $T = \int_0^t v\,dt$, we have

$$\frac{dK}{dT} = -2(K^2 G + K)^{1/2} K. \tag{96}$$

17.3. Average Geometrical Properties

This equation can be separated and integrated to give

$$F(KG) = C - 2G^{-1/2}T, \tag{97}$$

where

$$F(\kappa) = \int \frac{d\kappa}{\kappa^{3/2}(\kappa+1)^{1/2}}, \tag{98}$$

and C is a constant of integration which is related to the average curvature at time $t = 0$. The area is then given by

$$A = A(0) \exp\left[\int_0^T (K^2 G + K)^{1/2}(T')\, dT'\right], \tag{99}$$

and the volume fraction is

$$\alpha = \alpha(0) - s(0) \int_0^T \exp\left[\int_0^{T'} (K^2 G + K)^{1/2}(T'')\right] dT''. \tag{100}$$

2. *Area Coalescence with No Normal Motion.* With $\Phi_A \neq 0$, but with $v = 0$, we have

$$\frac{dA}{dt} = A\Phi_A, \tag{101}$$

$$\frac{dH}{dt} = -H\Phi_A, \tag{102}$$

$$\frac{dK}{dt} = -K\Phi_A. \tag{103}$$

Then

$$AH = h = \text{constant}, \tag{104}$$

and

$$AK = k = \text{constant}. \tag{105}$$

With reasonable assumptions about Φ_A, three situations can occur. If $A \to 0$ (either in finite time or infinite time), then H and K both $\to \infty$. If $A \to \infty$, then H and K both $\to 0$. If $A \to A_o$, a constant, then H and K both tend to constants. In this case, A_o must be a zero of Φ_A. The behavior in all cases can be determined by a phase plane analysis. If Φ_A has no zeros, then the interface will eventually disappear while becoming infinitely curved, or will fill large parts of the space with no curvature. The latter case seems to be impossible, possibly signaling a shortcoming of the model.

Let us examine a more specific model of interfacial area evolution. If the interfaces in the neighborhood of some point are moving (with zero average velocity), then there can exist surface coalescences near junctions of interfaces, as indicated in Figure 17.3. If we assume that the coalescences happen with a probability proportional to A^2, we see that we lose interfacial area at a rate

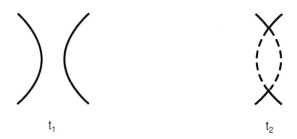

FIGURE 17.3. Surface coalescence.

proportional to A^2. Thus, we take $A\Phi_A = -\beta A^2$, where β is a constant. In this case,

$$A = \frac{1}{1/A(0) + \beta t}, \tag{106}$$

and we note that since both $A(0)$ and β are positive, $A \to 0$.

3. *Area Coalescence with Normal Motion.* Let us now examine the combined effects of normal motion and coalescence. Let $\Phi_A = -\beta A$ and $v = $ constant. We have

$$\frac{dA}{dt} = vHA - \beta A^2, \tag{107}$$

$$\frac{dH}{dt} = -(2H^2 - K)v + \beta AH, \tag{108}$$

$$\frac{dK}{dt} = -2HKv + \beta AK. \tag{109}$$

a. *"Planar" Evolution.* First, if $K(0) = 0$, then $K(t) \equiv 0$. If the exact Gaussian curvature $K = 0$, then one of the principal radii of curvature vanishes. Although the analogy does not necessarily follow for the averaged Gaussian curvature, we shall refer to this case as "planar." In this case, $H = H(A)$ is given by

$$\frac{dH}{dA} = \frac{-2H^2 + \hat{\beta} AH}{HA - \hat{\beta} A^2}. \tag{110}$$

With the substitution $\eta = H/A$, the equation is solvable by quadrature. The solution is

$$\frac{\eta}{(-3\eta + 2\hat{\beta})^{1/3}} = \frac{C}{A^2}, \tag{111}$$

where C is a constant of integration. From (111) we see that $\eta \to 0$ as $A \to \infty$. This implies that (107) will not allow $A \to \infty$. From an examination of the A–η phase plane, it is evident that as $t \to \infty$, $A \to 0$, and $\eta \to 2\hat{\beta}/3$. The latter implies that $H \to 0$ as $t \to \infty$.

b. General Evolution. Let us consider the case with $K(0) \neq 0$. If we let $\tau = \nu t$ and $\sigma = \hat{\beta} A$, the system becomes

$$\frac{d\sigma}{d\tau} = H\sigma - \sigma^2, \quad (112)$$

$$\frac{dH}{d\tau} = -(2H^2 - K) + \sigma H, \quad (113)$$

$$\frac{dK}{d\tau} = -2HK + \sigma K. \quad (114)$$

Combining (113) and (114) yields

$$\frac{H}{K} = \tau + b, \quad (115)$$

where $b = H(0)/K(0)$. Solving (115) for K and substituting into (113) leads to

$$\frac{dH}{d\tau} = -2H^2 + \frac{H}{\tau + b} + \sigma H. \quad (116)$$

With $\eta = H/\sigma$, the equations become

$$\frac{d\sigma}{d\tau} = \sigma^2(\eta - \hat{\beta}), \quad (117)$$

$$\frac{1}{\sigma}\frac{d\eta}{d\tau} = -3\eta^2 + 2\hat{\beta}\eta + \frac{\eta/\sigma}{\tau + b}. \quad (118)$$

If $b > 0$ and $H(\tau)$ remains finite, then for τ sufficiently large,

$$H(\tau)/(\tau + b) \to 0,$$

and the situation is analogous to that for $K \equiv 0$. Thus, $\sigma \to 0$ as $\tau \to \infty$. We see that in order for σ to grow, $\eta > \hat{\beta}$. If we let $\tau + b = \tau'$, we see that η changes sign at $\tau' = 0$. If $b < 0$, then $\tau' = 0$ occurs at a finite positive time $\tau = -b$. If $\eta(0) > 0$, then $\eta < 0$ for $\tau > -b$, and $\sigma \to 0$ in a finite time τ_0. If $\eta(0) < 0$, then for $\tau > -b$, we have $\eta/(\tau + b) > 0$. In this case, and $\sigma \to 0$ as $\tau \to \infty$.

Thus, with this model, $\sigma \to 0$ under all conditions. If, ultimately $H > 0$, then there will be some small amount of area present ($\sigma > 0$) at all times; whereas, if $H < 0$ for sufficiently large times, then the area will disappear in a finite time.

17.4. A Coalescence and Breakup Model and Geometry

The direction taken here is that if we supply a model for the coalescence and breakup terms, then that model will suggest constitutive equations for Φ_A, Φ_H, and Φ_K. We note that

$$n = \left(\frac{4\pi r^2}{4\pi r^2}\right) n = \frac{1}{4\pi r^2} A = \frac{1}{4\pi} KA, \quad (119)$$

where r is the average radius of the bubbles. We wish to "model" the right hand side of (16) by a functional dependence represented by the forms of the coalescence and breakup terms. We assume that the breakup rate is proportional to the number of bubbles, and that the coalescence rate is proportional to the number of bubbles that are statistically "inside" other bubbles. Then

$$\frac{dn}{dt} = -C\alpha_d n + Bn. \qquad (120)$$

We multiply (15) by u and integrate to obtain

$$\frac{d\alpha_d}{dt} = \frac{d}{dt}\int_0^\infty uf(u,t)\,du = 0. \qquad (121)$$

With this result, the solution of (120) for n is given by

$$n(t) = n(0)\exp(-C\alpha_d + B)t. \qquad (122)$$

Note that this solution monotonically approaches ∞, monotonically approaches 0, or stays constant, according to whether

$$B < C\alpha_d.$$

In order to derive results about the system (87)–(90), let us eliminate n for A and K. We have

$$A = 4\pi n r^2, \qquad (123)$$

so that

$$AK = 4\pi n. \qquad (124)$$

Thus,

$$\begin{aligned}\frac{dKA}{dt} &= A\Phi_K = 4\pi\frac{dn}{dt} \\ &= 4\pi(-C\alpha_d n + Bn) \\ &= (-C\alpha_d + B)AK.\end{aligned} \qquad (125)$$

The other moments of f do not lead to closed-form equations such as (120) and (121). We shall derive the moment equations leading to the quantities analogous to (88) and (89), and make assumptions to close these moment equations in terms of α, A, H, and K.

If we multiply (15) by $A_m u^m$, and integrate from 0 to ∞, we will be able to derive moment equations for f. If $m = \frac{1}{3}$ and $A_m = 4\pi\left(\frac{3}{4\pi}\right)^{1/3}$, the moment will be

$$AH = 4\pi\int_0^\infty \left(\frac{3u}{4\pi}\right)^{1/3} f(u,t)\,du, \qquad (126)$$

while if $m = \frac{2}{3}$ and $A_m = 4\pi\,(3/4\pi)^{2/3}$, the moment will be

$$A = 4\pi\int_0^\infty \left(\frac{3u}{4\pi}\right)^{2/3} f(u,t)\,du. \qquad (127)$$

17.4. A Coalescence and Breakup Model and Geometry

Denoting the mth moment by

$$F_m = \int_0^\infty u^m f(u,t)\,du, \tag{128}$$

we have

$$\frac{dF_m}{dt} = \frac{C}{2}\int_0^\infty \int_0^\infty (u'+u'')^{m+1} f(u',t) f(u'',t)\,du'\,du''$$
$$- C[\alpha_d F_m + n F_{m+1}] + B F_{m+1}\left(\frac{1}{2^{m+1}} - 1\right). \tag{129}$$

The desired moments are those for which $m = \frac{1}{3}$ or $m = \frac{2}{3}$. The equations for these moments involve F_{m+1} and the integral term. Neither can be expressed in terms of F_m and the other quantities directly. For F_{m+1}, note that

$$F_{m+1} = \int_0^\infty u^{m+1} f(u,t)\,du = \int_0^\infty u u^m f(u,t)\,du$$
$$= \bar{u}_m \int_0^\infty u^m f(u,t)\,du, \tag{130}$$

where $\bar{u}_m$ is the F_m-weighted average of the volume per particle. We approximate this by

$$\bar{u}_m \approx \frac{\alpha_d}{n} = \frac{4\pi \alpha_d}{AK}. \tag{131}$$

Thus,

$$F_{m+1} \approx \frac{4\pi \alpha_d F_m}{AK}. \tag{132}$$

For the integral, note that if we supply an approximation for $(u'+u'')^{m+1}$ in terms of $(u')^j$ and $(u'')^k$, and use (132), we have a closure for this term. For convenience, we use

$$(u'+u'')^{m+1} \approx (u')^{m+1} + (u'')^{m+1}. \tag{133}$$

This does have the property that it has the proper behavior when either u' or u'' is small. Then the integral becomes

$$\int_0^\infty \int_0^\infty (u'+u'')^{m+1} f(u',t) f(u'',t)\,du'\,du''$$
$$\approx \int_0^\infty \int_0^\infty [(u')^{m+1} + (u'')^{m+1}] f(u',t) f(u'',t)\,du'\,du''$$
$$= 2 F_{m+1} n. \tag{134}$$

Thus, the approximate equations for A and AH are

$$\frac{dA}{dt} = A\Phi_A = -C\alpha_d A + 4\pi B \frac{\alpha_d}{K}\left(2^{1/3} - 1\right), \tag{135}$$

and

$$\frac{dAH}{dt} = A\Phi_H = -C\alpha_d AH + 4\pi B \frac{\alpha_d H}{K}\left(2^{2/3} - 1\right). \quad (136)$$

17.5. Conclusion

The model represented by (87)–(90) can represent the evolution of complex geometry and is capable of predicting a variety of phenomena. The simplest version of this model, neglecting all effects arising due to averaging, is given by (91) and (92) and leads to infinite area per unit volume. Including a term representing a loss of surface area due to the coalescence of surfaces leads to a model that predicts the evolution to no remaining surface area. A simple model is proposed that accounts for the coalescence and breakup of bubbles.

18
Equations of Motion for Dilute Flow

18.1. Constitutive Equations

Some general properties of constitutive equations were discussed in Section 14.3. It is clear that any properly formulated theory of multicomponent fluids will be complex, and even in the simplest case will involve a large number of constants. Thus, the accustomed process for, e.g., finding the viscosity of a linearly viscous fluid by simple experimentation, must be superceded by a much more complex process. The current state of the art involves exploration of numerical solutions, extrapolation, exploration of analogies, and often not a small amount of guessing. For instance, additional guidance in formulating realistic constitutive equations can be obtained by assuming that the quantities computed from the "exact" solutions for single large particles in a flow can be taken as forms for constitutive equations. For example, the dynamics of the irrotational flow of an inviscid fluid about a sphere (see Section 15.3) can be used to compute the average force on a sphere and the average interfacial pressure. Constitutive equations arrived at in this way should reduce to the appropriate limit for dilute flows. As is evident from the state of our abilities to calculate the flow around assemblages of bodies of arbitrary shape, we cannot hope to obtain solutions that can be averaged to give appropriate constitutive equations for general motions. The calculated forms from "exact" solutions provide a guide to the appropriate forms for empirical testing. In order to attempt to obtain a system of equations that will predict the flow of a dispersed multicomponent material, it is helpful to find experiments that isolate appropriate phenomena and give (relatively) direct measurements of unknown coefficients involved therein. For best results, the physical experiment should be simple, and should correspond to a simple solution of the equations that depends mostly on the particular constitutive form being evaluated. Examples of such flows are sometimes called "viscometric," or "separate effects" experiments. We give some of the solutions in Part V. It is clear that simple exact solutions are desirable.

Three important fundamental motions of multicomponent materials are sedimentation, shearing flow, and wave motion. Solutions in each of these three phenomena are given in Part V. The simplicity of the solutions to these problems, plus the easy ability to analyze the flow fields experimentally, make them useful to assess the adequacy of a set of constitutive equations or to guide the selection of one set in favor of another.

Our goal here is to make full use of all of the techniques above to find a full set of useful field equations. Putative values for every function and constant are given. In doing so, necessarily the equations or coefficients will come from different sources, may have different underlying assumptions, and may even hold in different flow regimes. The desire here is completeness and usefulness in the context of the current state of what must be called an art. Many of the terms we give are debated and improved constantly in the literature. Thus, what we give is a starting point, not a definitive treatment.

18.1.1. Stress

We first discuss the "microscale stress." In order for a mixture to flow, one of the materials present must be fluid-like. We envision this as filling a connected region of space at each time, and refer to this material as material c, for "continuous" component. In particulate flows, the other component is solid; in gas–liquid flows, both are fluids. In both of these cases, we refer to this component as component d, for dispersed.

We shall first discuss constitutive equations for the pressure in both a fluid material and a particulate component. We then discuss the shear, or extra stresses in a viscous fluid component. Finally, we discuss the extra stress in the particle component of a particle–fluid flow.

In addition to the bulk stresses in both materials it is convenient to have expressions for the interfacial averaged values of the stresses, broken up into the interfacial pressure and the interfacial shear stress. We write

$$\mathbf{T}_{ki} = -p_{ki}\mathbf{I} + \tau_{ki}, \qquad (1)$$

where k has the values c and d, and i denotes interface. The utility of this relation will become more evident when we discuss the interfacial force later in this section.

Pressure

The pressure in a compressible fluid is related to the temperature and density of the fluid through the equation of state. When a component is incompressible, additional assumptions are required to give closure. These extra assumptions also are constitutive equations. If one component is finely dispersed throughout the

18.1. Constitutive Equations

other, the pressure in the dispersed component should not differ much from its value at the interface. Thus for dispersed flow, we take

$$p_{di} = p_d. \tag{2}$$

With constant surface tension and spherical bubbles or drops, we use the classical Laplace form:

$$p_{ci} = p_{di} - H\sigma, \tag{3}$$

where H is the mean curvature of the interface and σ is the surface tension. If the dispersed component is spherical, $H = 2/r_d$, where r_d is the radius of the dispersed component elements (bubbles or drops).

The "Bernoulli theorem" for the variation of pressure in a flowing (inviscid) fluid suggests that the interfacial pressure should be given by

$$p_{ci} = p_c - \xi \rho_c |\mathbf{v}_c - \mathbf{v}_d|^2. \tag{4}$$

Here we use this form even if the continuous phase is viscous. The calculations for the averaged fields for the flow of an inviscid fluid around a single sphere given in Section 15.3 suggest that for dilute flows, we have $\xi = \frac{1}{4}$ when the boundary layer remains attached to the spherical particle. For low Reynolds number flows, the calculation of the averaged fields indicates that $\xi = -\frac{9}{32}$. We speculate that since (4) holds in both limits of the inviscid and very viscous flows, then we might expect a similar form to hold in somewhat more generality. Thus we use

$$\xi = \xi(\alpha_k, \text{Re}_k), \tag{5}$$

where Re_k is the Reynolds number.

It is more difficult to interpret the "pressure" of the solid component. In the flow of a particle–fluid mixture, the solid particles have a variety of roles in transmitting momentum. First, the particles act as "stress transmitters," allowing the pressure and the shear stress in the fluid to be carried and concentrated. The second means of transporting momentum is through particle contacts. If the particles are essentially touching, they behave as a solid, in that they can transmit stresses elastically. Finally, they can transport momentum by undergoing random motions about the mean motion. In this mode, collisions are usually important, and the model will resemble the models used in the kinetic theory of gases. Here the particles are analogous to the molecules of the kinetic theory. In some sense, contact stresses and collisional stresses are not different, although we shall use quite different models for them.

The microscale pressure is assumed to be

$$p_{di} = p_d. \tag{6}$$

If the particles are so concentrated that they form a coherent solid material with spaces or pores where the fluid is, then the behavior is more appropriately modeled using the concepts of elastic and plastic deformation used by the soil mechanics community. A simple model of this type ignores the plastic behavior, and assumes that if the particles are at sufficiently high concentration, then the pressure is related to the volume fraction [85]. Thus, we assume that the solid component experiences a stress which depends on volume fraction

$$p_d = \begin{cases} 0 & \text{for } \alpha_d < \alpha_{\text{cont}}, \\ p_{\text{cont}}(\alpha_d) & \text{for } \alpha_d \geq \alpha_{\text{cont}}, \end{cases} \quad (7)$$

where $p_{\text{cont}}(\alpha_d)$ can be determined experimentally by compacting the particles statically and measuring the force necessary to change the volume of the sample.

The most complicated case is when collisions occur between the particles. In this case, stress is transmitted through the particles, and the fluctuation motions of the particles themselves transport momentum. We shall discuss only the model for the Reynolds stress, which is due to the velocity fluctuations.

The pressure is the spherical part of the momentum tensor, and is given in terms of the fluctuation kinetic energy as

$$p_d^{Re} = R\alpha_d \rho_d u_d^{Re} \quad (8)$$

where $R = \frac{2}{3}$ for frictionless collisions.

Shear Stress

Viscous Fluid Stress

Few flows of practical importance in vapor-liquid systems are sufficiently slow that viscous forces dominate turbulent stresses. However, many flows involve a boundary, which is the origin of flow resistance. Viscous forces can give a significant contribution to flow resistance, often through a (two-component) turbulent boundary layer having a single-component laminar sublayer. The effects are also the cause of component distribution [27],[28] which is important in determining profile parameters, key elements in one-dimensional modeling. In dilute flows, the stress in the dispersed component can differ little from its value on the interface. Hence,

$$\tau_{di} = \tau_d. \quad (9)$$

The assumption of no Marangoni effects, such as those caused by temperature-driven surface tension gradients, results in

$$\tau_{ci} = \tau_{di}. \quad (10)$$

A fairly general assumption for the form of the continuous component stress tensor is

$$\tau_c = \sum_j \mu_{cj} \mathbf{D}_j + \sum_j \sum_k \sum_l \gamma_{cjkl} \left[\nabla \alpha_j (\mathbf{v}_k - \mathbf{v}_l) + (\mathbf{v}_k - \mathbf{v}_l) \nabla \alpha_j \right]$$
$$+ \sum_j \sum_k \sum_l \sum_m a_{cjklm} \left[(\mathbf{v}_j - \mathbf{v}_k)(\mathbf{v}_l - \mathbf{v}_m) + (\mathbf{v}_l - \mathbf{v}_m)(\mathbf{v}_j - \mathbf{v}_k) \right],$$

$$(11)$$

where $\mu_{cc}(\alpha_d)$ is the effective fluid viscosity of the fluid, and $\mu_{cd}(\alpha_d)$ is the fluid cross viscosity. Note that we obtain Lundgren's [60] expression for the effective viscosity if we take $\mu_{cc} = \mu_f$, $\mu_{cd} = \alpha_d/(1-\alpha_d)\mu_f$, and $\gamma_{cddc} = \mu_f$.

Solid Component Stress

In the case of a flow of a dilute suspension of solid particles in a fluid, the particles act as transmitters for the fluid stress. If we examine the solution in Chapter 15.4, we see that the fluid (viscous) shear stress is "concentrated" by the presence of the particles, resulting in the "effective" viscosity of a suspension being significantly larger than that of the clear fluid. A simple model for this that may be appropriate is that the interfacial averaged fluid stress is a factor β of the averaged fluid stress. This gives

$$\tau_{ci} = \beta \tau_c. \tag{12}$$

Arguments in Chapter 15 suggest $\beta = \frac{5}{2}$ is valid for low concentrations of spheres.

The shear stress in a compacted bed of particles with fluid in the pores is often modeled by an elastic stress–strain law for small shears, and by a plastic model for stresses reaching some yield stress for the particles [99].

In a situation where the particle momentum transport is dominated by collisions, the stress can be related to the fluctuation kinetic energy u_d^{Re}. In Chapter 16, we suggest that

$$\tau_d = \gamma_d \rho_d a (u_d^{Re})^{1/2} \operatorname{sym} \nabla \mathbf{v}_d, \tag{13}$$

where a is the radius of the particles.

18.1.2. Reynolds Stress

The relative motion of two or more components gives rise to motions that deviate from the average motions. In addition, fluids often experience turbulent motion whether or not they carry other materials. These two concepts, taken together, lead one to the conclusion that it is necessary to give constitutive relations for the transport of momentum by velocity fluctuations.

The same assumption for the form of the Reynolds stress tensor for the continuous component is

$$\tau_c^{Re} = \sum_j \mu_{cj}^{Re} \mathbf{D}_j + \sum_j \sum_k \sum_l \gamma_{cjkl}^{Re} \left[\nabla \alpha_j (\mathbf{v}_k - \mathbf{v}_l) + (\mathbf{v}_k - \mathbf{v}_l) \nabla \alpha_j \right]$$
$$+ \sum_j \sum_k \sum_l \sum_m a_{cjklm}^{Re} \left[(\mathbf{v}_j - \mathbf{v}_k)(\mathbf{v}_l - \mathbf{v}_m) + (\mathbf{v}_l - \mathbf{v}_m)(\mathbf{v}_j - \mathbf{v}_k) \right], \tag{14}$$

where $\mu_{cc}^{Re}(\alpha_d)$ is the eddy viscosity of the fluid turbulence.

The analysis of Section 15.3 for an inviscid fluid indicates that

$$a_{cdcdc}^{Re} = -\frac{1}{20}, \tag{15}$$

and that the pressure generated by the turbulence has the form

$$p_c^{Re} = \frac{3}{20}\alpha_d \rho_c |\mathbf{v}_d - \mathbf{v}_c|^2. \tag{16}$$

18.1.3. Interfacial Force

It has become customary to write [47]

$$\mathbf{M}_k = p_{ki}\nabla \alpha_k - \tau_{ki} \cdot \nabla \alpha_k + \mathbf{M}_k', \tag{17}$$

where p_{ki} and τ_{ki} are the interfacially averaged pressure and the shear stress of component k. We shall assume

$$\mathbf{M}_k' = \mathbf{M}_k'\left(\alpha_k, \nabla \alpha_k, \mathbf{v}_d, \mathbf{v}_c, \frac{\partial \mathbf{v}_d}{\partial t}, \frac{\partial \mathbf{v}_c}{\partial t}, \nabla \mathbf{v}_d, \nabla \mathbf{v}_c, \ldots\right), \tag{18}$$

where ... includes scalar properties, such as the viscosities and the densities of the components, as well as intrinsic parameters, such as droplet or bubble radius and interfacial area density.

We represent $\mathbf{M}_d'$ as a sum of forces associated with drag, virtual mass, turbulent dispersion, and unsteady viscous effects, etc.,

$$\mathbf{M}_d' = \sum \mathbf{F}_n, \tag{19}$$

Drag

The drag force on a single particle or drop or bubble is the force felt by that object as it moves steadily through the surrounding fluid. The concept is clouded when an array of particles moves through a fluid, but the drag force is still attributed to steady effects. The drag force is usually given in terms of a dimensionless drag coefficient C_D, where the drag force is defined as

$$\mathbf{F}_D = \frac{3}{8}\frac{\alpha_d \rho_c C_D}{r_d}|\mathbf{v}_c - \mathbf{v}_d|(\mathbf{v}_c - \mathbf{v}_d). \tag{20}$$

Here, the drag coefficient C_D is assumed to be a function of α_d and

$$Re_d = \frac{r_d|\mathbf{v}_d - \mathbf{v}_c|}{\nu_c},$$

where ν_c is the fluid viscosity. The dependence of C_D on α_d and on Re_d for bubbly, droplet, and particle flows has been studied by Ishii and Zuber [46], among others. Some results are summarized here.

For low Reynolds number bubbly flows, in the so-called Stokes regime,

$$C_D = \frac{24}{Re_{2\phi}}, \tag{21}$$

where

$$Re_{2\phi} = \frac{2r_d \rho_c |\mathbf{v}_d - \mathbf{v}_c|}{\mu_m}, \tag{22}$$

18.1. Constitutive Equations

and the mixture viscosity is given by

$$\mu_m = \mu_c \left(\frac{1-\alpha_d}{\alpha_{dm}}\right)^{-2.5\alpha_{dm}(\mu_d+0.4\mu_c)/(\mu_d+\mu_c)}. \tag{23}$$

In the mixture viscosity the parameter α_{km} is the volume fraction of component k at its maximum packing (e.g., $\alpha_{dm} = 1.0$ for the vapor component). For Reynolds numbers greater than 1, the flow is in the undistorted particle regime and the drag coefficient becomes

$$C_D = \frac{24}{Re_{2\phi}} \left(1 + 0.1 Re_{2\phi}^{0.75}\right). \tag{24}$$

For larger bubbles, in the size range

$$r_d^* = r_d \left(\frac{\rho_c(\rho_c - \rho_d)g}{\mu_c^2}\right)^{1/3} > 34.65, \tag{25}$$

the drag coefficient is

$$C_D = 0.45 \left(\frac{1 + 17.67[f(\alpha_d)]^{6/7}}{18.67 f(\alpha_d)}\right)^2, \tag{26}$$

where f depends on the viscosities also,

$$f(\alpha_d) = (1 - \alpha_d)^{1/2} \left(\frac{\mu_c}{\mu_m}\right). \tag{27}$$

When

$$\frac{\mu_c}{\left[\rho_c \sigma \left(\frac{\sigma}{g(\rho_c - \rho_d)}\right)^{1/2}\right]^{1/2}} > 0.11 + \frac{1+\Psi}{\Psi^{8/3}}, \tag{28}$$

where

$$\Psi = 0.55\left[\left(1 + 0.08(r_d^*)^3\right)^{4/7} - 1\right]^{0.75}, \tag{29}$$

the flow is in the distorted particle regime and

$$C_D = \frac{4}{3} r_d \left(\frac{(\rho_c - \rho_d)g}{\sigma}\right)^{1/2} \left(\frac{1 + 17.67[f(\alpha_d)]^{6/7}}{18.67 f(\alpha_d)}\right)^2. \tag{30}$$

In the churn-turbulent regime, Ishii and Zuber [46] give

$$C_D = \frac{8}{3}(1 - \alpha_d)^2. \tag{31}$$

Virtual Mass and Lift

The virtual mass force is the force exerted on a moving object when it accelerates. If an object is immersed in fluid and accelerated, it must accelerate some of the surrounding fluid. This results in an interaction force on the object. This force is explicitly calculated for a single sphere in Section 15.3. We take this force to be

$$\mathbf{F}_{vm} = C_{vm}\alpha_d\rho_c \left[\left(\frac{\partial \mathbf{v}_c}{\partial t} + \mathbf{v}_c \cdot \nabla\mathbf{v}_c\right) - \left(\frac{\partial \mathbf{v}_d}{\partial t} + \mathbf{v}_d \cdot \nabla\mathbf{v}_d\right)\right]. \quad (32)$$

The parameter C_{vm} is called the virtual mass coefficient (sometimes the virtual volume coefficient). It is taken to be

$$C_{vm} = \frac{1}{2} \quad (33)$$

for dilute suspensions of spheres in a fluid because that is the calculated value for a single sphere in an infinite fluid.

In addition, if a particle moves through a fluid that is in shearing motion, then the particle experiences a force transverse to the direction of motion. Such a force is often called a "lift force." If the object is spherical, the force is calculated in Section 15.3. The lift force is

$$\mathbf{F}_L = -C_L\alpha_d\rho_c (\mathbf{v}_c - \mathbf{v}_d) \wedge (\nabla \wedge \mathbf{v}_c). \quad (34)$$

The coefficient C_L is called the lift coefficient. It has the value

$$C_L = \frac{1}{4} \quad (35)$$

for dilute flows of spheres.

Furthermore, if we examine (15.51.51), we see that there is a term proportional to $(\mathbf{v}_c - \mathbf{v}_d) \wedge (\nabla \wedge \mathbf{v}_d)$. This results in a part of the interfacial force density called the rotational force

$$\mathbf{F}_{\text{rot}} = -C_{\text{rot}} (\mathbf{v}_c - \mathbf{v}_d) \wedge (\nabla \wedge \mathbf{v}_d). \quad (36)$$

The coefficient C_{rot} is called the rotation coefficient. It has the value

$$C_{\text{rot}} = \frac{1}{4} \quad (37)$$

for dilute flows of spheres.

Note that these virtual mass and lift forces as proposed above are not frame indifferent separately. However, if we take the calculated values $C_{vm} = \frac{1}{2}, C_L = \frac{1}{4}$, and $C_{\text{rot}} = \frac{1}{4}$, the sum $\mathbf{F}_{vm} + \mathbf{F}_L + \mathbf{F}_{\text{rot}}$ *is frame indifferent* and agrees with calculations for the force on a single sphere in an inviscid fluid undergoing strain. The principle of objectivity requires that

$$C_L + C_{\text{rot}} = C_{vm}. \quad (38)$$

The virtual mass coefficient C_{vm} is believed to be a function of α_d and also Re_d. For the high Reynolds number (nearly inviscid) flow of a dilute suspension of spheres, Zuber [100] obtains

$$C_{vm} \approx \frac{1}{2} + \frac{3}{2}\alpha_d. \quad (39)$$

Turbulent Dispersion

An important force that is not captured by averaging the forces on a single sphere is that due to turbulent dispersion. The form of this force is

$$\mathbf{F}_{TD} = \hat{C}_{TD} \frac{3}{8} \frac{\alpha_d \rho_c C_D}{r_d} |\mathbf{v}_c - \mathbf{v}_d| \mathbf{T}_c \cdot \nabla \alpha_d, \tag{40}$$

where $\hat{C}_{TD}$ is called the turbulent dispersion coefficient. It depends on the particle size and the eddy time scale. For small particles, $\hat{C}_{TD} = 1$ is a good approximation. It is inconvenient to carry the full drag coefficient in the force due to turbulent dispersion. We adopt the notation

$$C_{TD}\rho_c = \hat{C}_{TD} \frac{3}{8} \frac{\alpha_d \rho_c C_D}{r_d} |\mathbf{v}_c - \mathbf{v}_d|. \tag{41}$$

Other Forces

Other forces can be calculated from the flow of an inviscid fluid around a single sphere. These forces are not usually included in multicomponent flow models. We include them here for completeness, and because some of them are sufficient to satisfy the second law of thermodynamics. We write

$$\begin{aligned}\mathbf{F}_O = &\, C_r \rho_c \left(\frac{\partial \alpha_d}{\partial t} + \mathbf{v}_d \cdot \nabla \alpha_d \right) (\mathbf{v}_d - \mathbf{v}_c) \\ &+ C_c \alpha_d \rho_c (\nabla \cdot (\mathbf{v}_d - \mathbf{v}_c))(\mathbf{v}_d - \mathbf{v}_c) + C_e \alpha_d \rho_c (\mathbf{v}_d - \mathbf{v}_c) \cdot \nabla(\mathbf{v}_d - \mathbf{v}_c) \\ &+ C_g \alpha_d \rho_c (\mathbf{v}_d - \mathbf{v}_c) \cdot (\nabla(\mathbf{v}_d - \mathbf{v}_c) + (\nabla(\mathbf{v}_d - \mathbf{v}_c))^T) \\ &+ C_j \rho_c |\mathbf{v}_d - \mathbf{v}_c|^2 \nabla \alpha_d + C_n \rho_c (\mathbf{v}_d - \mathbf{v}_c) \cdot \nabla \alpha_d (\mathbf{v}_d - \mathbf{v}_c).\end{aligned} \tag{42}$$

Arnold [4] gives the values $C_r = \frac{1}{2}$, $C_c = \frac{1}{20}$, $C_e = \frac{5}{4}$, $C_g = \frac{-9}{20}$, $C_j = \frac{2}{5}$, and $C_n = \frac{-9}{20}$.

Other forces which are sometimes included in the interfacial force are the Faxén and Basset forces. Both forces arise from viscous effects in the continuous component. The Faxén force is taken to be [14]

$$\mathbf{F}_F = C_F \alpha_d \mu_c \nabla^2 \mathbf{v}_c. \tag{43}$$

Single-sphere calculations indicate $C_F = \frac{3}{4}$. This value may be expected to be valid for low concentrations.

The Basset force is an unsteady force associated with a viscous wake or boundary layer formation. If a flat plate is impulsively started from rest moving parallel to itself, it experiences a force proportional to $t^{-1/2}$. Basset [7] calculated the analogous force on a single sphere. This force can be expressed as

$$\mathbf{F}_B = \alpha_d \frac{9}{2r_d} \left(\frac{\rho_c \mu_c}{\pi} \right)^{1/2} \int_{-\infty}^{t} \frac{\mathbf{a}(\mathbf{x}, \tau) \, d\tau}{\sqrt{(t-\tau)}}, \tag{44}$$

where the acceleration $\mathbf{a}$ is here taken to be the frame-indifferent quantity

$$\mathbf{a} = \left(\frac{\partial \mathbf{v}_c}{\partial t} + \mathbf{v}_c \cdot \nabla \mathbf{v}_c \right) - \left(\frac{\partial \mathbf{v}_d}{\partial t} + \mathbf{v}_d \cdot \nabla \mathbf{v}_d \right) - (\mathbf{v}_c - \mathbf{v}_d) \wedge (\nabla \wedge \mathbf{v}_c). \tag{45}$$

18.1.4. Momentum Sources

Momentum Source from Surface Tension

Ishii [45] gives the net momentum source due to surface tension as

$$\mathbf{m} = H\sigma \nabla \alpha_d. \tag{46}$$

The interfacial stress model gives

$$\mathbf{M}_d + \mathbf{M}_c = \nabla \cdot \mathbf{T}_i, \tag{47}$$

where the stress transmitted across the interface is calculated to be

$$\mathbf{T}_i = \alpha_d(\overline{p}_{di} - \overline{p}_{ci})\mathbf{I} + a_i\alpha_d\rho_c\left[b_i|\mathbf{v}_d - \mathbf{v}_c|^2\mathbf{I} + a_i(\mathbf{v}_d - \mathbf{v}_c)(\mathbf{v}_d - \mathbf{v}_c)\right], \tag{48}$$

where $b_i = \frac{3}{20}$, and $a_i = -\frac{9}{20}$.

18.1.5. Heat Flux

Conductive Heat Flux

Averaging the exact heat flux $\mathbf{q} = -k_k\nabla\theta$ yields

$$\mathbf{q}_k = -k_k\nabla\theta_k^m + k_k\frac{\gamma_k(\alpha_k)}{2\alpha_k}\nabla\alpha_k(\theta_i - \theta_k^m), \tag{49}$$

where $\gamma_k(\alpha_k)$ is the "thermal mobility" of component k, and θ_i is the interfacial average temperature.

Turbulent Heat Flux

Turbulent heat flux can be modeled in analogy with (14) for the Reynolds stress:

$$\hat{\mathbf{q}}_c^{Re} = d_c\alpha_d\rho_c(\mathbf{v}_c - \mathbf{v}_d)(\theta_c - \theta_d) + k_c^{Re}\nabla\theta_c^M, \tag{50}$$

$$\hat{\mathbf{q}}_d^{Re} = d_d\alpha_d\rho_d(\mathbf{v}_c - \mathbf{v}_d)(\theta_c - \theta_d) + k_d^{Re}\nabla\theta_d^M, \tag{51}$$

where d_k is a parameter and k_k^{Re} and c_c are the specific heats of the vapor and liquid, respectively, at constant volume.

Turbulent Kinetic Energy Flux

The turbulent kinetic energy flux is assumed to be

$$\mathbf{q}_c^K = f_c\alpha_d\rho_c|\mathbf{v}_c - \mathbf{v}_d|^2(\mathbf{v}_c - \mathbf{v}_d), \tag{52}$$

$$\mathbf{q}_d^K = f_d\alpha_d\rho_d|\mathbf{v}_c - \mathbf{v}_d|^2(\mathbf{v}_c - \mathbf{v}_d). \tag{53}$$

A calculation similar to that of Section 15.3.5 gives $f_c = \frac{1}{40}$.

Interfacial Energy Source

We write

$$E_k = \mathbf{q}_{ki} \cdot \nabla \alpha_k + E'_k, \tag{54}$$

where $\mathbf{q}_{ki}$ is the interfacial average energy flux, which is usually ignored. We propose

$$\mathbf{q}_{di} = \mathbf{q}_d, \tag{55}$$

$$\mathbf{q}_{ci} = \mathbf{q}_{di} + h_{cd}\Gamma_d, \tag{56}$$

where h_{cd} is the latent heat.

The term E'_k is usually taken to be

$$E'_k = H_k(\theta_i - \theta_k^m), \tag{57}$$

where θ_i is the interfacial temperature. The corresponding heat transfer coefficient H_k depends on α_k, r_d, and possibly other parameters.

We assume [45]

$$\epsilon = \theta_i \left(\frac{d\sigma}{d\theta_i}\right)\left(\frac{\partial H}{\partial t} + \mathbf{v}_i \cdot \nabla H\right) + H\sigma\frac{\partial \alpha_d}{\partial t} + \epsilon'; \tag{58}$$

ϵ' is usually neglected.

Interfacial Work

The interfacial work is related to the interfacial force by

$$W'_k = C_w \alpha_d \rho_c \mathbf{a} \cdot (\mathbf{v}_d - \mathbf{v}_c) + W_p p_c \left(\frac{\partial \alpha_d}{\partial t} + \mathbf{v}_d \cdot \nabla \alpha_d\right)$$
$$+ H_w p_c (\mathbf{v}_d - \mathbf{v}_c) \cdot \nabla \alpha_d + J_w \rho_c |\mathbf{v}_d - \mathbf{v}_c|^2 (\mathbf{v}_d - \mathbf{v}_c) \cdot \nabla \alpha_d$$
$$+ K_w \rho_c |\mathbf{v}_d - \mathbf{v}_c|^2 \left(\frac{\partial \alpha_d}{\partial t} + \mathbf{v}_d \cdot \nabla \alpha_d\right)$$
$$+ B_w \alpha_d \rho_c (\mathbf{v}_d - \mathbf{v}_c)(\mathbf{v}_d - \mathbf{v}_c) : \mathbf{D}_c + Q_w \alpha_d \rho_c (\mathbf{v}_d - \mathbf{v}_c)(\mathbf{v}_d - \mathbf{v}_c) : \mathbf{D}_d$$
$$+ R_w \alpha_d \rho_c |\mathbf{v}_d - \mathbf{v}_c|^2 (\nabla \cdot (\mathbf{v}_d - \mathbf{v}_c)) + P_w \alpha_d \rho_c |\mathbf{v}_d - \mathbf{v}_c|^2 (\nabla \cdot \mathbf{v}_c). \tag{59}$$

The coefficients in (59) have been calculated by Arnold [4] and are given by $C_w = \frac{1}{2}$, $W_p = 1$, $H_w = 1$, $J_w = \frac{1}{5}$, $K_w = \frac{1}{4}$, $B_w = -\frac{9}{20}$, $Q_w = \frac{-1}{10}$, $R_w = \frac{1}{5}$, and $P_w = \frac{3}{20}$.

18.2. Dispersed Flow Equations of Motion

In this section, we give equations of motion which include many different effects. We assume that $\Gamma_k = 0$. Then, using the relations between constitutive equations for the stresses and the interfacial force density, the equations of motion for a

18. Equations of Motion for Dilute Flow

particle–fluid mixture become

$$\frac{\partial \alpha_d \rho_d}{\partial t} + \nabla \cdot \alpha_d \rho_d \mathbf{v}_d = 0, \tag{60}$$

$$\frac{\partial \alpha_c \rho_c}{\partial t} + \nabla \cdot \alpha_c \rho_c \mathbf{v}_c = 0, \tag{61}$$

$$\begin{aligned}
\alpha_d \rho_d \left(\frac{\partial \mathbf{v}_d}{\partial t} + \mathbf{v}_d \cdot \nabla \mathbf{v}_d \right) &= -\alpha_d \nabla p_c + \alpha_d \rho_c \frac{3}{8 r_d} C_D |\mathbf{v}_c - \mathbf{v}_d|(\mathbf{v}_c - \mathbf{v}_d) \\
&+ \frac{1}{2} \alpha_d \rho_c \left[\left(\frac{\partial \mathbf{v}_c}{\partial t} + \mathbf{v}_c \cdot \nabla \mathbf{v}_c \right) - \left(\frac{\partial \mathbf{v}_d}{\partial t} + \mathbf{v}_d \cdot \nabla \mathbf{v}_d \right) \right] \\
&- \frac{1}{2} \rho_c (\mathbf{v}_d - \mathbf{v}_c) \left[\frac{\partial \alpha_d}{\partial t} + \mathbf{v}_d \cdot \nabla \alpha_d + \alpha_d \nabla \cdot (\mathbf{v}_d - \mathbf{v}_c) \right] \\
&+ \frac{1}{4} \alpha_d \rho_c (\mathbf{v}_d - \mathbf{v}_c) \wedge \nabla \wedge \mathbf{v}_d + \frac{1}{4} \alpha_d \rho_c (\mathbf{v}_d - \mathbf{v}_c) \wedge \nabla \wedge \mathbf{v}_c \\
&- \frac{3}{8 r_d} C_D |\mathbf{v}_c - \mathbf{v}_d| \frac{3}{20} \alpha_d \rho_c |\mathbf{v}_d - \mathbf{v}_c|^2 \nabla \alpha_d \\
&+ \alpha_d \nabla \cdot \left(\beta \mu_c [\nabla \mathbf{v}_v + (\nabla \mathbf{v}_v)^T] \right) + \nabla \cdot (\alpha_d \mu_d^{Re} [\nabla \mathbf{v}_d + (\nabla \mathbf{v}_d)^T]) \\
&+ \alpha_d \rho_d \mathbf{g},
\end{aligned} \tag{62}$$

$$\begin{aligned}
\alpha_c \rho_c \left(\frac{\partial \mathbf{v}_c}{\partial t} + \mathbf{v}_c \cdot \nabla \mathbf{v}_c \right) &= -\alpha_c \nabla p_c + \alpha_d \rho_c \frac{3}{8 r_d} C_D |\mathbf{v}_c - \mathbf{v}_d|(\mathbf{v}_d - \mathbf{v}_c) \\
&- \frac{1}{2} \alpha_d \rho_c \left[\left(\frac{\partial \mathbf{v}_c}{\partial t} + \mathbf{v}_c \cdot \nabla \mathbf{v}_c \right) - \left(\frac{\partial \mathbf{v}_d}{\partial t} + \mathbf{v}_d \cdot \nabla \mathbf{v}_d \right) \right] \\
&+ \frac{1}{2} \rho_c (\mathbf{v}_d - \mathbf{v}_c) \left[\frac{\partial \alpha_d}{\partial t} + \mathbf{v}_d \cdot \nabla \alpha_d + \alpha_d \nabla \cdot (\mathbf{v}_d - \mathbf{v}_c) \right] \\
&- \frac{1}{4} \alpha_d \rho_c (\mathbf{v}_d - \mathbf{v}_c) \wedge \nabla \wedge \mathbf{v}_d - \frac{1}{4} \alpha_d \rho_c (\mathbf{v}_d - \mathbf{v}_c) \wedge \nabla \wedge \mathbf{v}_c \\
&+ \frac{1}{2} \nabla \cdot \alpha_d \rho_c (\mathbf{v}_d - \mathbf{v}_c)(\mathbf{v}_d - \mathbf{v}_c) - \frac{1}{2} \alpha_d \rho_c (\mathbf{v}_d - \mathbf{v}_c) \cdot \nabla (\mathbf{v}_d - \mathbf{v}_c)^T \\
&+ \frac{1}{4} \rho_c |\mathbf{v}_c - \mathbf{v}_d|^2 \nabla \alpha_d + \frac{3}{8 r_d} C_D |\mathbf{v}_c - \mathbf{v}_d| \frac{3}{20} \alpha_d \rho_c |\mathbf{v}_d - \mathbf{v}_c|^2 \nabla \alpha_d \\
&+ \nabla \cdot \alpha_c \mu_c [\nabla \mathbf{v}_v + (\nabla \mathbf{v}_v)^T] + \nabla \cdot \alpha_c \mu_c^{Re} [\nabla \mathbf{v}_f + (\nabla \mathbf{v}_f)^T] \\
&+ \alpha_c \rho_c \mathbf{g},
\end{aligned} \tag{63}$$

where $\mathbf{v}_v = \alpha_d \mathbf{v}_d + \alpha_c \mathbf{v}_c$.

The equations of motion for the laminar flow of a dilute mixture can be obtained by adding (60) and (61) to obtain the equation of balance of mass for the mixture, and by adding (62) and (63) and neglecting the turbulent stresses, to obtain the equation of balance of momentum. The results are

$$\nabla \cdot (\alpha_d \mathbf{v}_d + \alpha_c \mathbf{v}_c) = 0, \tag{64}$$

18.2. Dispersed Flow Equations of Motion

$$\frac{\partial}{\partial t}(\alpha_c \rho_c \mathbf{v}_c + \alpha_d \rho_d \mathbf{v}_d) + \nabla \cdot [\alpha_c \rho_c \mathbf{v}_c \mathbf{v}_c + \alpha_d \rho_d \mathbf{v}_d \mathbf{v}_d + \frac{1}{2}\alpha_d \rho_c (\mathbf{v}_d - \mathbf{v}_c)(\mathbf{v}_d - \mathbf{v}_c)]$$
$$= -\nabla(\alpha_c p_c + \alpha_d p_d)$$
$$+ \nabla \cdot [\alpha_d \beta + \alpha_c]\mu_c[\nabla \mathbf{v}_v + (\nabla \mathbf{v}_v)^\mathsf{T}]. \tag{65}$$

To complete the model for nonlaminar flows, the Reynolds stress and the Reynolds heat flux must be specified. There is no consensus in the single-phase turbulence community regarding models for the specification of the Reynolds stress. Models which relate the eddy viscosity to the turbulent kinetic energy and dissipation have performed well in some flows. Similar models have been used for turbulent two-phase flows [1]. The results are sensitive to parameter values in the model. This presents both a theoretical and a practical problem.

Part V

Consequences

19
Nature of the Equations

It should be recognized that any system of equations that is expected to describe the behavior of a physical system is a *model*, and will, at best, describe the subset of phenomena that falls under the limitations of the model. These limitations are often unwritten and, unfortunately, are often unrecognized. As an example in a classical context, the equations of (linear) elasticity provide an excellent description of a large body of phenomena. However, the model fails to describe such common phenomena as permanent bending, crack propagation, and shear bands. For dispersed multicomponent flows, if the initial conditions, or the evolving fields, predict a high concentration of dispersed phase units in some region, we must consider the likely possibility that the predictions are not valid. In this situation, it is sometimes difficult to decide whether the model is incorrect, or whether the solution method led to an approximation that is invalid. Thus, we consider some general properties of the equations in order to understand the behavior in more complicated situations.

Faced with the task of solving a formulation of governing equations, we must reasonably be assured that the formulation reproduces the essential physics of the problem, so that the appearance of anomalous behavior can be attributed to the solution method. To this end, we next consider what can be understood about the nature of a formulation before attempts are made to solve it.

We give our attention to the model with constant densities of the components, and omit the Reynolds stresses. Thermal effects are ignored, except for some considerations in the acoustic model of Section 22.2. We examine the possibility that the equations reduce to the trajectories of a single sphere moving in a fluid, and give some insight why this is so. We present a model that retains only the viscous effects. Finally, we give a diffusive model for the particle component.

238   19. Nature of the Equations

We consider the general question of whether the system of equations is well-posed. Finally, we consider the characteristics of a number of models derived here and, in other places, to determine whether they are well-posed.

In the final two chapters in this Part V, we give several solutions to the equations for dilute multicomponent flow. Exact solutions, when obtainable for limiting cases and special geometries, give confidence in the equations and help to resolve problems associated with an approximate (numerical) solution of the complete model.

19.1. Special Cases of the Equations

19.1.1. Dispersed Equations

Consider the one-dimensional situation pictured in Figure 19.1. Momentum balance for the mixture (18.2.62) suggests that the rate of change of momentum for the mixture should be balanced by stresses applied at x and $x + \Delta x$. Two of these stresses are represented by the pressure in the fluid and the "pressure" inside the spheres. If the spheres are rigid solids, this "pressure" is the resultant of a distribution of fluid pressure over the exterior of the sphere. In addition, there is an extra "stress" due to the momentum transferred by velocity fluctuations in the fluid, which are the result of the relative motion of the spheres to the fluid. This is the Reynolds stress in the fluid, and is present in the equations an the left-hand side.

The continuum model for the particles between x and $x + \Delta x$ gives the rate of change of the momentum of the particles *and parts of particles* between x and $x+\Delta x$, denoted by $\dot{\mathbf{P}}_d(x, x + \Delta x)$, as the stress force transmitted to the particles by the particle parts outside the interval, denoted by $(\alpha_d \mathbf{i} \cdot \mathbf{T}_d)|_x + (\alpha_d(-\mathbf{i}) \cdot \mathbf{T}_d)|_{x+\Delta x}$, plus the force transmitted to the particles through their interface, denoted by $\mathbf{M}_d \Delta x$. Thus,

$$\dot{\mathbf{P}}_d(x, x + \Delta x) = (\alpha_d \mathbf{i} \cdot \mathbf{T}_d)|_x + (\alpha_d(-\mathbf{i}) \cdot \mathbf{T}_d)|_{x+\Delta x} + \mathbf{M}_d \Delta x . \quad (1)$$

The sum of the forces on all particles with their centers in the interval from x to $x + \Delta x$ is equal to the sum of the pressure forces on all the particles involved. This is denoted by $\sum \int p\mathbf{n}\, ds$. This is equal to the rate of change of the momentum of all the particles, denoted by $\dot{\mathbf{P}}_p$. Thus,

$$\dot{\mathbf{P}}_p = \sum \int p\mathbf{n}\, ds . \quad (2)$$

We note that (1) and (2) differ in the way they treat the particles being cut by the surfaces at x and $x + \Delta x$. The relation is that

$$\dot{\mathbf{P}}_d(x, x + \Delta x) = \dot{\mathbf{P}}_p + (\alpha_d \mathbf{i} \cdot \mathbf{T}_d)|_x + (\alpha_d(-\mathbf{i}) \cdot \mathbf{T}_d)|_{x+\Delta x}$$
$$- \sum_{\text{cut,in}} \int p\mathbf{n}\, ds|_x + \sum_{\text{cut,out}} \int p\mathbf{n}\, ds|_x$$

19.1. Special Cases of the Equations

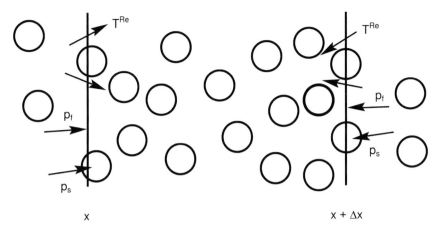

FIGURE 19.1. Momentum balance for mixture between x and $x + \Delta x$.

$$-\sum_{\text{cut,in}} \int p\mathbf{n}\,ds|_{x+\Delta x} + \sum_{\text{cut,out}} \int p\mathbf{n}\,ds|_{x+\Delta x}. \quad (3)$$

Here

$$\sum_{\text{cut,in}} \int p\mathbf{n}\,ds|_x \quad (4)$$

is the sum of the pressure forces on the surfaces of the cut particles at x whose centers are inside the interval from x to $x + \Delta x$,

$$\sum_{\text{cut,out}} \int p\mathbf{n}\,ds|_x \quad (5)$$

is the sum of the pressure forces on the surfaces of the cut particles at x whose centers are outside the interval from x to $x + \Delta x$. A similar interpretation is valid for the cut particles at $x + \Delta x$.

The terms on the right-hand side of (3) represent the resultants of forces on cut particles. Since there is a net force on the particles, and consequently a net force on the cut particles, these resultants do not necessarily add up to zero. However, the net contributions from the cut particles averages to zero.

The implication of this observation on the general dispersed phase momentum equation,

$$\frac{\partial \alpha_d \rho_d \mathbf{v}_d}{\partial t} + \nabla \cdot \alpha_d \rho_d \mathbf{v}_d \mathbf{v}_d = \nabla \cdot \alpha_d (\mathbf{T}_d + \mathbf{T}_d^{\text{Re}}) + \alpha_d \rho_d \mathbf{g} + \mathbf{M}_d, \quad (6)$$

when no collisions or particle dispersion are present in the dispersed component is that, if $\mathbf{F}_d$ is the force on a single sphere in a fluid, then

$$\nabla \cdot \alpha_d (\mathbf{T}_d + \mathbf{T}_d^{\text{Re}}) + \mathbf{M}_d = \alpha_d \mathbf{F}_d. \quad (7)$$

Then $\mathbf{F}_d$ should contain no "collective" effects; in particular, it should not contain terms proportional to $\nabla \alpha_d$. Then

$$\mathbf{M}_d = \mathbf{M}_d^0 - (\mathbf{T}_d + \mathbf{T}_d^{\text{Re}}) \cdot \nabla \alpha_d ,$$

whence (7) becomes

$$\alpha_d \nabla \cdot (\mathbf{T}_d + \mathbf{T}_d^{\text{Re}}) + \mathbf{M}_d^0 = \alpha_d \mathbf{F}_d + \rho_d \mathbf{g} ,$$

and the equation of balance of momentum for the dispersed component becomes

$$\rho_d \frac{D\mathbf{v}_d}{Dt} = \mathbf{F}_d . \tag{8}$$

19.1.2. The Force on a Sphere Due to Inertia

The force on a sphere of radius a at $\mathbf{x}$ in the straining flow of an inviscid fluid is computed by integrating $-p\mathbf{n}$ over the surface of the sphere. Thus

$$\mathbf{F}_p(\mathbf{x}, t) = \int_{\Omega(a)} \mathbf{n} \left(p_0 - \rho_c \left[\frac{1}{2} |\nabla \phi|^2 + \frac{\partial \phi}{\partial t} \right] \right) d\Omega , \tag{9}$$

where the integration is over the variable $\mathbf{x}'$, with the sphere centered at $\mathbf{x}$, and $|\mathbf{x}'| = a$. Substituting the pressure from Bernoulli's equation (15.33) and performing the integration results in

$$\mathbf{F}_p(\mathbf{z}) = \frac{4}{3}\pi a^3 \rho_p \left(\frac{\partial \mathbf{v}_f}{\partial t} + \mathbf{v}_f \cdot \nabla \mathbf{v}_f + \frac{1}{2} \left[\frac{\partial \mathbf{v}_f}{\partial t} - \frac{\partial \mathbf{v}_p}{\partial t} + \mathbf{v}_f \cdot \nabla \mathbf{v}_f \right] \right). \tag{10}$$

Note that this force agrees with Taylor's [84] calculation of the force necessary to hold a sphere at rest in an accelerating stream. Furthermore, it is equal to the product of the number density with the *volume averaged* interfacial force given by (15.3.114).

Another force that has been calculated for a stationary sphere in a rotating fluid is given by (15.3.90). This force is given by

$$\mathbf{F}_p^L = \frac{4}{3}\pi a^3 \frac{1}{4} \rho_c (\mathbf{v}_f - \mathbf{v}_p) \wedge \nabla \mathbf{v}_f . \tag{11}$$

From (18.2.64) and (18.2.65), it is evident that the mixture behaves as an inviscid fluid in the limit as $\alpha_d \to 0$. Then,

$$\nabla \cdot \mathbf{v}_c = 0 , \tag{12}$$

$$\frac{\partial}{\partial t} \rho_c \mathbf{v}_c + \nabla \cdot \rho_c \mathbf{v}_c \mathbf{v}_c = -\nabla p_c . \tag{13}$$

The dispersed component momentum equation then reduces to

$$\left(\rho_d + \frac{1}{2}\rho_c \right) \left(\frac{\partial \mathbf{v}_d}{\partial t} + \mathbf{v}_d \cdot \nabla \mathbf{v}_d \right) = \frac{3}{2} \rho_c \left(\frac{\partial \mathbf{v}_c}{\partial t} + \mathbf{v}_c \cdot \nabla \mathbf{v}_c \right)$$
$$- \frac{1}{4}\rho_c (\mathbf{v}_d - \mathbf{v}_c) \wedge \nabla \wedge \mathbf{v}_d - \frac{1}{4}\rho_c (\mathbf{v}_d - \mathbf{v}_c) \wedge \nabla \wedge \mathbf{v}_c . \tag{14}$$

This equation shows that for dilute flows of spheres, the equations reduce to Taylor's result [84] for an accelerating sphere, or for a stationary sphere in an accelerating fluid. Moreover, it also includes a lift force that reduces to (11) if the spheres are stationary. It also includes a lift force that gives a transverse force if the spheres are translating and rotating as a unit. Note that this is not a rotation of each sphere.

19.1.3. Viscous Model

If we ignore the inertia terms in (18.2.62) and (18.2.63), we have

$$0 = -\alpha_d \nabla p_c + \alpha_d \rho_c \frac{3}{8r_d} C_D |\mathbf{v}_c - \mathbf{v}_d|(\mathbf{v}_c - \mathbf{v}_d)$$
$$+ \alpha_d \nabla \cdot \left(\beta \mu_c [\nabla \mathbf{v}_v + (\nabla \mathbf{v}_v)^T]\right) + \nabla \cdot \alpha_d \mu_d^{Re}[\nabla \mathbf{v}_d + (\nabla \mathbf{v}_d)^T]$$
$$+ \alpha_d \rho_d \mathbf{g}, \qquad (15)$$

$$0 = -\alpha_c \nabla p_c + \alpha_d \rho_c \frac{3}{8r_d} C_D |\mathbf{v}_c - \mathbf{v}_d|(\mathbf{v}_d - \mathbf{v}_c)$$
$$+ \nabla \cdot \alpha_c \mu_c [\nabla \mathbf{v}_v + (\nabla \mathbf{v}_v)^T] + \nabla \cdot \alpha_c \mu_c^{Re}[\nabla \mathbf{v}_c + (\nabla \mathbf{v}_c)^T] + \alpha_c \rho_c \mathbf{g}. \qquad (16)$$

19.1.4. Diffusion Model

If we assume that the drag, turbulent dispersion, and the particle Reynolds stress dominate the particle momentum equation (18.2.62), we have

$$\mathbf{v}_d = \mathbf{v}_c - C_{TD}\rho_c |\mathbf{v}_d - \mathbf{v}_c|^2 \nabla \alpha_d + \frac{1}{\beta} \nabla \cdot \alpha_d \mathbf{T}_d^{Re}, \qquad (17)$$

where

$$\beta = \frac{3}{8r_d} C_D |\mathbf{v}_c - \mathbf{v}_d|.$$

Substituting this into the equation of balance of mass for the particle component gives

$$\frac{\partial \alpha_d}{\partial t} + \nabla \cdot \alpha_d \mathbf{v}_c = \nabla \cdot D_p \nabla \alpha_d + \frac{1}{\beta} \nabla \cdot \alpha_d \mathbf{T}_d^{Re}, \qquad (18)$$

where the particle diffusivity is defined as

$$D_p = C_{TD} \rho_c |\mathbf{v}_d - \mathbf{v}_c|^2.$$

Thus, we see that the turbulent dispersion term gives rise to a term that appears as diffusion in the equation of balance of particle mass. The second term on the right-hand side of (18) has been called *turbophoresis*.

19.1.5. Sedimentation/Terminal Rise

The simplest flow of a two-component mixture is a vertical flow where a balance is achieved between drag and buoyancy forces, to produce a flow with constant velocities and volume fractions. If we assume

$$\alpha_d = \breve{\alpha}, \qquad \alpha_c = 1 - \breve{\alpha}, \qquad \mathbf{v}_k = \breve{w}_k \mathbf{e}_z, \qquad (19)$$

19. Nature of the Equations

the equations of balance of mass (18.60) and (18.61) are satisfied identically, while the equations of balance of momentum (18.62) and (18.63) reduce to

$$0 = -\check{\alpha}\frac{\partial p_c}{\partial z} + \check{\alpha}\rho_c\frac{3}{8r_d}C_D|\check{w}_c - \check{w}_d|(\check{w}_c - \check{w}_d) - \check{\alpha}\rho_d g, \tag{20}$$

$$0 = -(1-\check{\alpha})\frac{\partial p_c}{\partial z} + \check{\alpha}\rho_c\frac{3}{8r_d}C_D|\check{w}_c - \check{w}_d|(\check{w}_d - \check{w}_c) - (1-\check{\alpha})\rho_c g. \tag{21}$$

Adding the two momentum equations (20) and (21) gives

$$\frac{\partial p_c}{\partial z} - [\check{\alpha}\rho_d + (1-\check{\alpha})\rho_c]g. \tag{22}$$

Thus, the pressure gradient is hydrostatic, with the effective density for the fluid being the mixture density, $\rho_m = \check{\alpha}\rho_d + (1-\check{\alpha})\rho_c$.

Using (22) to eliminate the pressure gradient, the dispersed component momentum equation reduces to

$$\rho_c\frac{3}{8r_d}C_D|\check{w}_c - \check{w}_d|(\check{w}_c - \check{w}_d) = -(1-\check{\alpha})(\rho_c - \rho_d)g. \tag{23}$$

Note that if $\rho_d < \rho_c$, the dispersed component rises relative to the continuous component, while if $\rho_d > \rho_c$, the dispersed component sinks relative to the continuous component. The latter process is sometimes called sedimentation. Obtaining the velocities from this relation (23) is complicated by the dependence of C_D on the Reynolds number, and hence, on the relative velocity. The velocities of the individual components also depend on the initial and boundary conditions appropriate to the specific application.

We also note that the appropriate force balances involve only the forces included here. Specifically, the pressure gradient force is intimately involved in a final force balance that has a buoyant force on the dispersed component. Models that do not have the forces included here do exist in the literature. While those models may describe some flow situations adequately, caution should be exercised in applying them to general flows.

20
Well-Posedness

20.1. Formulation

A model that is not properly formulated mathematically cannot describe physical phenomena correctly.[1] Our confidence that our equations, along with our boundary and initial conditions, can correspond with physical experience, is increased if their solutions satisfy the following three prerequisites:

- the solutions must exist;
- the solutions must be uniquely determined; and
- the solutions must depend in a continuous fashion on the initial and boundary data.

If a formulation satisfies these prerequisites, it is said to be well posed [25]. However, because of the absence of general theorems, it is usually only possible to investigate existence and uniqueness for special cases of the equations.

The reader who is familiar with the large body of work in "ill-posed" problems will recognize that the physically meaningful ill-posed problems occur in a context different from that intended here. An ill-posed problem that occurs in the inviscid limit of some well-posed problem deserves study. In many situations, multiply valued steady states occur, but the initial-value problem that governs the evolution to them should be well-posed.

---

[1]While mathematical correctness does not imply physical validity, the latter cannot be obtained without the former.

20.2. Characteristic Values

One aspect of the requirement of stability can be quantified by examining the characteristic values of the governing equations in the following way.

The one-dimensional model equations can be rearranged to the matrix form

$$\mathbf{A}\frac{\partial \boldsymbol{\Psi}}{\partial t} + \mathbf{B}\frac{\partial \boldsymbol{\Psi}}{\partial z} = \mathbf{C} \tag{1}$$

where

$$\boldsymbol{\Psi} = [\alpha, v_d, v_c, p_c, p_d, e_c, e_d]^T, \tag{2}$$

and $\mathbf{A}(\boldsymbol{\Psi})$ and $\mathbf{B}(\boldsymbol{\Psi})$ are 7×7 square matrices; and $\mathbf{C}(\boldsymbol{\Psi})$ is a column vector, none of which contains derivatives of $\boldsymbol{\Psi}$. To investigate the behavior of this set of equations, suppose that arbitrary data for $\boldsymbol{\Psi}$ are specified at all points (z, t) along a curve $\mathcal{C}_1$, which is expressed as $z = z(t)$. A solution for $\boldsymbol{\Psi}$ in the neighborhood of $\mathcal{C}_1$ can be obtained by means of a Taylor series expansion about points on $\mathcal{C}_1$ provided that the first and higher derivatives of $\boldsymbol{\Psi}$ can be calculated. The calculation is simplified by a transformation to normal $n(z, t)$ and tangential $\tau(z, t)$ variables (see Figure 20.1), whence (1) becomes

$$\left[\mathbf{A}(\boldsymbol{\Psi})\frac{\partial n}{\partial t} + \mathbf{B}(\boldsymbol{\Psi})\frac{\partial n}{\partial z}\right]\frac{\partial \boldsymbol{\Psi}}{\partial n} = \mathbf{C}(\boldsymbol{\Psi}) - \left[\mathbf{A}(\boldsymbol{\Psi})\frac{\partial \tau}{\partial t} + \mathbf{B}(\boldsymbol{\Psi})\frac{\partial \tau}{\partial z}\right]\frac{\partial \boldsymbol{\Psi}}{\partial \tau}. \tag{3}$$

Since $\boldsymbol{\Psi}$ is given by the initial data as a function of τ along the curve, the right-hand side of (3) is known, and the derivatives $\partial \boldsymbol{\Psi}/\partial n$ will be uniquely determined,

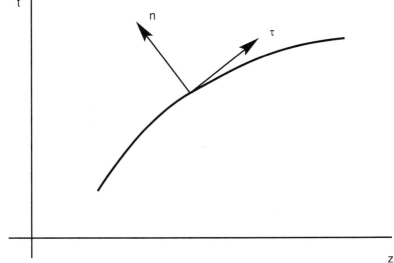

FIGURE 20.1. Normal and tangential coordinates.

if the coefficient matrix

$$\mathbf{A}(\boldsymbol{\Psi})\frac{\partial n}{\partial t} + \mathbf{B}(\boldsymbol{\Psi})\frac{\partial n}{\partial z} \tag{4}$$

is nonsingular, that is, if

$$\det\left(\mathbf{A}(\boldsymbol{\Psi})\frac{\partial n}{\partial t} + \mathbf{B}(\boldsymbol{\Psi})\frac{\partial n}{\partial z}\right) \neq 0. \tag{5}$$

Combining the solution from (3) with the results of successive differentiations of the governing equation (1) then provides all the derivatives needed for the Taylor series expansion; hence, the solution can be developed from the data given on $\mathcal{C}_1$.

When the determinant of the coefficient matrix is zero, the curve $\mathcal{C}_1$ is called a *characteristic curve*. In this case, the existence of a solution for $\partial\boldsymbol{\Psi}/\partial n$ requires that each determinant that appears as a numerator in the solution for each element of $\partial\boldsymbol{\Psi}/\partial n$, must also vanish. The resulting equations become compatibility conditions. Since terms from the right side of (3) are involved in the compatibility conditions, an immediate consequence is that the initial data cannot be arbitrarily prescribed on a characteristic curve. We also find that not all of the compatibility conditions are independent, so that there exist infinite combinations of $\partial\boldsymbol{\Psi}/\partial n$ and $\partial\boldsymbol{\Psi}/\partial \tau$, or $\partial\boldsymbol{\Psi}/\partial z$ and $\partial\boldsymbol{\Psi}/\partial t$, that will satisfy the problem. Finally, it can be shown that higher-order normal derivatives can no longer be determined from the differential equation; hence, the solution at points in the neighborhood of a characteristic curve cannot be obtained from data on the curve. For this reason, a solution is said to be "transported" along a characteristic curve, and the solution at a neighboring point "arrives" there on the characteristic curve that passes through that point.

The dependence of the solution on prescribed initial data can thus be reduced to an investigation of the equation

$$\det\left(\mathbf{A}(\boldsymbol{\Psi})\lambda - \mathbf{B}(\boldsymbol{\Psi})\right) = 0, \tag{6}$$

in which we have introduced the characteristic values λ defined as

$$\lambda = -\frac{\partial n}{\partial t} \bigg/ \frac{\partial n}{\partial z}. \tag{7}$$

If the values of λ that satisfy (6) are complex, the coefficient matrix in (3) is nonsingular and the solution is uniquely determined from the data given along the curve $\mathcal{C}_1$. In this case, system (1) is called *elliptic*.

On the other hand, if the characteristic values of (6) are real, then $\mathcal{C}_1$ is a characteristic curve, and the solution acquires the special properties just described. In this case, system (1) is called *hyperbolic*. Based on the orthogonality of the local tangent vector $dt\hat{\mathbf{e}}_t + dz\hat{\mathbf{e}}_z$ with the normal

$$\frac{\partial n}{\partial t}\hat{\mathbf{e}}_t + \frac{\partial n}{\partial z}\hat{\mathbf{e}}_z$$

to curve C_1 it follows that

$$\lambda = -\frac{\partial n/\partial t}{\partial n/\partial z} = \frac{dz}{dt}. \tag{8}$$

Thus, real characteristic values are the velocities of the characteristic curves.

It is significant to note [25] that boundary-value problems are associated with elliptic equations, whose solutions depend on data on the boundary; whereas initial-value problems are typically represented by hyperbolic equations, whose solutions depend on the initial and boundary data prescribed on restricted domains.

The classical example of the type of "ill-posedness" that arises from complex characteristics can be seen in the Hadamard problem, where Laplace's equation is used for an initial-value problem. The initial-value problem for Laplace's equation can be written as

$$\frac{\partial u}{\partial t} - \frac{\partial v}{\partial x} = 0,$$
$$\frac{\partial v}{\partial t} + \frac{\partial u}{\partial x} = 0, \tag{9}$$

for $-\infty < x < \infty$, and $0 < t < \infty$, with initial conditions

$$u(x, 0) = f(x),$$
$$v(x, 0) = g(x), \tag{10}$$

for $\infty < x < \infty$.

Suppose $u_0(x, t)$ and $v_0(x, t)$ give a solution of this problem. Consider the problem given by

$$\frac{\partial u'}{\partial t} - \frac{\partial v'}{\partial x} = 0,$$
$$\frac{\partial v'}{\partial t} + \frac{\partial u'}{\partial x} = 0, \tag{11}$$

with initial conditions

$$u'(x, 0) = \epsilon^2 \sin(x/\epsilon),$$
$$v'(x, 0) = -\epsilon \cos(x/\epsilon), \tag{12}$$

for $\infty < x < \infty$. The solution of this second problem is easily seen to be

$$u'(x, t) = \epsilon^2 e^{t/\epsilon} \sin(x/\epsilon),$$
$$v'(x, t) = -\epsilon e^{t/\epsilon} \cos(x/\epsilon). \tag{13}$$

Then the solution of the problem

$$\frac{\partial u}{\partial t} - \frac{\partial v}{\partial x} = 0,$$
$$\frac{\partial v}{\partial t} + \frac{\partial u}{\partial x} = 0, \tag{14}$$

with initial conditions

$$u(x, 0) = f(x) + \epsilon^2 \sin(x/\epsilon),$$
$$v(x, 0) = g(x) - \epsilon \sin(x/\epsilon), \quad (15)$$

is

$$u_0(x, t) + u'(x, t),$$
$$v_0(x, t) + v'(x, t), \quad (16)$$

for any ϵ. For small ϵ, the error made in the initial conditions can be made as small as desired. However, the error in the solution can be made to exceed any given value at any time $t > 0$ by choosing ϵ small enough. This is an unhealthy state of affairs; no matter how well the correct initial condition is approximated for numerical purposes, there is no assurance that the computed solution will be close to the desired one.

The problem is generic to complex characteristics. If we consider the characteristics in one complex space variable $z = x + iy$ and in one real time variable, we see that we need to extend the (real) initial-values for Ψ from the real axis $y = 0$ to a neighborhood of the real axis. The most reasonable way to do this is to extend the initial conditions as analytic functions (i.e., functions expressible in a convergent power series in the complex variable z). If an error proportional to $\sin(x/\epsilon)$ is incurred in the initial conditions, this extends analytically as $\sin(x/\epsilon)\cosh(y/\epsilon)$. This complex initial condition error grows exponentially away from the real axis. Complex characteristics entering a point (x, t) in the real plane come from both sides (i.e., $y > 0$ and $y < 0$) to the point (x, t). As t increases, these complex characteristics come from farther away from the real (x, t) plane, and consequently can pick up larger and larger errors from a small error proportional to $\sin(x/\epsilon)$.

A physically realizable two-phase flow evolves from some initial conditions, subject to the physically imposed boundary conditions. Depending on the phenomena being modeled, the model should have either a parabolic or hyperbolic nature. Hyperbolic systems, those having real characteristic values, may be expected to transport either continuous or discontinuous initial data as well as to develop embedded discontinuities, either stationary or moving. A parabolic model has the ability to describe the formation and evolution of regions of sharp change that provide a better model than the discontinuities inherent in hyperbolic models.

Elliptic behavior, when one of the variables is time, is unacceptable. Nonetheless, there are some subtle issues in requiring real characteristics. It seems clear that a formulation which produces complex characteristic values is physically unsound and in need of revision to incorporate missing physics. Consider the classical Taylor–Lewis instability associated with the acceleration of a denser fluid into a lighter fluid apparently derives from a nominally ill-posed formulation. This appears to be a real physical instability that is described by an ill-posed problem. On closer examination, we require the rapid evolution at the smallest scales be described by a model that includes capillary forces.

In subsequent sections, we consider whether, and in what sense, that the instabilities associated with "ill-posed" problems in multicomponent flow may sometimes be real and observable. Whatever the case, further theoretical and experimental investigations will be needed to resolve this issue.

20.3. The Simplest Model

The simplest model for "multiphase flow" is obtained by assuming that the phasic pressures p_k, $k = 1, 2$, are both equal and that $\mathbf{M}_d$ is given as an algebraic combination of the averaged fields, but that it contains no derivatives of these averaged fields. The most commonly used form is $\mathbf{M}_d = \alpha C_D \rho_c / r_d |\mathbf{v}_c - \mathbf{v}_d|(\mathbf{v}_c - \mathbf{v}_d)$. The equations of motion in one dimension then become

$$\frac{\partial \alpha}{\partial t} + \frac{\partial \alpha v_d}{\partial z} = 0, \tag{17}$$

$$\frac{\partial (1 - \alpha)}{\partial t} + \frac{\partial (1 - \alpha) v_c}{\partial z} = 0, \tag{18}$$

$$\alpha \rho_d \left(\frac{\partial v_d}{\partial t} + v_d \frac{\partial v_d}{\partial z} \right) = -\alpha \frac{\partial p}{\partial z} + M_d, \tag{19}$$

$$(1 - \alpha) \rho_c \left(\frac{\partial v_c}{\partial t} + v_c \frac{\partial v_c}{\partial z} \right) = -(1 - \alpha) \frac{\partial p}{\partial z} - M_d, \tag{20}$$

where M_d involves terms not having derivatives.

The matrices $\mathbf{A}$ and $\mathbf{B}$ are

$$\mathbf{A} = \begin{bmatrix} 1 & 0 & 0 & 0 \\ -1 & 0 & 0 & 0 \\ 0 & \alpha \rho_d & 0 & 0 \\ 0 & 0 & (1-\alpha) \rho_c & 0 \end{bmatrix}, \tag{21}$$

$$\mathbf{B} = \begin{bmatrix} v_d & \alpha & 0 & 0 \\ -v_c & 0 & 1-\alpha & 0 \\ 0 & \alpha \rho_d v_d & 0 & \alpha \\ 0 & 0 & (1-\alpha) \rho_c v_c & 1-\alpha \end{bmatrix}. \tag{22}$$

The characteristics are given by

$$\lambda = \left\{ \infty, \infty, \frac{1}{d} \left(r \pm s^{1/2} \right) \right\}, \tag{23}$$

where

$$r = (1-\alpha)\rho_c v_c + (1-\alpha)^2 \rho_d v_d ,$$
$$s = -\alpha(1-\alpha)^2 \rho_c^2 (v_d - v_c)^2 , \quad (24)$$
$$d = (1-\alpha)^2 \rho_d + \alpha(1-\alpha)\rho_c .$$

The characteristics are complex, unless $v_c = v_d$. Thus, the simplest model is not physically realistic to describe the evolution of multicomponent flows.

20.4. Effect of Viscosity

A model including the effects of viscosity in the form of a higher-order derivative of the fluid velocity can be put in the form

$$\frac{\partial \alpha}{\partial t} + \frac{\partial \alpha v_d}{\partial z} = 0 , \quad (25)$$

$$\frac{\partial (1-\alpha)}{\partial t} + \frac{\partial (1-\alpha) v_c}{\partial z} = 0 , \quad (26)$$

$$\alpha \rho_d \left(\frac{\partial v_d}{\partial t} + v_d \frac{\partial v_d}{\partial z} \right) = -\alpha \frac{\partial p}{\partial z} + M_d , \quad (27)$$

$$(1-\alpha)\rho_c \left(\frac{\partial v_c}{\partial t} + v_c \frac{\partial v_c}{\partial z} \right) = -(1-\alpha)\frac{\partial p}{\partial z} + \frac{\partial}{\partial z}(1-\alpha)\tau_c - M_d , \quad (28)$$

$$\mu_c \frac{\partial v_c}{\partial z} = \tau_c , \quad (29)$$

where the last equation gives the definition of the viscous stress in terms of the derivative of the continuous phase velocity.

The vector of unknowns becomes $(\alpha, v_d, v_c, p, \tau_c)^T$, and the matrices of the formulation with viscosity ($\mu_c > 0$) become

$$\hat{A} = \begin{bmatrix} & & & & 0 \\ & & & & 0 \\ & A & & & 0 \\ & & & & 0 \\ 0 & 0 & 0 & 0 & 0 \end{bmatrix} , \quad (30)$$

$$\hat{\mathbf{B}} = \begin{bmatrix} & & & 0 \\ & \mathbf{B} & & 0 \\ & & & 0 \\ & & & -1 \\ 0 & 0 & (1-\alpha) & 0 & 0 \end{bmatrix}. \tag{31}$$

The characteristics are

$$\lambda = \{\infty, \infty, \infty, \infty, v_c\}; \tag{32}$$

the characteristic which is finite is always real. Thus, the characteristics of the system with fluid viscosity ($\mu_c > 0$) and the characteristics of the system for an inviscid fluid ($\mu_c = 0$) are unrelated. If the flow is sufficiently viscous that the appropriate model should contain the viscosity terms, then no problems with complex characteristics will be evident.

On the other hand, when the flow is nearly inviscid, we expect to be able to use the equations for the inviscid fluid, possibly with a drag model to account for the effects of viscosity in the boundary layer around the dispersed component. In this case, the roles of inertia, virtual mass, and other first-derivative terms become dominant in determining whether the characteristics are real, that is, whether the problem is well-posed.

If the characteristics of the system for an inviscid fluid are complex, an infinitesimally small disturbance in the initial data will grow large in a short time. A small amount of viscosity can alleviate the instability. To understand this, consider the linear stability of a steady-state solution Ψ_0 to the formulation represented by (25)–(28). A perturbation Ψ' of the form $\hat{\Psi} e^{\sigma t} e^{ikx}$ will be a nontrivial solution, if

$$\det\left(\frac{\sigma}{ik}\mathbf{A}(\Psi_o) + \mathbf{B}(\Psi_0) + \frac{1}{ik}\mathbf{D}(\Psi_o)\right) = 0. \tag{33}$$

Here $\mathbf{A}$ and $\mathbf{B}$ are the same matrices encountered in the inviscid characteristics problem and $\mathbf{D} = \partial \mathbf{C}/\partial \Psi$. As $k \to \infty$, $-\sigma/ik \to \lambda$, which is the characteristic slope at Ψ_0. Suppose that $\lambda = \mu + i\nu$ and $\lambda^* = \mu - i\nu$ are complex conjugate characteristic values, where $\nu \neq 0$. The two corresponding growth rates of the perturbation are $\Re\sigma = \pm k\nu$. Since $\nu \neq 0$, one of the growth rates must be positive; in fact, as $k \to \infty$, the positive growth rate becomes infinite.

For the viscous system, direct substitution of the perturbation in Ψ and linearization yield

$$\det\left(\frac{\sigma}{ik}\mathbf{A}(\Psi_o) + \mathbf{B}(\Psi_0) + ik\mathbf{B}'(\Psi_0) + \frac{1}{ik}\mathbf{D}(\Psi_o)\right) = 0 \tag{34}$$

as the condition for a nontrivial solution, where

$$\mathbf{B}' = \mu_c(1 - \alpha_o) \begin{bmatrix} 0 & 0 & 0 & 0 \\ 0 & 0 & 0 & 0 \\ 0 & 0 & 0 & 0 \\ 0 & 0 & 0 & -1 \end{bmatrix}. \tag{35}$$

If $\mu_c k/\rho_c v_o \ll 1$, the values of σ are close to those for the system with no viscosity. Thus, for small values of viscosity, there will exist disturbances whose wavenumbers k satisfy this criterion, and we conclude that complex characteristics can also taint viscous models in the inviscid limit. In this section we consider the consequences of several modeling assumptions to the characteristics of the associated formulations.

20.5. Inertial Effects

The simplest model in this chapter includes the inertia effects of the "left-hand side" terms, but, as derived in Section 15.3, there are many other inertial effects that can be included in the dilute multiphase flow equations. The model given in (18.2.60)–(18.2.63), specialized to one dimension, can be written in the form

$$\frac{\partial \alpha}{\partial t} + \frac{\partial \alpha v_d}{\partial z} = 0, \qquad (36)$$

$$\frac{\partial (1-\alpha)}{\partial t} + \frac{\partial (1-\alpha) v_c}{\partial z} = 0, \qquad (37)$$

$$\alpha \rho_d \left(\frac{\partial v_d}{\partial t} + v_d \frac{\partial v_d}{\partial z} \right) = -\alpha \frac{\partial p}{\partial z}$$
$$+ \alpha C_{vm} \rho_c \left[\left(\frac{\partial v_c}{\partial t} + v_c \frac{\partial v_c}{\partial z} \right) - \left(\frac{\partial v_d}{\partial t} + v_d \frac{\partial v_d}{\partial z} \right) \right]$$
$$+ \alpha C_{gdd} \rho_c (v_d - v_c) \frac{\partial v_d}{\partial z} + \alpha C_{gdc} \rho_c (v_d - v_c) \frac{\partial v_c}{\partial z}$$
$$+ C^*_{vm} \rho_c (v_d - v_c) \left(\frac{\partial \alpha}{\partial t} + v_d \frac{\partial \alpha}{\partial z} \right)$$
$$+ (C_{ad} - C_{TD}) \rho_c (v_d - v_c)^2 \frac{\partial \alpha}{\partial z} + F_D, \qquad (38)$$

$$(1-\alpha) \rho_c \left(\frac{\partial v_c}{\partial t} + v_c \frac{\partial v_c}{\partial z} \right) = -(1-\alpha) \frac{\partial p}{\partial z}$$
$$+ \alpha C_{vm} \rho_c \left[\left(\frac{\partial v_d}{\partial t} + v_d \frac{\partial v_d}{\partial z} \right) - \left(\frac{\partial v_c}{\partial t} + v_c \frac{\partial v_c}{\partial z} \right) \right]$$
$$+ \alpha C_{gcd} \rho_c (v_d - v_c) \frac{\partial v_d}{\partial z} + \alpha C_{gcc} \rho_c (v_d - v_c) \frac{\partial v_c}{\partial z}$$
$$- C^*_{vm} \rho_c (v_d - v_c) \left(\frac{\partial \alpha}{\partial t} + v_d \frac{\partial \alpha}{\partial z} \right)$$
$$+ (C_{ac} + C_{TD}) \rho_c (v_d - v_c)^2 \frac{\partial \alpha}{\partial z} - F_D. \qquad (39)$$

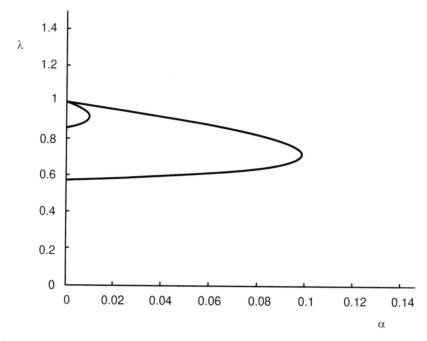

FIGURE 20.2. Characteristic values: The smaller loop corresponds to $C_{TD} = 0.5$, and the larger one corresponds to $C_{TD} = 1.0$.

where the calculation for an inviscid fluid gives $C_{vm} = C_{vm}^* = \frac{1}{2}$, $C_{ad} = \frac{11}{20}$, $C_{gc} = -\frac{19}{20}$, $C_{ac} = \frac{17}{20}$, and $C_{gc} = \frac{7}{20}$. The turbulent dispersion term is left as a parameter. Here, F_D is the drag force, and involves terms not having derivatives. The inviscid system with $C_{ad} = C_{gc} = C_{ac} = C_{gc} = C_{TD} = 0$ was studied by Pauchon and Banerjee [70], [71].

The "simplest" two-fluid model with fluids of no viscosity is obtained for $C_{vm} = 0$, $a_c + b_c = 0$, $C_{ad} = C_{gc} = C_{ac} = C_{gc} = C_{TD} = 0$, and $\xi = 0$; its characteristics are complex, unless $u = v$. For $C_{vm} = \frac{1}{2}$, $\xi = \frac{1}{4}$, $a_c + b_c = \frac{1}{5}$, $C_{ad} = \frac{11}{20}$, $C_{gc} = -\frac{19}{20}$, $C_{ac} = \frac{17}{20}$, $C_{gc} = \frac{7}{20}$, and $C_{TD} = 0$, the characteristics are also complex unless $u = v$. This corresponds to a model for the dispersed component that is at "absolute zero," that is, there are no effects of velocity fluctuations. One effect of velocity fluctuations is the presence of turbulent dispersion. The characteristics for different values of C_{TD} are shown in Figure 20.2.

20.6. Summary Observations

The problem of complex characteristics seems to arise from the coupling between the two momentum equations, since it does not appear in models, which have only one momentum equation. The argument has been advanced that viscosity will change the type of system of partial differential equations, and the problem of complex characteristics will be irrelevant. While it is true that a viscous system has real characteristics, in the limit of vanishing viscosity, the complex characteristics of the inviscid system give rise to small-scale instabilities which are artifacts of the model, and not physically real.

Thus, effort should be devoted to understanding the inviscid, incompressible, two-fluid momentum equations. The alternative of including differential terms to force real characteristics, either in the name of engineering or by invoking continuum mechanics principles, is unsatisfactory. We believe that the systematic inclusion of all terms arising in the averaged momentum equations, each soundly based on physics, will yield an appropriate working model.

21
Solutions for Shearing Flows

It is common physical experience that when a nondilute multicomponent fluid is placed in a viscometric testing device, the flow that is produced is not a viscometric flow, that is, the velocity gradient is not necessarily constant, and the concentration is not uniform. Instead, relatively thin layers of fluid with relatively few suspended particles accumulate near the boundaries. These layers exhibit high shear rates. The suspended particles accumulate at some distance from the boundaries, giving a high local viscosity and a low shear rate. These phenomena are so mathematically robust that they are exhibited by a large number of theories for suspensions. Here, we consider two properly invariant theories of multicomponent fluids. We investigate the simplest type of viscometric test, steady flow between parallel plates with one plate stationary and the other plate moving parallel to it at constant speed. We also consider steady flow in a channel, with the flow forced by a pressure gradient. For one theory, it is possible to demonstrate exact solutions to the field equations. For the other, plausible approximate solutions are found. Both types of solutions exhibit the phenomenon of a core of concentrated suspension surrounded by a layer of clear fluid.

The state of experiment for multicomponent mixtures is less than optimal. We know of elegant experimental work which agrees qualitatively with the solutions given here. However, we know of no precise measurements of full velocity and volume fraction fields appropriate for comparison with theory. At present, the best choice appears to be to continue to refine theory and solution techniques, and attempt to assimilate experimental results into the theories as they become available.

21.1. Field Equations

The field equations are of the following types: equations of conservation of mass for each of the constituents, equations of balance of momentum for each of the constituents, constitutive equations for the stress of each of the constituents, a constitutive equation for the forces of interaction between the constituents, and an "equation of state" (actually just another constitutive equation) for the solid.

21.1.1. Balance Equations

Conservation equations for mass of the constituents, assuming the fluid to be incompressible and the solid to have incompressible grains, are

$$\frac{\partial \alpha}{\partial t} + \nabla \cdot \alpha \mathbf{v}_s = 0, \qquad (1)$$

$$\frac{\partial (1-\alpha)}{\partial t} + \nabla \cdot (1-\alpha)\mathbf{v}_f = 0. \qquad (2)$$

The most general suspension we deal with is a mixture of an incompressible fluid with incompressible particles. Thus the equations of momentum are of the form

$$\alpha \rho_s \left(\frac{\partial \mathbf{v}_s}{\partial t} + \mathbf{v}_s \cdot \nabla \mathbf{v}_s \right) = \nabla \cdot \alpha \mathbf{T}_s - \alpha \rho_s \mathbf{g} + \mathbf{M}_s, \qquad (3)$$

$$(1-\alpha)\rho_f \left(\frac{\partial \mathbf{v}_f}{\partial t} + \mathbf{v}_f \cdot \nabla \mathbf{v}_f \right) = \nabla \cdot (1-\alpha)\mathbf{T}_f - (1-\alpha)\rho_f \mathbf{g} - \mathbf{M}_s. \qquad (4)$$

21.1.2. Constitutive Equations

In the balance principles (3) and (4), constitutive equations are required for the stresses of the constituents and the momentum exchange. Those are considered here. In addition, "equations of state" are needed in theories of materials with compressible components. Those are given in sections appropriate to the particular materials.

Stresses

The arguments in Chapter 8 indicate that the stresses for the components should have two parts, a "microscale" stress and a "Reynolds" stress:

$$\mathbf{T}_s = \mathbf{T}_s^m + \mathbf{T}_s^{\text{Re}},$$
$$\mathbf{T}_f = \mathbf{T}_f^m + \mathbf{T}_f^{\text{Re}}. \qquad (5)$$

In addition, it is often convenient to give an explicit notation for a hydrostatic part of some of the stresses,

$$\mathbf{T}_s = -p_s \mathbf{I} + \tau_s,$$
$$\mathbf{T}_f = -p_f \mathbf{I} + \tau_f, \qquad (6)$$

Microscale Stresses

For the purposes of this chapter it suffices to assume the microscale stresses have linear viscosity for each constituent with each viscosity depending on α,

$$\alpha \mathbf{T}_s^m = -\alpha p_s^m \mathbf{I} + \mu_{ss} \mathbf{D}_s + \mu_{sf} \mathbf{D}_f$$
$$+ \mu_{s\nabla}(\alpha) \left[\nabla \alpha (\mathbf{v}_f - \mathbf{v}_s) + (\mathbf{v}_f - \mathbf{v}_s) \nabla \alpha \right], \qquad (7)$$

$$(1-\alpha) \mathbf{T}_f^m = -(1-\alpha) p_f^m \mathbf{I} + \mu_{fs} \mathbf{D}_s + \mu_{ff} \mathbf{D}_f$$
$$+ \mu_{f\nabla}(\alpha) \left[\nabla \alpha (\mathbf{v}_f - \mathbf{v}_s) + (\mathbf{v}_f - \mathbf{v}_s) \nabla \alpha \right], \qquad (8)$$

where all of the constitutive coefficients μ can depend on volume fraction α. Included are the possibility of "cross-viscosities" $\mu_{sf}(\alpha)$ and $\mu_{fs}(\alpha)$ because preliminary calculations both in kinetic theory and in continuum theory indicate that such quantities should exist (Section 15.4).

One model for viscosities, consistent with that of Joseph and Lundgren [51] and Lundgren [60], is of particular interest. This model for the fluid stress is given by (15.4.153), and can be expressed as

$$(1-\alpha) \tau_f^m = \alpha \mu_f \mathbf{D}_s + (1-\alpha) \mu_f \mathbf{D}_f$$
$$- \mu_f(\alpha) \left[\nabla \alpha (\mathbf{v}_f - \mathbf{v}_s) + (\mathbf{v}_f - \mathbf{v}_s) \nabla \alpha \right], \qquad (9)$$

where μ_f is the viscosity of the pure fluid, taken as a constant. Then

$$\mu_{ff}(\alpha) = (1-\alpha) \mu_f, \qquad (10)$$
$$\mu_{fs}(\alpha) = \alpha \mu_f, \qquad (11)$$
$$\mu_{f\nabla}(\alpha) = -\mu_f. \qquad (12)$$

The values for μ_{ss} or μ_{sf} can be obtained as follows. We assume that there is a function $\beta(\alpha)$ with the properties

$$\mu_{ss} = \alpha \beta \mu_{fs}, \qquad (13)$$
$$\mu_{sf} = \alpha \beta \mu_{ff}. \qquad (14)$$

The function β is called an "amplification factor" for the stress. Motivation from averaging theory indicates that it arises because of the rigidity of the individual particles. Then (15.4.145) gives the "effective viscosity" for a suspension. That is, we should have

$$\mu_f (1 + \alpha \beta) = \mu_{\text{eff}}. \qquad (15)$$

This reduces to the Einstein expression for viscosity if $\beta = \frac{5}{2}$. It also allows other expressions to be used for β.

While this set of assumptions has some pleasing features, we need not restrict matters this much at this stage. We retain the generality allowed by (7) and (8) with the cross-viscosities. In our subsequent calculations it will be possible to set them equal to zero with no danger of division by zero. On the other hand, the algebraic gain from this simplification will be marginal at best. It is plausible to expect the viscosities to dominate the cross-viscosities so that

$$\mu_{ss}(\alpha) \gg \mu_{sf}(\alpha), \tag{16}$$

$$\mu_{ff}(\alpha) \gg \mu_{fs}(\alpha), \tag{17}$$

however, it is not obvious that such results can be derived in any standard way, e.g., from thermodynamics.

Reynolds Stresses

The representation theory in Section 14.3 indicates that the "Reynolds" stresses should depend on deformation rates and forces of interaction. Thus the form

$$\alpha \mathbf{T}_s^{Re} = -\alpha p_s^{Re} \mathbf{I} + \mu_{ss}^{Re}(\alpha) \mathbf{D}_s + \mu_{sf}^{Re}(\alpha) \mathbf{D}_f$$
$$+ \alpha \rho_s \left[a_s |\mathbf{v}_f - \mathbf{v}_s|^2 \mathbf{I} + b_s (\mathbf{v}_f - \mathbf{v}_s)(\mathbf{v}_f - \mathbf{v}_s) \right],$$
$$(1-\alpha) \mathbf{T}_f^{Re} = -(1-\alpha) p_f^{Re} \mathbf{I} + \mu_{fs}^{Re}(\alpha) \mathbf{D}_s + \mu_{ff}^{Re}(\alpha) \mathbf{D}_f$$
$$+ \alpha \rho_f \left[a_f |\mathbf{v}_f - \mathbf{v}_s|^2 \mathbf{I} + b_f (\mathbf{v}_f - \mathbf{v}_s)(\mathbf{v}_f - \mathbf{v}_s) \right], \tag{18}$$

is plausible.

In our constitutive equations for the stresses, the two types of pressures play intrinsically different roles. The pressures p_f^m and p_s^m are reactions to the constraint of incompressibility of the fluid and the solid, and are determined as part of the solution to a boundary-value problem. The pressures $p_s^{Re}(\alpha)$ and $p_f^{Re}(\alpha)$, on the other hand, can arise due to the fluctuation motions of the particles and the fluid turbulence, and are given by constitutive equations. Generally it would be necessary to specify such an equation precisely to solve boundary-value problems. However, in the particular structure presented here, it suffices that p_s^{Re} depend on α only.

Momentum Exchange

In a sense, the essence of the study of the mechanics of multicomponent fluids is the structure of the constitutive equation for the momentum exchange $\mathbf{M}_s$. It is the subject for very substantial debate. We wish to elucidate some physics that arises from some part of the momentum exchange in the next section. However, we do not take the most general expression for it here. We consider forces of interaction for drag, buoyancy, the Faxén force, virtual mass, and lift. The form

$$\mathbf{M}_s = p_{si} \nabla \alpha + \alpha S(\mathbf{v}_f - \mathbf{v}_s) + \alpha k \mu_f \nabla^2 \mathbf{v}_f - \nabla \alpha \cdot \tau_{si}$$
$$+ \alpha \rho_f C_{vm} \left[\frac{\partial \mathbf{v}_f}{\partial t} + \mathbf{v}_f \cdot \nabla \mathbf{v}_f - \frac{\partial \mathbf{v}_s}{\partial t} - \mathbf{v}_s \cdot \nabla \mathbf{v}_s + (\mathbf{v}_s - \mathbf{v}_f) \wedge (\nabla \wedge \mathbf{v}_f) \right], \tag{19}$$

contains all of these. Note that we have combined the virtual mass and lift forces. The individual terms are not objective but the sum is.

Henceforth we assume no body forces, so that $\mathbf{b}_p = \mathbf{b}_f = \mathbf{g} = 0$. No buoyancy of the solid in the fluid is a special case of this assumption.

21.2. Kinematics and Dynamics of Shearing Flow

We wish to investigate the behavior of different constitutive models in the same flow apparatus. Thus it is economical to carry the solutions through as far as is possible in as much generality as possible. For classical simple materials, it is known that in viscometric flows only a small portion of very complex constitutive behavior is exhibited [24]. A consequence is simple velocity fields, independent of all or most of the constitutive complexity of the particular material. For example, in a rectilinear shearing viscometer, all incompressible simple fluids exhibit a rectilinear velocity profile.

No results quite so strong are known for suspensions. In fact, we exhibit in this chapter a rich and complex variety of behavior for slow flow of suspensions in rectilinear shearing viscometers. Nonetheless, it is true that the assumptions of boundary conditions and basic kinematics appropriate to viscometric flows yields problems in analysis that sometimes are tractable. It is also true that viscometers are often used to test suspensions, so the mathematical analysis is physically interesting.

Choose a set of right-handed Cartesian coordinates $\{x, y, z\}$. Define a *rectilinear shearing flow* by

$$\mathbf{v}_k = \hat{u}_k(y)\mathbf{e}_x,$$
$$\alpha = \hat{\alpha}(y), \qquad (20)$$

where, for example, $\mathbf{v}_k$ means both of the velocities $\mathbf{v}_s$ and $\mathbf{v}_f$, and $\mathbf{e}_x$ is the unit vector in the positive x-direction. Thus

$$\nabla \alpha \sim [0, d\hat{\alpha}/dy, 0], \qquad (21)$$

and

$$\nabla \mathbf{v}_k \sim \begin{bmatrix} 0 & d\hat{u}_k/dy & 0 \\ 0 & 0 & 0 \\ 0 & 0 & 0 \end{bmatrix}. \qquad (22)$$

Then

$$2\mathbf{D}_k \sim \begin{bmatrix} 0 & d\hat{u}_k/dy & 0 \\ d\hat{u}_k/dy & 0 & 0 \\ 0 & 0 & 0 \end{bmatrix}. \qquad (23)$$

21.2. Kinematics and Dynamics of Shearing Flow

The notation $\sim$ denotes that the vector or tensor to its left has the coordinate representation given by the array to its right.

It is an immediate consequence of (20) that flows of this nature are accelerationless, i.e.,

$$\frac{\partial \mathbf{v}_k}{\partial t} + \mathbf{v}_k \cdot \nabla \mathbf{v}_k = 0 \,. \tag{24}$$

Therefore, in particular, the virtual mass force is identically zero.

There are two different steady rectilinear shearing flows of interest. One is flow between parallel plates with one plate stationary and the other plate moving parallel to it. This flow is called plane Couette flow. A second is steady flow between parallel plates at rest. The flow is driven by a pressure gradient in the axial direction. This flow is called plane Poiseuille flow.

We use a spatial Cartesian coordinate system with origin on the lower plate, x-axis along the lower plate, with positive direction in the direction of movement of the upper plate, and y-axis directed toward the upper plate. The plates are a distance $2L$ apart. For the problem of plane Couette flow, the top plate moves at constant speed $2V$.

The boundary conditions appropriate for rectilinear shearing flow are dependent on the material of which the walls are constructed, whether it is smooth, or bumpy, or sticky. In at least some situations, observations indicate that α vanishes or nearly does so at boundaries. Moreover, the average value of α across a cross-section defines the average concentration of solids in the material, and may be specified at the outset. Thus we have the conditions

$$\begin{aligned}&(1) \quad \hat{\alpha}(0) = 0 \,, \\ &(2) \quad \hat{\alpha}(2L) = 0 \,, \\ &(3) \quad \int_0^{2L} \hat{\alpha} \, dy = 2L \langle \alpha \rangle \,.\end{aligned} \tag{25}$$

We require an adequate number of boundary conditions for the velocities. For the fluid and solid, we assume the classical boundary condition, that is, the velocity of each component at a solid boundary matches the velocity of the boundary.

We perform three separate analyses, one to account for complicated expressions for the viscous terms, one to assess the effect of nonuniform particle concentration, and one to ascertain the role of the interfacial terms in determining the particle concentration. The first analysis shows the role of the fluid and particle viscosity terms in the shearing of the mixture. It is exact. The second analysis, also exact, shows the effects of separation of the components with clear layers in the flow. The third analysis involves an approximate solution which treats the dynamics of the separation process in more detail, and shows that the interaction terms are instrumental in separating the components into particle-laden and particle-free zones. That analysis requires the viscosity of the particle-laden region to become infinite, so that the core undergoes a rigid motion.

21.2.1. Plane Poiseuille Flow

In this analysis, we assume that the solid component volume fraction α_s is constant. We also assume that the pressure gradient in (15) and (16) is constant. The boundary conditions are

$$
\begin{aligned}
&(1) \quad \hat{u}_f(0) = 0, \\
&(2) \quad \hat{u}_s(0) = 0, \\
&(3) \quad \hat{u}_f(2L) = 0, \\
&(4) \quad \hat{u}_s(2L) = 0.
\end{aligned}
\tag{26}
$$

In accord with (2) and (4) for the pressures, we take

$$p_s^m = p_{si} = p_{fi} = p_f^m - C_p \rho_f |\mathbf{v}_f - \mathbf{v}_s|^2. \tag{27}$$

For the stress terms, we take

$$\tau_s^m = \tau_{si}^m = \tau_{fi}^m. \tag{28}$$

The stress transmitted by the particles corresponds to the extra stress needed to make the mixture more viscous. To account for this effect, we take

$$\tau_{si} = \beta(\alpha)\tau_f^m. \tag{29}$$

For the viscosities of the fluid, we take

$$
\begin{aligned}
\mu_{ss} &= 0, \\
\mu_{fs} &= 0, \\
\mu_{ff} &= (1-\alpha)\mu_f, \\
\mu_{sf} &= \beta\alpha\mu_f.
\end{aligned}
\tag{30}
$$

We further ignore the Faxén force, so that $k = 0$. Substituting these relations and (20) for the shearing flow of a particle–fluid mixture, the equations of conservation of mass are satisfied identically. The equations of conservation of momentum in the axial direction become

$$\hat{\alpha}S(\hat{u}_f - \hat{u}_s) + \hat{\alpha}\mu_f \frac{d}{dy}\left[\beta\frac{d\hat{u}_f}{dy}\right] = \hat{\alpha}\mathcal{P}, \tag{31}$$

$$\hat{\alpha}S(\hat{u}_s - \hat{u}_f) + \mu_f\beta\frac{d\hat{\alpha}}{dy}\frac{d\hat{u}_f}{dy} + \mu_f\frac{d}{dy}\left[(1-\hat{\alpha})\frac{d\hat{u}_f}{dy}\right] = (1-\hat{\alpha})\mathcal{P}, \tag{32}$$

where $\mathcal{P}$ is the pressure gradient.

In this section, we shall ignore the transverse (y-direction) momentum equations, and instead, assume that $\hat{\alpha}$ is constant.

Adding (31) and (32) results in

$$[\hat{\alpha}\beta + (1-\hat{\alpha})]\mu_f \frac{d^2\hat{u}_f}{dy^2} = \mathcal{P}. \tag{33}$$

Then the solution satisfying the boundary conditions is

$$\hat{u}_f = -\frac{\mathcal{P}}{2\left[\hat{\alpha}\beta + (1-\hat{\alpha})\right]\mu_f} y(2L-y). \tag{34}$$

Thus, the fluid velocity profile is parabolic. Note that for flow in the positive x-direction, we must have $\mathcal{P} < 0$, so that pressure decreases in the positive x-direction. Substituting this into the solid component momentum equation gives

$$\hat{u}_s = \hat{u}_f + \frac{(1-\hat{\alpha})(\beta-1)}{S\left[\hat{\alpha}\beta + (1-\hat{\alpha})\right]}\mathcal{P}. \tag{35}$$

Note that if $\beta > 1$, then the particles lag the fluid (i.e., $\hat{u}_s < \hat{u}_f$), but if $\beta < 1$, the particles lead the fluid.

Also note that the particle component boundary conditions on the walls cannot be satisfied with this solution. This prompts the search for solutions with nonuniform $\hat{\alpha}$.

21.2.2. Model Emphasizing Viscosity

In this section, we emphasize the effects of viscosity. Thus we retain only sufficient forces of interaction for plausible physics. In the subsequent section, we study the transition layer between the clear fluid layer and the particle-laden layer. Here, we study more thoroughly the shearing motions of both layers, in order to see the effects of the division of the viscous terms.

We assume sufficient symmetry so that conditions at the center line replace conditions at the top boundary $y = 2L$. These symmetry conditions are that the velocity of the fluid *and* of the solid at the centerline both are half of the velocity of the top boundary. Thus

$$\begin{array}{ll}(1) & \hat{u}_f(0) = 0, \\ (2) & \hat{u}_s(0) = 0, \\ (3) & \hat{u}_f(L) = U, \\ (4) & \hat{u}_s(L) = U.\end{array} \tag{36}$$

This situation is depicted in Figure 21.1.

The equations are obtained by making the following nonunique choices in (3) and (4). For pressures,

$$\begin{aligned} \alpha p_s^{\text{Re}} &= \mathcal{P}_s(\alpha), \\ (1-\alpha)p_f^{\text{Re}} &= 0, \\ \alpha p_s^m &= 0, \\ (1-\alpha)p_f^m &= \mathcal{P}_f, \end{aligned} \tag{37}$$

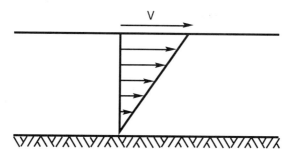

FIGURE 21.1. Simple shearing flow.

for the Reynolds stresses,

$$\tau_d^{Re} = 0,$$
$$\tau_c^{Re} = 0, \qquad (38)$$

for coefficients in the force of interaction,

$$\alpha S = \lambda_1(\alpha),$$
$$p_{si} = \lambda_2(\alpha),$$
$$p_{fi} = \lambda_2(\alpha),$$
$$C_{vm} = 0,$$
$$k = 0, \qquad (39)$$

for interfacial shear stresses,

$$\tau_{di} = \tau_{ci} = 0. \qquad (40)$$

We assume that $\mu_{ss}(\alpha)$, $\mu_{sf}(\alpha)$, $\mu_{fs}(\alpha)$, and $\mu_{ff}(\alpha)$ are known. The virtual mass, lift, and Faxén forces are assumed to vanish. In the particular flows studied here, in fact, the virtual mass forces vanish identically, and the Faxén force has no qualitative effect on the flow, so the only significant force that is ignored is the lift force.

It is convenient to use the notation

$$\mu_S = \mu_{sf} + \mu_{ss},$$
$$\mu_F = \mu_{ff} + \mu_{fs}. \qquad (41)$$

We substitute assumptions (20) into the field equations, and assume that $\mathcal{P}_f$ is a function of y only. Equations (1) and (2) for conservation of mass are satisfied identically. The remaining field equations come from substituting the constitutive equations (18) for stress, and the constitutive equations (19) for momentum interaction into the balance of momentum (3) and (4). The resulting equations are most easily dealt with by using their sum and one of the individual equations. Here, we choose to do that, using the sum and the equation for the momentum of the fluid.

21.2. Kinematics and Dynamics of Shearing Flow

The resulting system is

$$\frac{d}{dy}\left[\mu_F \frac{d\hat{u}_f}{dy} + \mu_S \frac{d\hat{u}_s}{dy}\right] = 0, \tag{42}$$

$$\frac{d}{dy}[\mathcal{P}_f + \mathcal{P}_s] = 0, \tag{43}$$

and

$$\frac{d}{dy}\left[\mu_{ff}\frac{d\hat{u}_f}{dy} + \mu_{fs}\frac{d\hat{u}_s}{dy}\right] + \lambda_1(\hat{u}_s - \hat{u}_f) = 0, \tag{44}$$

$$\frac{d\mathcal{P}_f}{dy} + \lambda_2 \frac{d\alpha}{dy} = 0. \tag{45}$$

To begin solving this system, let us consider (43) and (45). Notice that this part of the system is decoupled from the rest in the sense that it involves no dependence on the velocities. Then (45) becomes

$$\left[\frac{d\mathcal{P}_s}{d\hat{\alpha}} + \lambda_2(\hat{\alpha})\right]\frac{d\hat{\alpha}}{dy} = 0, \tag{46}$$

so either there is a relation between the two independent constitutive functions $\mathcal{P}_s(\cdot)$ and $\lambda_2(\cdot)$ or else $d\hat{\alpha}/dy = 0$. We consider the latter to be the only realistic possibility. If we assume that the function $\hat{\alpha}(y)$ is continuous, there is no solution that agrees with the boundary condition (25). This difficulty can be overcome by considering a larger class of solutions, for example, solutions $\hat{\alpha}(y)$ that are piecewise continuous. Then there is a solution of the form

$$\hat{\alpha} = \begin{cases} 0, & y \in [0, h), \\ \alpha_0, & y \in (h, 2L - h), \\ 0, & y \in (2L - h, 2L], \end{cases} \tag{47}$$

for some number $h < L$, and with $\hat{\alpha}$ undefined at $y = h$ and $y = 2L - h$. This makes the constitutive functions $\lambda_1, \lambda_2, \mu_{ff}, \mu_{fs}, \mu_{sf}$, and μ_{ss} constant also. Here, α_0 and h are quantities to be determined. Not all of the parameters α_0, $\langle \alpha \rangle$, and h are independent. Suppose, for example, that the steady shearing flow evolves at long time from a homogeneous flow with uniform volume fraction $\langle \alpha \rangle$. Then the total fraction of the solid constituent in the cross section is given by

$$\langle \alpha \rangle L = \alpha_0 (L - h). \tag{48}$$

Equation (47) is not unique among all the piecewise constant functions satisfying all of the conditions we have assumed. However, we shall study the consequences of taking the function $\hat{\alpha}(y)$ to have this form.

To solve for the velocity field, note that (47) implies that near the walls, that is, for $y \in [0, h)$ and $y \in (2L - h, 2L]$, there is no solid constituent. Thus the fluid may be considered to be an ordinary linearly viscous fluid, governed by the Navier–Stokes equations. The problem has sufficient symmetry so that a solution in the region $y \in [0, L]$ defines the solution in the whole region $y \in [0, 2L]$. Thus in the region $y \in [0, h]$ we have

$$\hat{u}_f = \hat{A}y + \hat{B}, \qquad (49)$$

and $\hat{u}_s$ not defined. In the region $y \in (h, L)$ the situation is more complicated. There, the governing equations are

$$\frac{d}{dy}\left[\mu_F \frac{d\hat{u}_f}{dy} + \mu_S \frac{d\hat{u}_s}{dy}\right] = 0, \qquad (50)$$

$$\frac{d}{dy}\left[\mu_{ff} \frac{d\hat{u}_f}{dy} + \mu_{fs} \frac{d\hat{u}_s}{dy}\right] + \lambda_1(\hat{u}_s - \hat{u}_f) = 0. \qquad (51)$$

Equation (50) may be integrated twice to give

$$\mu_F \hat{u}_f + \mu_S \hat{u}_s = Ay + B. \qquad (52)$$

Our principal interest is in $\hat{u}_f$, which is somewhat easier to observe than $\hat{u}_s$. Thus we solve the above result for $\hat{u}_s$ and substitute into (51). This results in

$$\frac{d}{dy}\left[\left(\mu_{ff} - \frac{\mu_F \mu_{fs}}{\mu_S}\right)\frac{d\hat{u}_f}{dy}\right] + \lambda_1\left(\frac{Ay+B}{\mu_S} - \frac{\mu_F}{\mu_S} - 1\right)\hat{u}_f = 0. \qquad (53)$$

The coefficient of $d\hat{u}_f/dy$ in (53) has a significant role in subsequent arguments, and it is worthwhile to consider it further. Let

$$\bar{\mu}_f = \frac{\mu_F \mu_{fs}}{\mu_S}, \qquad (54)$$

then substitution of (41) into this relation gives

$$\bar{\mu}_f = \frac{\mu_{ff}(\mu_{sf} + \mu_{ss} - \mu_{fs}) - \mu_{fs}^2}{\mu_{sf} + \mu_{ss}}. \qquad (55)$$

Notice that in the case of separation of components, where $\mu_{sf} = \mu_{fs} = 0$, we have $\bar{\mu}_f = \mu_f > 0$. It is obvious that there is a neighborhood of $\mu_{sf} = \mu_{fs} = 0$ in which $\mu_{sf} \neq 0$ and $\mu_{fs} \neq 0$ but for which $\bar{\mu}_f > 0$. Another combination of viscosities of interest is

$$\mathsf{T} = \frac{\mu_F}{\mu_S} + 1. \qquad (56)$$

In general

$$\mathsf{T} = \frac{\mu_{ff} + \mu_{fs} + \mu_{ss} + \mu_{sf}}{\mu_{sf} + \mu_{ss}}, \qquad (57)$$

and in the case of separated phases

$$\mathsf{T} = \frac{\mu_{ff}}{\mu_{ss}} + 1. \qquad (58)$$

Thus, there is a neighborhood of the case of separated phases where $\mathsf{T} > 0$. Note that $\bar{\mu}_f$ has the dimensions of viscosity and T is dimensionless.

With the above notation, (53) becomes

$$\bar{\mu}_f \frac{d^2 \hat{u}_f}{dy^2} + \lambda_1 \mathsf{T} \hat{u}_f = -\frac{\lambda_1}{\mu_S}(Ay + B). \qquad (59)$$

The particular solution of this equation is

$$\overset{P}{u}_f = \frac{1}{\mathsf{T}\mu_s}(Ay + B). \tag{60}$$

The form of the homogeneous solution of (59) depends on the signs of its coefficients. Since the coefficient λ_1 in (51) represents a drag force, it must be positive. Then the definition

$$\beta^2 = \frac{\mathsf{T}\lambda_1}{\bar{\mu}_f} \tag{61}$$

makes sense, at least in the neighborhood of separated phases. The solution is

$$\overset{H}{u}_f = C\cosh\beta y + D\sinh\beta y. \tag{62}$$

The velocity fields for the two materials are then

$$\hat{u}_f = \hat{A}y + \hat{B}, \tag{63}$$

for $y \in [0, h)$, and

$$\hat{u}_f = \frac{1}{\mathsf{T}\mu_s}(Ay + B) + (C\cosh\beta y + D\sinh\beta y),$$

$$\hat{v}_s = \frac{1}{\mathsf{T}\mu_s}(Ay + B) - \frac{\mu_F}{\mu_S}(C\cosh\beta y + D\sinh\beta y), \tag{64}$$

for $y \in (h, L]$. Notice that in the central "mixed material" region, both materials undergo a basic simple shearing. There is no reason to think that the shearing in the central region is the same as in the edge "clear fluid" region. Superposed on the simple shearing is a nonlinear flow profile caused by the drag of the fluid on the solid. The profile for the solid constituent is adjusted from the profile for the solid constituent by the negative of a viscosity ratio.

To complete the formal solution, we require an adequate number of boundary conditions to specify A, B, C, D, $\hat{A}$, and $\hat{B}$. For the clear fluid, we assume the classical boundary condition, that is, the velocity at a solid boundary matches the velocity of the boundary. In this particular boundary-value problem, there is no solid boundary for the mixed material regime. Thus only interface conditions, not boundary conditions, are needed. Our assumption is that the clear fluid at the interface affects only the fluid in the mixture at the interface, and the solid at the interface acts as if it were at a traction-free boundary. We assume sufficient symmetry so that conditions at the centerline replace conditions at the interface $y = 2L - h$ and at the top boundary $y = 2L$. These symmetry conditions are that the velocity of the fluid *and* of the solid at the interface are both half of the velocity of the top boundary. Thus

(1) $\hat{u}_f(0) = 0$,

(2) $\lim\limits_{y \to h^+} \hat{u}_f(y) = \lim\limits_{y \to h^-} \hat{u}_f(y)$,

(3) $\lim\limits_{y \to h^+} \left(\mu_{ff} \dfrac{\hat{u}_f(y)}{dy} + \mu_{fs} \dfrac{\hat{u}_s(y)}{dy} \right) = \lim\limits_{y \to h^-} \mu_f \dfrac{\hat{u}_f(y)}{dy}$, (65)

(4) $\lim\limits_{y \to h^+} \left(\mu_{sf} \dfrac{d\hat{u}_f(y)}{dy} + \mu_{ss} \dfrac{\hat{u}_s(y)}{dy} \right) = 0$,

(5) $\hat{u}_f(L) = U$,

(6) $\hat{u}_s(L) = U$,

where μ_f is the viscosity of the clear fluid.[1] These boundary conditions, when applied to (64) give a nonsingular set of six linear algebraic equations in the six unknown constants A, B, C, D, $\hat{A}$, and $\hat{B}$.

Applying boundary conditions (1), (5), and (6) gives

$$\hat{u}_f = \hat{A} y \qquad (66)$$

for $y \in [0, h)$, and

$$\hat{u}_f = U + A(y - L) + C \left(\frac{\sinh \beta(y - L)}{\sinh \beta L} \right),$$

$$\hat{u}_s = U + A(y - L) - C \frac{\mu_F}{\mu_S} \left(\frac{\sinh \beta(y - L)}{\sinh \beta L} \right), \qquad (67)$$

for $y \in (h, L]$. The remaining three constants, $\hat{A}$, A, and C are obtained from the interface conditions (2), (3), and (4), and are somewhat complicated to write out. Let

$$\Delta = \beta[(L - h)\mu_f(\mu_{ss}\mu_S - \mu_{fs}\mu_F)$$
$$+ h(\mu_S + \mu_F)(\mu_{ss}\mu_{ff} - \mu_{fs}\mu_{sf})] \cosh \beta(L - h)$$
$$+ \mu_f \mu_F \mu_S \sinh \beta(L - h). \qquad (68)$$

In the case of separated phases, this expression has the simpler form

$$\Delta = \beta \left[(L - h)\mu_f \mu_{ss}^2 + h(\mu_{ss} + \mu_{ff})\mu_{ss}\mu_{ff} \right] \cosh \beta(L - h)$$
$$+ \mu_f \mu_{ff} \mu_{ss} \sinh \beta(L - h). \qquad (69)$$

These expressions are the determinants resulting from solving three linear algebraic equations in three unknowns. For the case of separated phases, it is clear that the determinant cannot vanish. For the general case, the possibility of its vanishing exists only if the cross-viscosities μ_{sf} and μ_{fs} dominate the viscosities μ_{ss} and

---

[1] It is plausible to assume $\mu_{ff} = (1 - \alpha)\mu_f$. The arguments here are independent of that choice.

21.2. Kinematics and Dynamics of Shearing Flow   267

μ_{ff}. For $\hat{A}$, A, and C, then

$$\hat{A} = \frac{\beta(\mu_S + \mu_F)(\mu_{ss}\mu_{ff} - \mu_{fs}\mu_{sf})\cosh\beta(L-h)}{\Delta}U, \qquad (70)$$

while

$$A = \frac{\beta\mu_f(\mu_{ss}^2 + \mu_{sf}\mu_{ss} - \mu_{fs}\mu_{sf} - \mu_{ff}\mu_{sf})\cosh\beta(L-h)}{\Delta}U, \qquad (71)$$

and

$$C = -\frac{\mu_f(\mu_{fs} + \mu_{ff})(\mu_{ss} + \mu_{sf})\sinh\beta L}{\Delta}U. \qquad (72)$$

For the case of separated phases, these expressions simplify thus:

$$\frac{1}{\mu_F + \mu_S}\hat{A} = \frac{\beta(\mu_{ss} + \mu_{ff})(\mu_{ss}\mu_{ff})\cosh\beta(L-h)}{\Delta}U, \qquad (73)$$

$$A = \frac{\beta\mu_f\mu_{ss}^2\cosh\beta(L-h)}{\Delta}U, \qquad (74)$$

and

$$C = -\frac{\mu_f\mu_{ff}\mu_{ss}\sinh\beta L}{\Delta}U. \qquad (75)$$

The quantity $\hat{A}$ is the shear rate in the clear fluid. The situation is slightly more complicated in the mixed material because the velocity profiles are nonlinear. Nonetheless, the quantity A is essentially the shear rate in the mixed material.

21.2.3. Separation Processes

In this section, we consider the forces that could cause and maintain the geometry appropriate to the clear fluid and concentrated layers needed in the previous section. Our field equations are sufficiently complex so that we are not able to demonstrate exact solutions. However, approximate solutions are obtained rather easily. The process of obtaining them includes arguments that they are internally consistent. It is found that for the same geometry as considered in the previous section, there are three types of layers: clear fluid layers near the boundaries, layers of rigid motion far away from the boundaries, and transition layers.

Substituting (27), (28), (29), (30), and (20) for the shearing flow of a particle-fluid mixture, the equations of balance of mass are again satisfied identically. The equations of balance of momentum become

$$\hat{\alpha}S(\hat{u}_s - \hat{u}_f) + \hat{\alpha}\mu_f\frac{d}{dy}\left[\beta\frac{d\hat{u}_f}{dy}\right] + \hat{\alpha}k\mu_f\frac{d^2\hat{u}_f}{dy^2} = 0,$$

$$\hat{\alpha}S(\hat{u}_f - \hat{u}_s) + \mu_f\beta\frac{d\hat{\alpha}}{dy}\frac{d\hat{u}_f}{dy} + \mu_f\frac{d}{dy}\left[(1-\hat{\alpha})\frac{d\hat{u}_f}{dy}\right] + \hat{\alpha}k\mu_f\frac{d^2\hat{u}_f}{dy^2} = 0,$$

21. Solutions for Shearing Flows

$$-\hat{\alpha}\frac{\partial p_f}{\partial y} + 2\hat{\alpha}C_p(\hat{u}_s - \hat{u}_f)\frac{d}{dy}(\hat{u}_s - \hat{u}_f)$$

$$+\hat{\alpha}\rho_f C_{vm}(\hat{u}_s - \hat{u}_f)\frac{d\hat{u}_f}{dy} + \frac{d}{dy}\hat{\alpha}\rho_s a_s(\hat{u}_f - \hat{u}_s)^2 = 0,$$

$$-(1-\hat{\alpha})\frac{\partial p_f}{\partial y} + C_p(\hat{u}_s - \hat{u}_f)^2\frac{d\hat{\alpha}}{dy}$$

$$+\hat{\alpha}\rho_f C_{vm}(\hat{u}_f - \hat{u}_s)\frac{d\hat{u}_f}{dy} + \frac{d}{dy}\hat{\alpha}\rho_f a_f(\hat{u}_f - \hat{u}_s)^2 = 0. \quad (76)$$

The relative motion between the spheres and the fluid can be obtained from either of the remaining momentum equations in the x-direction. We divide (76)$_1$ by $\hat{\alpha}S$ to obtain

$$\hat{u}_s - \hat{u}_f = \frac{\mu_f}{S}\frac{d}{dy}\left[(\beta + k)\frac{d\hat{u}_f}{dy}\right]. \quad (77)$$

The momentum equations in the y-direction are used to determine the distribution of spheres across the channel. These equations involve the transverse pressure gradient dp_c/dy. This pressure gradient can be eliminated by subtracting the product of $\hat{\alpha}/(1-\hat{\alpha})$ with (76)$_4$ from (76)$_3$. The result is

$$0 = \frac{\hat{\alpha}}{1-\hat{\alpha}}\rho_f C_{vm}(\hat{u}_s - \hat{u}_f)\frac{d\hat{u}_f}{dy}$$

$$+ \left[C_p + a_s - a_f\frac{\hat{\alpha}}{1-\hat{\alpha}}\right]2\hat{\alpha}\rho_f(\hat{u}_s - \hat{u}_f)\frac{d(\hat{u}_s - \hat{u}_f)}{dy}$$

$$+ \left[-(C_p + a_f)\frac{\hat{\alpha}}{1-\hat{\alpha}} + a_s\right]\rho_f(\hat{u}_s - \hat{u}_f)^2\frac{d\hat{\alpha}}{dy}. \quad (78)$$

We can further eliminate $\hat{u}_s - \hat{u}_f$ by using (77). The result is

$$0 = \hat{\alpha}\left[C_p + a_s - a_f\frac{\hat{\alpha}}{1-\hat{\alpha}}\right]\frac{2\mu_f}{S}\frac{d^2}{dy^2}(\beta + k)\frac{d\hat{u}_f}{dy} - \frac{\hat{\alpha}C_{vm}}{1-\hat{\alpha}}\frac{d\hat{u}_f}{dy}$$

$$+ \left[-(C_p + a_f)\frac{\hat{\alpha}}{1-\hat{\alpha}} + a_s\right]\left[\frac{\mu_f}{S}\frac{d}{dy}(\beta + k)\frac{d\hat{u}_f}{dy}\right]\frac{d\hat{\alpha}}{dy}. \quad (79)$$

We nondimensionalize the problem by

$$w = \frac{\hat{u}_f(y)}{U}, \quad (80)$$

$$\zeta = \frac{y}{C_{vm}}. \quad (81)$$

The equations become

$$[1 + (\beta - 1)\hat{\alpha}]\frac{dw}{d\zeta} = \Lambda, \quad (82)$$

21.2. Kinematics and Dynamics of Shearing Flow

$$0 = 2\hat{\alpha}\left[C_p + a_s - a_f\frac{\hat{\alpha}}{1-\hat{\alpha}}\right]\frac{d^2}{d\zeta^2}(\beta+k)\frac{dw}{d\zeta} - \text{St}\frac{\hat{\alpha}C_{vm}}{1-\hat{\alpha}}\frac{dw}{d\zeta}$$
$$+ \left[a_s - (C_p + a_f)\frac{\hat{\alpha}}{1-\hat{\alpha}}\right]\left[\frac{d}{d\zeta}(\beta+k)\frac{dw}{d\zeta}\right]\frac{d\hat{\alpha}}{d\zeta}, \tag{83}$$

where Λ is the dimensionless mixture shear rate, and

$$\text{St} = \frac{Sh^2}{\mu_f}, \tag{84}$$

is the Stokes number, a measure of the drag. Note that if we assume Stokes flow around a single sphere of radius a, then $S = 9\mu_f/2a^2$, and therefore

$$\text{St} = \frac{9}{2}\left(\frac{h}{a}\right)^2.$$

The boundary conditions are

$$w(0) = 0, \tag{85}$$

$$w(1) = 1, \tag{86}$$

$$\int_0^1 \hat{\alpha}(\zeta)\,d\zeta = \langle\alpha\rangle. \tag{87}$$

Let us seek a solution for St large. This makes the reasonable assumption that the channel is many sphere diameters wide. If St is large, we see that we must have β large, and either $dw/d\zeta$ is small, or $\hat{\alpha}$ is small. From (82), if $\Lambda = O(1)$ and $dw/d\zeta$ is small, β must be large. With $dw/d\zeta = \text{St}^{-q}g$, and $\beta = \text{St}^q b'$, we have $\hat{\alpha} = \alpha_m + \text{St}^{-q}\hat{\alpha}'$ and $b' = -9\alpha_m/8\hat{\alpha}'$. In order to obtain a balance in (83), we must have $q = 1$. Thus, the motion suggested by this balance is a region of high concentration with $\hat{\alpha}$ nearly equal to α_m, with small shearing due to the high viscosity there. The rest of the flow is a region of small $\hat{\alpha}$, that is, nearly clear fluid. The shearing in these regions is not small.

Let us assume that the region with $\hat{\alpha}$ small is near the wall and the region with $dw/d\zeta$ small is at the center of the channel, in a region that we shall call the core. The resulting approximate concentration of spheres is then given by

$$\hat{\alpha} = \begin{cases} \alpha_m, & 0 < \zeta < h^*, \\ 0, & h^* < \zeta < 1, \end{cases} \tag{88}$$

where h^* is the location of the edge of the core.

In the clear fluid region, the fluid velocity must satisfy

$$\frac{dw}{d\zeta} = \Lambda, \tag{89}$$

so that

$$w = \Lambda\zeta. \tag{90}$$

At $\zeta = h^*$, we have
$$w(h^*) = \Lambda h^* = 1. \tag{91}$$
Thus, we see that
$$\Lambda = \frac{1}{h^*}. \tag{92}$$
If $h^* = O(1)$, then $\Lambda = O(1)$ and the approximation is consistent. Moreover, since $\langle\alpha\rangle = \alpha_m(1-h^*)$, we have
$$\Lambda = \frac{\langle\alpha\rangle}{\langle\alpha\rangle - \alpha_m}. \tag{93}$$
Note that Λ is large when $\langle\alpha\rangle$ is near α_m.

There must be a transition layer separating the core from the clear fluid. In this transition layer, the concentration of spheres goes from α_m to 0, while $dw/d\zeta$ goes from 0 to Λ. Note that finding the appropriate $\hat{\alpha}$ forces $dw/d\zeta$ to have the proper transition.

Let $\zeta = \zeta^* + \mathrm{St}^{-p}\zeta'$. Using this, we obtain a balance in (83) for $p = \tfrac{1}{2}$, and $dw/d\zeta'$ can be eliminated to give
$$\frac{d^2\hat{\alpha}}{d\zeta'^2} + f_1(\hat{\alpha})\left(\frac{d\hat{\alpha}}{d\zeta'}\right)^2 + f_2(\hat{\alpha}) = 0, \tag{94}$$
where
$$f_1(\hat{\alpha}) = \frac{f''(\hat{\alpha})}{f'(\hat{\alpha})} + \frac{[-(C_p + a_f)\hat{\alpha} + (1-\hat{\alpha})a_s]}{2\hat{\alpha}\left[(1-\hat{\alpha})(C_p + a_s) - a_f\hat{\alpha}\right]}, \tag{95}$$
$$f_2(\hat{\alpha}) = \frac{1}{(1+(\beta-1)\hat{\alpha})f'(\hat{\alpha})\left[(1-\hat{\alpha})(C_p + a_s) - a_f\hat{\alpha}\right]}, \tag{96}$$
$$f(\hat{\alpha}) = \frac{\beta + k}{1+(\beta-1)\hat{\alpha}}. \tag{97}$$

Numerical solutions to (94) are shown in Figure 21.2. The values of the parameters for the dotted curve are
$$a_f = -\frac{1}{5}, \quad a_s = -\frac{1}{5}, \quad C_p = -\frac{9}{32}, \tag{98}$$
and for the solid curve,
$$a_f = -\frac{1}{5}, \quad a_s = -1, \quad C_p = -\frac{1}{4}. \tag{99}$$
Solutions for a_s too small do not exist.

When $\zeta^* = 1 - o(1)$ the clear fluid layer no longer exists, and the wall lies in the transition region. In this case, Λ need not be $O(1)$. If we let $\zeta^* = 1$, we see that the same approximation holds in (82) and (83), leading to the same equation (94) for $\hat{\alpha}$. The difference is that the boundary conditions are not those of matching at $\zeta' \to \pm\infty$ but instead
$$w(0) = 0. \tag{100}$$

21.2. Kinematics and Dynamics of Shearing Flow

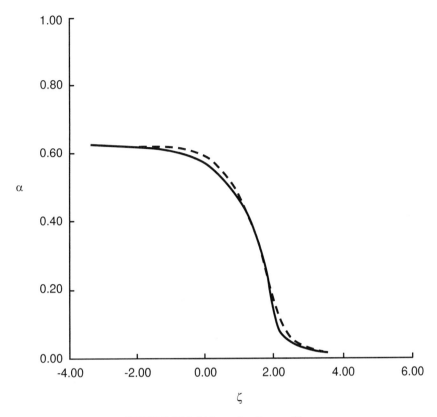

FIGURE 21.2. Volume fraction profile.

In addition, we have

$$\langle \alpha \rangle = \alpha_m + \int_{-\infty}^{0} \hat{\alpha}'(\zeta')\,d\zeta', \tag{101}$$

and

$$1 = w(-\infty) = -\int_{-\infty}^{0} \frac{dw}{d\zeta'}\,d\zeta' = \Lambda\,\mathrm{St}^{-1/2} \int_{-\infty}^{0} \frac{d\zeta'}{1+(\beta-1)\hat{\alpha}}. \tag{102}$$

Figure 21.3 shows the dependence of Λ on $\langle \alpha \rangle$. For $\langle \alpha \rangle$ near α_m the dependence on $\langle \alpha \rangle$ is derived through (94) using the boundary conditions (101) and (102). There is a weak dependence on St.

This analysis suggests a difficulty faced in experimental studies of such problems. If we begin with a uniformly particle-laden fluid and shear it, the concentration profile does not remain uniform. Rather it develops to one that is nonuniform. Generally, the particles move away from the boundary, so that the effective viscosity near the boundary is small and the local shear rate large. Away from the boundary, the converse is true. In the limiting case there are clear layers near the boundary with a concentrated core. Thus the type of gross measurements

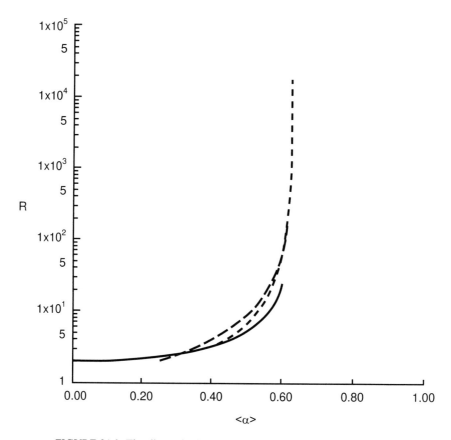

FIGURE 21.3. The dimensionless shear versus average volume fraction.

on boundaries done to measure the viscosity of a single fluid are without meaning for a multicomponent fluid. It is difficult to shear a uniform layer of particle-laden fluid to obtain the shear stress as a function of concentration. Measurements on the interior of the flow field must be made for a multicomponent fluid to find the local viscosity as a function of concentration.

22
Wave Dynamics

22.1. Introduction

The study of waves [73] represents a rich and useful branch of mechanics. For fluids, the simplest branch of this study is the investigation of infinitesimal waves in homogeneous compressible fluids with no surfaces. Assume one-dimensional motion. For such fluids, the governing equations are

$$\frac{\partial \rho}{\partial t} + \frac{\partial \rho v}{\partial z} = 0, \tag{1}$$

$$\rho \left(\frac{\partial v}{\partial t} + v \frac{\partial v}{\partial z} \right) = -\frac{\partial p}{\partial z}. \tag{2}$$

These are supplemented by an "equation of state" of the form

$$p = p(\rho). \tag{3}$$

The problem of linear infinitesimal waves is considered. To do so, drop the convective acceleration from the balance of momentum, and let the density consist of a constant large part and a variable small part $\rho = \rho_0 + \delta\rho$ and assume that the velocity is small, $v = \delta v$. Substituting these assumptions into the governing equations and doing some routine manipulations gives the linear wave equation

$$\frac{\partial^2 \delta\rho}{\partial t^2} = c^2 \frac{\partial^2 \delta\rho}{\partial z^2}, \tag{4}$$

where c is the *wave speed*, given by

$$c^2 = \frac{\partial p}{\partial \rho}(\rho_0).\tag{5}$$

Thus, there is one speed of propagation of linear waves, and the speed is independent of the wavelength, and is therefore a property of the material.

For multicomponent materials, the matter is considerably more complex, so much so that not even a fraction of all possible phenomena have been investigated. To make the investigation easier, we restrict our attention to two components, for the most part bubbly liquids, that is, bubbles of a compressible fluid suspended in another fluid. Some of the reasons for the complexity are obvious. There are two materials, so that the possibility of two wave speeds exists. This can be stated in physical terms also. When the medium transmitting the wave is heterogeneous, the particles, drops, or bubbles making up the dispersed component can act as scatterers, or absorbers, or means of faster or slower propagation. In addition, these materials are coupled mechanically and thermally, leading to the possibility of a wealth of phenomena to be explored.

22.2. Acoustic Propagation

Small amplitude pressure waves propagate through a heterogeneous medium by propagating through both media. The processes occurring at the interface between the two media are important in determining the propagation speed and the attenuation of pressure waves.

Some appreciation of the importance of the presence of bubbles on the propagation of acoustic waves can be gained by doing the following experiment. Fill identical glasses to the same depth, one with plain water, the other with carbonated water. Tap each glass with a spoon. Note that the plain water glass "rings" with a tone that takes a second or so to attenuate. The glass containing carbonated water yields a tone that sounds completely different. It is almost a noise rather than a tone and attenuates in a fraction of the time of attenuation of the glass containing the clear water. In this case the bubbles increase the ability of the medium to dissipate the energy in the acoustic wave, and change the dominant frequency of the system. The fact that this particular system includes a glass complicates the physics, but in fact this particular case happens to demonstrate the physics of the pure materials correctly.

Let us try to model the physics of such materials. For simplicity, let us consider variations in one spatial direction only, and let that direction coincide with that of gravity, which we assume to be the only body force. The governing equations are then

$$\frac{\partial \alpha \rho_d}{\partial t} + \frac{\partial \alpha \rho_d v_d}{\partial z} = 0,\tag{6}$$

22.2. Acoustic Propagation

$$\frac{\partial(1-\alpha)\rho_c}{\partial t} + \frac{\partial(1-\alpha)\rho_c v_c}{\partial z} = 0, \qquad (7)$$

$$\alpha\rho_d\left(\frac{\partial v_d}{\partial t} + v_d\frac{\partial v_d}{\partial z}\right) = -\alpha\frac{\partial p_d}{\partial z} - \frac{\partial}{\partial z}\alpha_d\rho_c C_i(v_d - v_c)^2$$
$$- M_c - \alpha\rho_d g, \qquad (8)$$

$$(1-\alpha)\rho_c\left(\frac{\partial v_c}{\partial t} + v_c\frac{\partial v_c}{\partial z}\right) = -(1-\alpha)\frac{\partial p_c}{\partial z} - \frac{\partial}{\partial z}\alpha\rho_c C_r(v_d - v_c)^2$$
$$- (p_{ci} - p_c)\frac{\partial \alpha}{\partial z} + M_c - (1-\alpha)\rho_c g, \qquad (9)$$

$$\alpha\rho_d\left(\frac{\partial h_d}{\partial t} + v_d\frac{\partial h_d}{\partial z}\right) = -\alpha\left(\frac{\partial p_d}{\partial t} + v_d\frac{\partial p_d}{\partial z}\right) + E_d, \qquad (10)$$

$$(1-\alpha)\rho_c\left(\frac{\partial h_c}{\partial t} + v_c\frac{\partial h_c}{\partial z}\right) = -(1-\alpha)\left(\frac{\partial p_c}{\partial t} + v_c\frac{\partial p_c}{\partial z}\right) - E_d. \qquad (11)$$

Here the equations of state are

$$h_d = h_d(\rho_d, \theta_d), \qquad (12)$$
$$h_c = h_c(\rho_c, \theta_c). \qquad (13)$$
$$p_d = p_d(\rho_d, \theta_d), \qquad (14)$$
$$p_c = p_c(\rho_c, \theta_c). \qquad (15)$$

The interfacial force density is assumed to consist of Stokes drag, virtual mass, a relative acceleration force, bubble dispersion, and a compressibility force [5], and is given by (18.1.17), where the various force densities in (18.1.19) are given by (18.1.20), (18.1.32), (18.1.34), and (18.1.42). Thus,

$$M_c = \alpha S(\alpha)(v_d - v_c)$$
$$+ \alpha C_{vm}\rho_c\left[\left(\frac{\partial v_d}{\partial t} + v_d\frac{\partial v_d}{\partial z}\right) - \left(\frac{\partial v_c}{\partial t} + v_c\frac{\partial v_c}{\partial z}\right)\right]$$
$$+ C_{m1}\alpha\rho_c(v_d - v_c)\frac{\partial v_d - v_c}{\partial z} + C_{m2}\rho_c(v_d - v_c)^2\frac{\partial \alpha}{\partial z}$$
$$+ \frac{3}{R_d}C_{vm}\rho_c(v_d - v_c)\left(\frac{\partial R_d}{\partial t} + v_d\frac{\partial R_d}{\partial z}\right). \qquad (16)$$

The pressure terms are assumed to be governed by the Rayleigh equation

$$p_d = p_{di}, \qquad (17)$$

$$p_{ci} = p_{di}, \qquad (18)$$

$$p_{ci} - p_c = C_p\rho_c|v_c - v_d|^2 + \rho_c\left[R_d\frac{d^2 R_d}{dt^2} + \frac{3}{2}\left(\frac{d R_d}{dt}\right)^2\right]. \qquad (19)$$

The model derived in Section 15.3 gives the coefficients the values

$$C_{vm} = \frac{1}{2}, \quad C_p = \frac{1}{4}, \quad C_r = \frac{1}{5},$$

$$C_i = \frac{3}{10}, \quad C_{m1} = -\frac{1}{10}, \quad C_{m2} = -\frac{3}{10}.$$

The interfacial energy transfer term is taken to be dependent on Joule heating

$$E_d = \alpha H(\theta_c - \theta_d). \tag{20}$$

Finally, the bubble radius and the volume fraction are related to the number density by

$$\alpha = n \frac{4}{3} \pi R_d^3, \tag{21}$$

where the condition that no bubbles break is given by

$$\frac{\partial n}{\partial t} + \frac{\partial n v_d}{\partial z} = 0. \tag{22}$$

The result is a system of equations that are a rather great generalization of the system governing a single compressible fluid. We study its behavior by generalizing the technique used for the single fluid. Thus we linearize the equations by assuming

$$\Psi = \Psi_0 + \delta \Psi, \tag{23}$$

where $\Psi = [\alpha, \rho_d, \rho_c, v_d, v_c, h_d, h_c, R_d]^T$. Here the 0 subscript denotes the steady state, and δ indicates a perturbation. We neglect products of δ-terms. We consider only situations where the rise velocity of the bubbles through the liquid is negligible. Typically this happens when the bubbles are small. In this situation it is reasonable to ignore the gravitational terms, and we do so. We further assume that $v_{d0} = v_{c0} = 0$ and that the other steady-state quantities are independent of z. Thus, we seek solutions of the form

$$\delta \Psi = \hat{\Psi} \exp(i\omega t - ikz), \tag{24}$$

where ω and k are the frequency and wavenumber, respectively.

If (24) is substituted in the system (6)–(11), an equation of the form

$$\mathbf{A}(k, \omega) \hat{\Psi} = 0, \tag{25}$$

results, where $\mathbf{A}(k, \omega)$ is an 8×8 matrix depending on k and ω. The *dispersion relation* is given by det $\mathbf{A}(k, \omega) = 0$. The vector(s) satisfying (25) give the relative amplitudes of the components of the propagating solution.

There are two different physical experiments that can be performed to elucidate the physics of this situation. If a transducer is placed at the bottom of the two-component flow region, the waves are excited at a given frequency (or superposition of several frequencies), the resultant wavenumber may be complex. In this situation, the combination $\Re(k)/\omega = 1/c$ is the *celerity*, defined as the reciprocal of the wave speed. The quantity $\Im(k)$ is the spatial attenuation. If, on the other hand, the physical experiment starts with a spatially distributed pressure variation as an initial condition, with homogeneous boundary conditions at the top and bottom of the flow region, the analysis uses the (complex) Fourier representation of the initial pressure and derivatives to get amplitudes and wavenumbers of the initial pressure. Then the dispersion relation gives the (complex) frequencies ω, and the speed of sound is given by $\Re(\omega)/k$, and $\Im(\omega)$ is the decay constant.

22.2.1. Special Cases

The general dispersion relation potentially gives nine "speeds of sound" [21]. Of these, all but two (one pair) have speeds close to the speeds of the components, which we have assumed to be zero here. Of the small speeds, many can be identified as convective velocities. Two of these small speeds are the so-called volumetric[1] wave speeds. These waves will be discussed more fully in the next subsections. In the next two sub-subsections we shall discuss the sound speeds and the effects of the modeling assumptions and the sizes of the parameters on them. All the results we shall discuss are based on the neglect of interfacial heat transfer, that is, we shall assume $E_d = -E_c = 0$. With this assumption, the energy equations become

$$\rho_{d0} c_{pd} \hat{\theta}_d = \hat{p}_d. \tag{26}$$

$$\rho_{c0} c_{pc} \hat{\theta}_c = \hat{p}_c, \tag{27}$$

where c_{pd} and c_{pc} are the specific heats of the bubbles and the liquid, respectively, at constant pressure. The equations of state are linearized by substituting $p_d = p_{d0} + \hat{p}_d$, $p_c = p_{c0} + \hat{p}_c$, $\theta_d = \theta_{d0} + \hat{\theta}_d$, and $\theta_c = \theta_{c0} + \hat{\theta}_c$, and neglecting products of the fluctuating quantities. We have

$$\rho_{d0} + \hat{\rho}_d = \rho_d(p_{d0} + \hat{p}_d, \theta_{d0} + \hat{\theta}_d)$$
$$\approx \rho_d(p_{d0}, \theta_{d0}) + \frac{\partial \rho_d}{\partial p_d}(p_{d0}, \theta_{d0})\hat{p}_d + \frac{\partial \rho_d}{\partial \theta_d}(p_{d0}, \theta_{d0})\hat{\theta}_d. \tag{28}$$

$$\rho_{c0} + \hat{\rho}_c = \rho_c(p_{c0} + \hat{p}_c, \theta_{c0} + \hat{\theta}_c)$$
$$\approx \rho_c(p_{c0}, \theta_{c0}) + \frac{\partial \rho_c}{\partial p_c}(p_{c0}, \theta_{c0})\hat{p}_c + \frac{\partial \rho_c}{\partial \theta_c}(p_{c0}, \theta_{c0})\hat{\theta}_c. \tag{29}$$

Substituting $\hat{\theta}_d$ and $\hat{\theta}_c$ in terms of $\hat{p}_d$ and $\hat{p}_c$ from (26) and (27) gives

$$\hat{\rho}_d = c_d^{-2}\hat{p}_d, \tag{30}$$

$$\hat{\rho}_c = c_c^{-2}\hat{p}_c, \tag{31}$$

where c_d and c_c are the isentropic speeds of sound for the dispersed and continuous components, respectively. They are defined by

$$c_d^{-2} = \left.\frac{\partial \rho_d}{\partial p_d}\right|_{s_d}$$

---

[1] Also called kinematic wave speeds.

$$= \left.\frac{\partial \rho_d}{\partial p_d}\right|_{\theta_d} + \frac{\partial \theta_d}{\partial p_d}\left.\frac{\partial \rho_d}{\partial \theta_d}\right|_{p_d}$$

$$= \frac{\partial \rho_d}{\partial p_d}(p_{d0}, \theta_{d0}) + \frac{1}{c_{pd}\rho_{d0}}\frac{\partial \rho_d}{\partial \theta_d}(p_{d0}, \theta_{d0}), \tag{32}$$

$$c_c^{-2} = \left.\frac{\partial \rho_c}{\partial p_c}\right|_{s_c}$$

$$= \left.\frac{\partial \rho_c}{\partial p_c}\right|_{\theta_c} + \frac{\partial \theta_c}{\partial p_c}\left.\frac{\partial \rho_c}{\partial \theta_c}\right|_{p_c}$$

$$= \frac{\partial \rho_c}{\partial p_c}(p_{c0}, \theta_{c0}) + \frac{1}{c_{pc}\rho_{c0}}\frac{\partial \rho_c}{\partial \theta_c}(p_{c0}, \theta_{c0}). \tag{33}$$

Low-Frequency Homogeneous Flow

Let us assume that there is no diffusion, that is, the velocities of the two components are the same. This is called the *homogeneous flow model*. Then the two momentum equations can be replaced by their sum.[2] We shall also assume that $E_d = -E_c = 0$. This assumption means that no heat transfer occurs across the interface. The physical rationalization of this is that the gas inside the bubbles is a poor conductor of heat. Clearly, at sufficiently low frequency this assumption will be violated. However, experience with physical systems indicates that it gives good results in the range of tens to hundreds of Hertz. Finally, we shall assume that $p_d = p_c = p$. This assumption neglects the effects of bubble expansion or contraction on the pressure. We use the term "low frequency" to mean the neglect of the right-hand side of the Rayleigh equation (19).

With $\hat{v}_d = \hat{v}_c = \hat{v}$ and $\hat{p}_d = \hat{p}_c = p$, the two conservation of mass equations become

$$\omega\hat{\alpha} - k\alpha_0\hat{v} + \alpha_0\omega\frac{\hat{\rho}_d}{\rho_{d0}} = 0, \tag{34}$$

$$-\omega\hat{\alpha} - k(1-\alpha_0)\hat{v} + (1-\alpha_0)\omega\frac{\hat{\rho}_c}{\rho_{c0}} = 0, \tag{35}$$

and the sum of the conservation of momentum equations yields the equation of conservation of mixture momentum

$$[\alpha_0\rho_{d0} + (1-\alpha_0)\rho_{c0}]\omega\hat{v} = k\hat{p}. \tag{36}$$

Adding the two mass equations and substituting $\hat{v}$, $\hat{\rho}_d$, and $\hat{\rho}_c$ in terms of $\hat{p}$ gives

$$\left\{-\frac{k^2}{\omega[\alpha_0\rho_{d0} + (1-\alpha_0)\rho_{c0}]} + \omega\left(\frac{\alpha_0}{\rho_{d0}c_d^2} + \frac{1-\alpha_0}{\rho_{c0}c_c^2}\right)\right\}\hat{p} = 0. \tag{37}$$

---

[2]This model can be rationalized by assuming a large value for the drag coefficient S.

Thus, for nontrivial pressure waves (i.e., $\hat{p} \neq 0$), we must have

$$\frac{\omega}{k} = \pm c_{2\phi},$$

where the homogeneous two-component wave speed is

$$c_{2\phi} = \left\{ [\alpha_0 \rho_{d0} + (1-\alpha_0)\rho_{c0}] \left(\frac{\alpha_0}{\rho_{d0} c_d^2} + \frac{1-\alpha_0}{\rho_{c0} c_c^2} \right) \right\}^{-1/2}. \tag{38}$$

These values are the propagation speeds for pressure waves in bubbly flow. The "speed of sound" is given by the positive value. The speed of sound is shown in Figure 22.1 for air–water flow for various values of α_0. Note that the two-component speed of sound is much lower than the speed of sound in either medium. Also note that there is no attenuation in this sound propagation model. If we assume that this speed is also the critical velocity for choked two-component flow, this low speed of sound explains why champagne spurting from the top of a newly opened bottle is flowing at the speed of sound (so-called "choked flow"), but not at the (potentially dangerous!) speed of sound for the liquid.

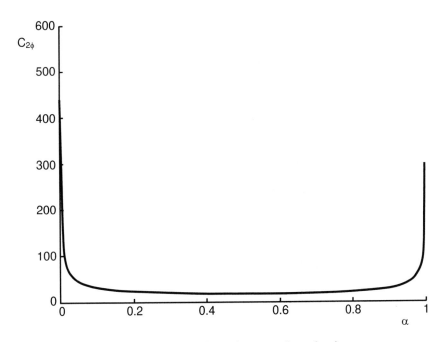

FIGURE 22.1. Speed of sound versus volume fraction.

Bubble Resonance

At higher frequencies, bubble resonance occurs. We shall derive a simplified model for this phenomenon. The linearized Rayleigh equation is

$$\hat{p}_d - \hat{p}_c = -\omega^2 \rho_{c0} R_{d0} \hat{R}_d . \tag{39}$$

If we assume that the density of the gas is given by

$$\hat{\rho}_d = c_d^{-2} \hat{p}_d , \tag{40}$$

and that the mass in a single bubble is independent of t and z, we have

$$m = \rho_d \frac{4}{3}\pi R_d^3 , \tag{41}$$

so that

$$\hat{\rho}_d \frac{4}{3}\pi R_{d0}^3 + \rho_{d0} 4\pi R_{d0}^2 \hat{R}_d = 0 . \tag{42}$$

Substituting (40) into (42), solving for $\hat{p}_d$, and substituting this into Rayleigh's equation (39), we have

$$\hat{p}_c = \left(\omega^2 \rho_{c0} R_{d0} - \frac{3\rho_{d0} c_d^2}{R_{d0}}\right) \hat{R}_d . \tag{43}$$

If we assume $\hat{p}_c = 0$, we have a nontrivial solution if

$$\omega = \pm \frac{c_d}{R_{d0}} \left(\frac{3\rho_{d0}}{\rho_{c0}}\right)^{1/2} . \tag{44}$$

This is the condition for bubble resonance in a nonpropagating situation, *i.e.*, when the wavelength of the disturbance is infinite.

For propagating acoustic waves, we couple the linearized Rayleigh equation (39) with the momentum equation

$$[\alpha_0 \rho_{d0} + (1 - \alpha_0)\rho_{c0}]\omega \hat{v} = k[\alpha_0 \hat{p}_d + (1 - \alpha_0)\hat{p}_c] , \tag{45}$$

and the two conservation of mass equations

$$\omega \hat{\alpha} - k\alpha_0 \hat{v} + \alpha_0 \omega \frac{\hat{p}_d}{\rho_{d0} c_d^2} = 0 , \tag{46}$$

$$-\omega \hat{\alpha} - k(1-\alpha_0)\hat{v} + (1-\alpha_0)\omega \frac{\hat{p}_c}{\rho_{c0} c_c^2} = 0 . \tag{47}$$

The relation among α, n, and R_d comes from linearizing (21) and (22), to obtain

$$\hat{\alpha} = \hat{n}\frac{4}{3}\pi R_{d0}^3 + n_0 4\pi R_{d0}^2 \hat{R}_d , \tag{48}$$

and

$$\omega \hat{n} - kn_0 \hat{v} = 0 . \tag{49}$$

Substituting $\hat{n}$ from (49) into (48) gives

$$\hat{\alpha} = \frac{k}{\omega}\hat{v} n_0 \frac{4}{3}\pi R_{d0}^3 + n_0 4\pi R_{d0}^2 \hat{R}_d = \frac{k}{\omega}\alpha_0 \hat{v} + \frac{3\alpha_0}{R_{d0}}\hat{R}_d . \tag{50}$$

22.2. Acoustic Propagation

Substituting this relation into the two equations of conservation of mass (46) and (47) yields

$$\frac{3\hat{R}_d}{R_{d0}} + \frac{\hat{p}_d}{\rho_{d0}c_d^2} = 0, \tag{51}$$

$$-\frac{3\omega\alpha_0}{R_{d0}}\hat{R}_d - k\hat{v} + (1-\alpha_0)\omega\frac{\hat{p}_c}{\rho_{c0}c_c^2} = 0. \tag{52}$$

If we now substitute (36) into (52) we have

$$-\frac{3\omega\alpha_0}{R_{d0}}\hat{R}_d - \frac{k^2[\alpha_0\hat{p}_d + (1-\alpha_0)\hat{p}_c]}{\omega[\alpha_0\rho_{d0} + (1-\alpha_0)\rho_{c0}]} + (1-\alpha_0)\omega\frac{\hat{p}_c}{\rho_{c0}c_c^2} = 0. \tag{53}$$

Equations (39), (51), and (53) are three linear equations in the three variables $\hat{p}_d$, $\hat{p}_c$, and $\hat{R}_d$. The condition for a solution gives the dispersion relation [67]

$$\left\{\frac{\omega^2}{k^2}\left[\frac{\alpha_0}{\rho_{d0}c_d^2} + \frac{(1-\alpha_0)}{\rho_{c0}c_c^2}\right] - \frac{1}{\alpha_0\rho_{d0} + (1-\alpha_0)\rho_{c0}}\right\}$$

$$+ \omega^2\frac{(1-\alpha_0)R_{d0}^2}{3\rho_{d0}c_d^2}\left[\frac{\rho_{c0}}{\alpha_0\rho_{d0} + (1-\alpha_0)\rho_{c0}} - \frac{\omega^2}{k^2}\frac{1}{c_c^2}\right] = 0. \tag{54}$$

Solving for ω/k as a function of ω gives

$$\frac{\omega^2}{k^2} = c_{2\phi}^2 \frac{1 - a_1\omega^2}{1 - a_2\omega^2}, \tag{55}$$

where

$$a_1 = \frac{(1-\alpha_0)c_{2\phi}^2 R_{d0}^2}{3\rho_{d0}c_d^2 c_c^2}\left(\frac{\alpha_0}{\rho_{d0}c_d^2} + \frac{1-\alpha_0}{\rho_{c0}c_c^2}\right)^{-1},$$

and

$$a_2 = \frac{(1-\alpha_0)\rho_{c0} R_{d0}^2}{3\rho_{d0}[\alpha_0\rho_{d0} + (1-\alpha_0)\rho_{c0}]c_d^2}\left(\frac{\alpha_0}{\rho_{d0}c_d^2} + \frac{1-\alpha_0}{\rho_{c0}c_c^2}\right)^{-1}.$$

Note that since $c_{2\phi} < c_c$, we have $a_1 > a_2$.

For ω sufficiently small, the speed of sound is approximately that derived in Section 22.2.1 above. There is no attenuation. As the frequency increases, the sound speed increases, until at $\omega^- = \sqrt{a_1}$, the sound speed is infinite. For larger frequencies, sound will be attenuated. With zero attenuation at ω^-. The attenuation increases until the frequency reaches $\omega^+ = \sqrt{a_2}$, where the attenuation becomes infinite, and propagation starts again. For frequencies greater than ω^+, the sound speed is less than $c_{2\phi}$, again with no attenuation. The situation is shown in Figures 22.2 and 22.3. The dashed curve indicates the model (55), and the solid curve shows a full solution of $\det \mathbf{A} = 0$.

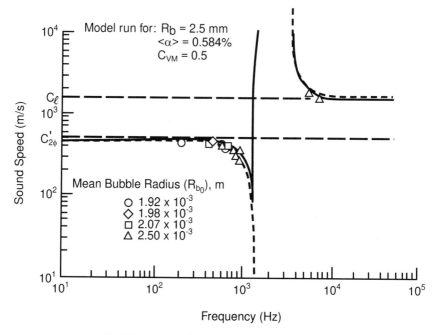

FIGURE 22.2. Speed of sound versus frequency.

22.3. Volumetric Waves

One motivation for the study of waves is the appearance of complex characteristics in the "simplest" model for multicomponent flow, discussed in Section 20.2. Complex characteristics manifest themselves in the inability of the model to predict the propagation of small disturbances.

22.3.1. Kinematic Waves

One means of making models capable of predicting wave motions and stability of multicomponent flows is to add in the missing physics that causes the simplest model to have complex characteristics. These models are discussed in Section 20.2. A second approach to the problem is to use an even simpler model to describe the waves. The model is analogous to the the drift-flux model. For our purposes, we take the drift-flux model to be that given by neglecting the inertia of each component, and assuming that the densities of each component are constant. If the densities are assumed to be constant, then (6) and (7) can be integrated to yield

$$\alpha v_d + (1 - \alpha)v_c = j_0 . \tag{56}$$

The remaining equations have the form

$$\frac{\partial \alpha}{\partial t} + \frac{\partial \alpha v_d}{\partial z} = 0, \tag{57}$$

22.3. Volumetric Waves 283

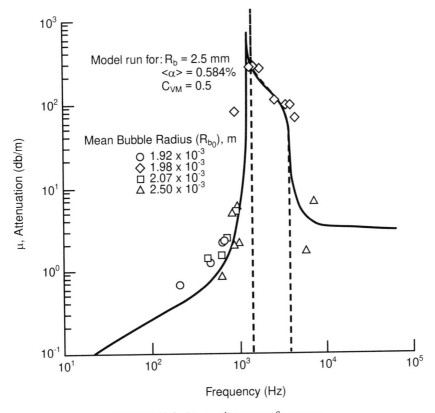

FIGURE 22.3. Attenuation versus frequency.

$$0 = -\alpha \frac{\partial p}{\partial z} + M_d - \alpha \rho_d g, \qquad (58)$$

$$0 = -(1-\alpha)\frac{\partial p}{\partial z} - M_d - (1-\alpha)\rho_c g, \qquad (59)$$

where the drag is given by

$$M_d = \alpha S(\alpha)(v_d - v_c). \qquad (60)$$

The volumetric flux parameter j_0 can be a function of time t; however, we shall assume that it is constant. Eliminating $\partial p/\partial z$ from (58) and (59), and solving for $v_d - v_c$ in terms of α gives

$$v_d - v_c = \frac{(1-\alpha)(\rho_c - \rho_d)g}{S(\alpha)}. \qquad (61)$$

Solving (56) for $v_d - v_c$ in terms of v_d and α gives

$$v_d = j_0 + (1-\alpha)(v_d - v_c) = j_0 + \frac{(1-\alpha)^2}{S(\alpha)}(\rho_c - \rho_d)g. \qquad (62)$$

Thus v_d is a function of α. Differentiating αv_d with respect to z, and substituting in (57) gives

$$\frac{\partial \alpha}{\partial t} + \left(v_d + \alpha \frac{\partial v_d}{\partial \alpha} \right) \frac{\partial \alpha}{\partial z} = 0. \tag{63}$$

The literature on this equation is rich; see, for example, Whitham [97]. Wallis [94] gives a discussion relevant to two-component flow. We summarize that discussion here.

If the initial data is smooth for $-\infty < z < \infty$, then for a short time, the solution will be given by the following procedure, from the method of characteristics. On a characteristic, given by

$$\frac{dz}{dt} = a = v_d + \alpha \frac{\partial v_d}{\partial \alpha}, \tag{64}$$

the value of α is constant. If the characteristics are "spreading," as shown in Figure 22.4, then the values of α over the varying part of the solution at any time t are getting farther apart. This sort of phenomenon is called a *rarefaction*. If, on the other hand, the characteristics are "converging," as shown in Figure 22.5, then the solution cannot stay smooth and single valued. Some call this *wave breaking*. Since the volume fraction cannot be multiple valued, we must relax the requirement that the solution be smooth. Specifically, we allow the solution to be discontinuous along some curve(s) called "shocks." We note here that the term "shock" has a very special meaning to some, namely, a discontinuity of density in gas dynamics. Here we use the term in its mathematical sense to describe a propagating discontinuity in the solution to a hyperbolic partial differential equation.

A shock will evolve out of an initial profile for $\alpha(z, 0) = \alpha_0(x)$ if there is a region where the change in α is such that the characteristics converge. If the region is such that characteristics move away from the wave, the region will become less steep. If the initial condition is a discontinuity, the resulting wave will not stay as a shock, but will immediately smooth out and become smoother (i.e., the change will be less abrupt). This is sometimes called a "fan," but we shall refer to it as a *rarefaction*. On the other hand, the discontinuous initial condition will remain a discontinuity if the characteristics approach the shock as time increases. A discontinuous initial condition will allow the construction of a large number of solutions that remain discontinuous along one (or more) curves in the x–t plane. However, not all of these constructions lead to physically realistic solutions.

The shock that does correspond to the physics will conserve the appropriate quantities across the shock.

22.4. Characteristics and Linear Stability

For more complicated systems, such as the 4×4 system given by (6)–(9) with constant densities, the description of shocks and rarefactions is considerably more complex than that given in Section 22.3.1.

22.4. Characteristics and Linear Stability

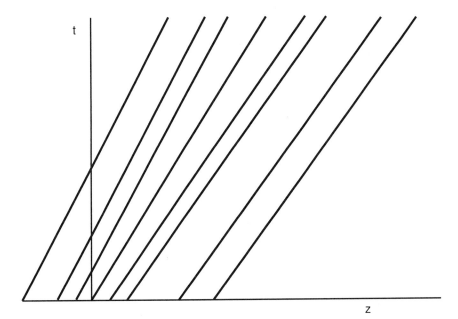

FIGURE 22.4. Spreading characteristics.

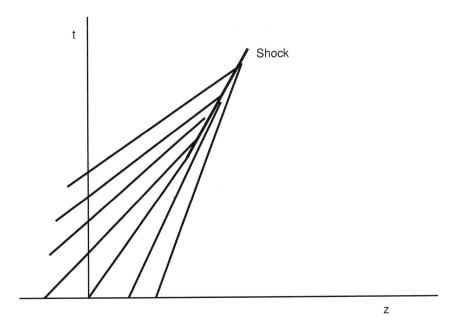

FIGURE 22.5. Converging characteristics, leading to a shock.

We shall consider the equations of motion with the inertia of the dispersed component neglected. A model of this type was studied by Pauchon and Banerjee [70], [71], and by Park [66]. The system can be written as

$$\frac{\partial(1-\alpha)}{\partial t} + \frac{\partial(1-\alpha)v_c}{\partial z} = 0, \tag{65}$$

$$\alpha v_d + (1-\alpha)v_c = j_0, \tag{66}$$

$$\alpha\rho_d\left(\frac{\partial v_d}{\partial t} + v_d\frac{\partial v_d}{\partial z}\right) = -\alpha\frac{\partial p_d}{\partial z} - \frac{\partial}{\partial z}\alpha_d\rho_c C_i(v_d - v_c)^2 + \alpha S(\alpha)(v_c - v_d)$$

$$+ \alpha C_{vm}\rho_c\left[\left(\frac{\partial v_c}{\partial t} + v_c\frac{\partial v_c}{\partial z}\right) - \left(\frac{\partial v_d}{\partial t} + v_d\frac{\partial v_d}{\partial z}\right)\right]$$

$$- C_{m1}\alpha\rho_c(v_d - v_c)\frac{\partial(v_d - v_c)}{\partial z}$$

$$- C_{m2}\rho_c(v_d - v_c)^2\frac{\partial\alpha}{\partial z} - \alpha\rho_d g, \tag{67}$$

$$(1-\alpha)\rho_c\left(\frac{\partial v_c}{\partial t} + v_c\frac{\partial v_c}{\partial z}\right) = -(1-\alpha)\frac{\partial p_c}{\partial z} - \frac{\partial}{\partial z}\alpha\rho_c C_r(v_d - v_c)^2$$

$$- (p_{ci} - p_c)\frac{\partial\alpha}{\partial z} + \alpha S(\alpha)(v_d - v_c)$$

$$+ \alpha C_{vm}\left[\left(\frac{\partial v_d}{\partial t} + v_d\frac{\partial v_d}{\partial z}\right) - \left(\frac{\partial v_c}{\partial t} + v_c\frac{\partial v_c}{\partial z}\right)\right]$$

$$+ C_{m1}\alpha\rho_c(v_d - v_c)\frac{\partial(v_d - v_c)}{\partial z}$$

$$+ C_{m2}\rho_c(v_d - v_c)^2\frac{\partial\alpha}{\partial z} - (1-\alpha)\rho_c g, \tag{68}$$

$$p_d = p_c - C_p\rho_c(v_d - v_c)^2. \tag{69}$$

This model includes the inertial effects known to be important in the two-component models of the type derived in Section 15.3.2. We also use the integrated form of the equation of balance of total volume (66) with the volumetric flow assumed to be constant.

This system can be written as

$$\frac{\partial\Psi}{\partial t} + \mathbf{B}\frac{\partial\Psi}{\partial z} = \mathbf{C}, \tag{70}$$

where

$$\Psi = \begin{bmatrix} 1-\alpha \\ v_c \end{bmatrix}, \tag{71}$$

22.4. Characteristics and Linear Stability

$$\mathbf{B} = \begin{bmatrix} v_c & 1-\alpha \\ (v_d - v_c)^2 K_1(\alpha) & v_c + (v_d - v_c) K_2(\alpha) \end{bmatrix}, \quad (72)$$

$$\mathbf{C} = \frac{\alpha}{\alpha(1-\alpha) + C_{vm}} \begin{bmatrix} 0 \\ -g\alpha(1-\alpha) + \alpha S(\alpha)(v_d - v_c) \end{bmatrix},$$

$$K_1(\alpha) = \frac{C_{m2} - C_{m1} + C_p(\alpha - 2) - \alpha C_r + C_i(1-\alpha) + C_{vm}}{C_{vm} + \alpha - \alpha^2}, \quad (73)$$

$$K_2(\alpha) = \frac{-2\alpha C_r + 2(\alpha - 1)C_p + 2C_i(1-\alpha) - C_{m1} + 2(1-\alpha)C_{vm}}{C_{vm} + \alpha - \alpha^2} \quad (74)$$

$$(v_d - v_c) = \frac{j_0 - v_c}{\alpha}. \quad (75)$$

As discussed in Section 20.2, this system has characteristics given by

$$\det(\mathbf{B} - \lambda \mathbf{I}) = 0. \quad (76)$$

If we nondimensionalize the characteristics with

$$\lambda^* = \frac{\lambda - v_c}{v_d - v_c}, \quad (77)$$

then $\lambda^* = \lambda^*(\alpha)$ turns out to be a function of α only. The expression for λ^* is

$$\lambda^*_{\pm} = \frac{1}{2} \left(K_2(\alpha) \pm \sqrt{K_2^2(\alpha) + 4(1-\alpha) K_1(\alpha)} \right). \quad (78)$$

The graph of λ^* versus α is that shown in Figure 20.2. Note that for each value of α, there are two values of λ^*, each corresponding to a curve in x–t space given by $dx/dt = \lambda$. Thus, the more complicated model has more waves than the kinematic wave model.

We note here that the eigenvectors can be expressed as

$$\mathbf{R} = \begin{bmatrix} 1 - \alpha \\ v_c - \lambda \end{bmatrix}. \quad (79)$$

22.4.1. Linear Stability

The relative role of the kinematic wave and the characteristics can be seen more clearly by considering the stability of the steady, uniform state. If we let $\alpha = \alpha_0 + \delta\alpha$ and $v_c = v_{c0} + \delta v_c$, and linearize by neglecting products of the δ terms, we can then eliminate δv_c and obtain the system

$$\frac{\partial \delta\alpha}{\partial t} + a_0 \frac{\partial \delta\alpha}{\partial z} + T \left(\frac{\partial}{\partial t} + \lambda_+ \frac{\partial}{\partial z} \right) \left(\frac{\partial}{\partial t} + \lambda_- \frac{\partial}{\partial z} \right) \delta\alpha = 0, \quad (80)$$

where a_0 is the kinematic wave speed given by (64) and λ_+ and λ_- are the two characteristic speeds, given by (78). All quantities are evaluated at $\alpha = \alpha_0$, $v_c =$

v_{c0}. Here T is a relaxation time given by

$$\frac{1}{T} = \frac{S(\alpha_0)}{\alpha_0(1-\alpha_0) + C_{vm}}. \tag{81}$$

This equation for $\delta\alpha$ possesses a "wave hierarchy" [57], [97], where for $T = 0$ the wave is the kinematic wave, while for $T = \infty$, there are two linear waves, each traveling at the characteristic speeds. In order to analyze this equation, we form the dispersion relation by letting $\delta\alpha = \alpha' e^{ikx} e^{-i\omega t}$. The relation between frequency ω and wavenumber k is given by

$$\omega - a_0 k + iT[\omega^2 - (\lambda_+ + \lambda_-)\omega k + \lambda_+\lambda_- k^2] = 0. \tag{82}$$

For real k and complex ω, we shall refer to $\Re(\omega/k)$ as the celerity, and $\Im(\omega)$ as the attenuation. If $\lambda_- < a_0 < \lambda_+$, the wave is stable, with the slowest decay for the lowest frequency waves, while if $a_0 > \lambda_+$, all the waves faster than the characteristic, that is, with $\Re(\omega/k) > \lambda_+$, are amplified. The wave having the most amplification is the one with $\Re(\omega/k) = \lambda_+$. If $\lambda_+ \to \lambda_-$, as it does at the nose of the curve of λ^* versus α, then $\Im(\omega) \to \infty$ as $k \to \infty$. Thus, the appearance of complex characteristics corresponds to infinite growth rates, and the frequencies involved become infinite. This is consistent with the example of Hadamard discussed in Section 20.2.

22.5. Inlet Step Response

We wish to study the behavior of the system (70) for $-0 < x < \infty, t > 0$, subject to the initial condition $\Psi(x, 0) = \Psi^+$ for $x > 0$, and the boundary condition $\Psi(0, t) = \Psi^-$ for $t > 0$. The physical situation that is usually cited for this initial-boundary-value problem corresponds to allowing the system to reach a steady state $\Psi = \Psi^+$ and then, at $t = 0$, changing the conditions at the inlet to $\Psi = \Psi^-$. This problem is an initial-boundary-value problem. However, the initial-value problem specified above is more easily studied. In general, we expect to see several waves emanate from the origin in the x–t plane.

The theory is relatively well developed for homogeneous, or unforced systems, with $\mathbf{C} = 0$, which are in conservation form. A system is said to be in conservation form if it is in the form

$$\frac{\partial \Upsilon}{\partial t} + \frac{\partial \mathbf{Q}(\Upsilon)}{\partial z} = 0. \tag{83}$$

The system (65) and (69) is not in this form; however, a transformation of dependent variables yields an equivalent system that is in conservation form.

We note that the equation of conservation of mass for the continuous component is in conservation form; consequently,

$$\Upsilon_1 = 1 - \alpha, \tag{84}$$

$$Q_1 = (1-\alpha)v_c. \tag{85}$$

22.5. Inlet Step Response

A second conserved quantity can be found by assuming that the conserved quantity is of the form $\Upsilon_2 = f(\alpha)(j_0 - v_c)$, where $f(\alpha)$ is to be determined. The flux can be determined by differentiating Υ_2 with respect to t, and using (70). We have

$$\frac{\partial}{\partial t}\Upsilon_2 = f'(\alpha)\frac{\partial \alpha}{\partial t}(j_0 - v_c) - f(\alpha)\frac{\partial v_c}{\partial t}$$

$$= f'(\alpha)(j_0 - v_c)\left(-v_c\frac{\partial \alpha}{\partial z} + (1-\alpha)\frac{\partial v_c}{\partial z}\right)$$

$$+ f(\alpha)\left[-\frac{K_1(j_0 - v_c)^2}{\alpha^2}\frac{\partial \alpha}{\partial z} - \left(v_c + \frac{K_2(j_0 - v_c)}{\alpha}\right)\frac{\partial v_c}{\partial z}\right]$$

$$= -\frac{\partial Q_2}{\partial \alpha}\frac{\partial \alpha}{\partial z} - \frac{\partial Q_2}{\partial v_c}\frac{\partial v_c}{\partial z}. \tag{86}$$

Thus,

$$\frac{\partial Q_2}{\partial v_c} = f(\alpha)\left(v_c + \frac{K_2(j_0 - v_c)}{\alpha}\right) - (1-\alpha)(j_0 - v_c)f'(\alpha). \tag{87}$$

Integrating gives

$$Q_2 = f(\alpha)\left(\frac{1}{2}v_c^2 + \frac{K_2(j_0 v_c - \frac{1}{2}v_c^2)}{\alpha}\right) - (1-\alpha)(j_0 v_c - \frac{1}{2}v_c^2)f'(\alpha) + g(\alpha). \tag{88}$$

Differentiating (88) with respect to α, we have

$$\frac{\partial Q_2}{\partial \alpha} = f'(\alpha)\left(\frac{1}{2}v_c^2 + \frac{K_2(j_0 v_c - \frac{1}{2}v_c^2)}{\alpha}\right) - f(\alpha)\frac{K_2(j_0 v_c - \frac{1}{2}v_c^2)}{\alpha^2}$$

$$- \frac{\partial}{\partial \alpha}\left[(1-\alpha)f'(\alpha)\right](j_0 v_c - \frac{1}{2}v_c^2) + g'(\alpha)$$

$$= f'(\alpha)(j_0 - v_c)v_c + f(\alpha)\frac{K_1(j_0 - v_c)^2}{\alpha^2}. \tag{89}$$

Thus,

$$g'(\alpha) = \frac{f(\alpha)K_1 j_0^2}{\alpha^2} \tag{90}$$

and

$$\frac{\partial}{\partial \alpha}\left[(1-\alpha)f'(\alpha) - \frac{f(\alpha)K_2}{\alpha}\right] - 2f'(\alpha) - \frac{2K_1 f(\alpha)}{\alpha^2} = 0. \tag{91}$$

This is a second-order ordinary differential equation for $f(\alpha)$ which has a regular singular point at $\alpha = 0$. The solution(s) that are acceptable are not singular at $\alpha = 0$. From the function $f(\alpha)$, the flux Q_2 can be obtained from (88). A typical graph of $f(\alpha)$ is shown in Figure 22.6.

The second conserved quantity is proportional to the relative velocity $v_d - v_c$. The "density" of the conserved quantity is a complicated function of α.

According to the theory for conservation laws for systems of dimension n in one space dimension [57], the wave consists of a total of n shocks and rarefactions. For

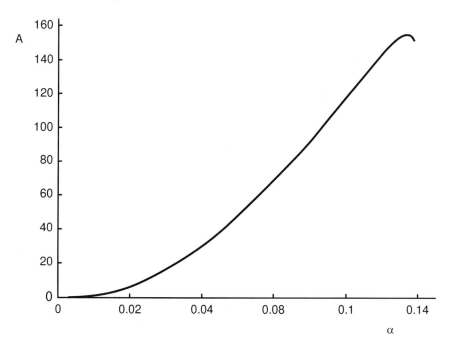

FIGURE 22.6. "Density" of the conserved momentum.

$n = 2$, there will be a shock followed by a rarefaction, or a rarefaction followed by a shock. To determine the structure of the wave, consider the characteristics emanating from behind the discontinuity at $t = 0$ in the x–t plane, with conditions given by Υ^-. Note that there are two sets of these, since there are two characteristics λ^* for each α. Similarly, there are two sets of characteristics emanating from just ahead of the discontinuity. According to the theory, one of these sets will come together, and one set will diverge. By shock stability arguments, the sets of characteristics that come together signal the appearance and propagation of a shock, while the sets that diverge from each other lead to a rarefaction. The intermediate state Υ^i must be compatible with both the shock and the rarefaction. We illustrate the wave in Figure 22.7, for the situation where the shock is "fast," and the rarefaction is "slow."

22.5.1. Shock

If the speed of the shock is s, conservation of Υ across the shock requires that

$$-s[\Upsilon] + [Q] = 0. \tag{92}$$

where [ ] denotes the jump across the shock. Thus,

$$s = \frac{[(1-\alpha)v_c]}{[(1-\alpha)]} = \frac{[Q_2]}{[\Upsilon_2]}. \tag{93}$$

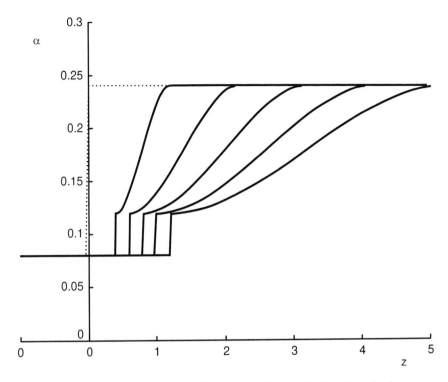

FIGURE 22.7. Evolution of an initial discontinuity as a shock and rarefaction.

For given α^- and v_c^-, (93) gives v_c^i as a function of α^i. Indeed, in the system described above, the equation for v_c^i is quadratic, where the coefficients involve the function $f(\alpha)$. If we treat α^i as a parameter, the curve $v_c^i(\alpha^i)$, called a shock locus, describes the possible intermediate states that can be reached from (α^-, v_c^-). Note that the point (α^-, v_c^-) lies on this curve. Several of these curves, for different (α^-, v_c^-), are shown in Figure 22.8.

22.5.2. Rarefaction

Since the "fan" emanates from the origin in the x–t plane, $\Upsilon = \hat{\Upsilon}(x/t)$, where $x/t = \sigma$ is the slope of the characteristic through the origin. Thus, on any characteristic having eigenvalue λ, we have $\sigma = \lambda$. If we differentiate $\hat{\Upsilon}(x/t)$ with respect to t and x, and substitute into (83), we have

$$\frac{\partial \Upsilon}{\partial \sigma}\left(-\frac{x}{t^2}\right) + \frac{\partial \mathbf{Q}}{\partial \Upsilon}\frac{\partial \Upsilon}{\partial \sigma}\left(\frac{1}{t}\right) = 0. \qquad (94)$$

Thus, $\partial \Upsilon/\partial \sigma$ is a right eigenvector of the matrix $\partial \mathbf{Q}/\partial \Upsilon$, and, as suggested above, $\sigma = x/t = \lambda$ is an eigenvalue.

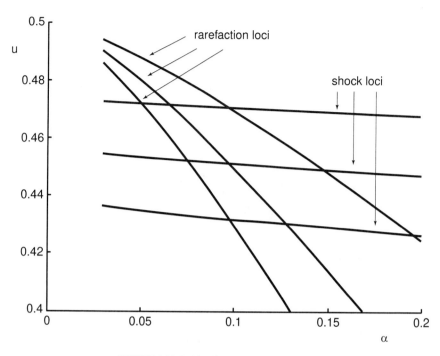

FIGURE 22.8. Shock and rarefaction loci.

If we write the system as

$$\frac{\partial \Upsilon}{\partial t} + \frac{\partial \mathbf{Q}}{\partial \Upsilon}\frac{\partial \Upsilon}{\partial z} = \frac{\partial \Upsilon}{\partial \Psi}\frac{\partial \Psi}{\partial t} + \frac{\partial \mathbf{Q}}{\partial \Upsilon}\frac{\partial \Upsilon}{\partial \Psi}\frac{\partial \Psi}{\partial z}, \qquad (95)$$

this equation can be recognized in terms of the original system (70) if

$$\mathbf{B} = \frac{\partial \Psi}{\partial \Upsilon}\frac{\partial \mathbf{Q}}{\partial \Upsilon}\frac{\partial \Upsilon}{\partial \Psi}. \qquad (96)$$

Then the eigenvalues of $\partial \mathbf{Q}/\partial \Upsilon$ are the eigenvalues of $\mathbf{B}$, and if $\mathbf{R}$ is a right eigenvector of $\partial \mathbf{Q}/\partial \Upsilon$, then $\partial \Psi/\partial \Upsilon \mathbf{R}$ is an eigenvector of $\mathbf{B}$. We note that since $\partial \Upsilon/\partial \sigma$ is a right eigenvector of $\partial \mathbf{Q}/\partial \Upsilon$, then

$$\frac{\partial \Psi}{\partial \sigma} = \frac{\partial \Psi}{\partial \Upsilon}\frac{\partial \Upsilon}{\partial \sigma} = \mathbf{R} \qquad (97)$$

is a right eigenvector of $\mathbf{B}$.

Since one right eigenvector is given by (79), we have

$$\frac{d(1-\alpha)}{d\sigma} = R_1 = 1 - \alpha, \qquad (98)$$

$$\frac{dv_c}{d\sigma} = R_2 = v_c - \lambda = \frac{\lambda^*(\alpha)}{\alpha}(j_0 - v_c). \qquad (99)$$

The rarefaction locus is given by

$$\frac{dv_c}{d\alpha} = \frac{R_2(\alpha, u)}{R_1(\alpha, v_c)}. \tag{100}$$

If a rarefaction connects (α^a, v_c^a) to (α^b, v_c^b), then v_c^b is the solution of the differential equation (100) at $\alpha = \alpha^b$, where the equation is solved using the initial conditions $v_c(\alpha^a) = v_c^a$.

We further note that the speeds of the various parts of the rarefaction are the eigenvalues of the system, evaluated at the conditions between (α^a, v_c^a) and (α^b, v_c^b) on the rarefaction curve. The eigenvalue can be found from the eigenvector by using (77). This gives

$$\lambda = v_c + (1 - \alpha)\frac{R_2}{R_1}. \tag{101}$$

Then (100) implies that

$$\lambda = v_c - (1 - \alpha)\frac{dv_c}{d\alpha} = -\frac{d(1-\alpha)v_c}{d(1-\alpha)}. \tag{102}$$

Thus, the speed of any part of the rarefaction is given by the negative of the slope of the rarefaction locus plotted in $(1 - \alpha)v_c - (1 - \alpha)$ space. The rarefaction locus in these coordinates is shown in Figure 22.8.

The possible intermediate states starting from (α^+, v_c^+) then lie on the integral curves given by (100). Several integral curves are shown in Figure 22.8. The intermediate state separating the shock and the rarefaction is then given by the intersection of the shock curves and the rarefaction curves.

An arbitrary discontinuity in α and u evolves into a shock and a rarefaction traveling at different speeds. The possible sets of shocks and rarefactions can be found by considering the α–u plane filled with shock loci and rarefaction loci, as indicated in Figure 22.8. Then if the conditions are $\alpha = \alpha^+$, $v_c = v_c^+$ ahead of the discontinuity, and $\alpha = \alpha^-$, $v_c = v_c^-$ behind the discontinuity, then either the shock or the rarefaction will have conditions ahead of $\alpha_a = \alpha^+$, $v_c^a = v_c^+$, and the other will have conditions behind $\alpha_b = \alpha^-$, $v_c^b = v_c^-$. They must be connected by an intermediate state, α^i, v_c^i, which is the state behind the leading wave (rarefaction or shock) and ahead of the trailing wave. This state must lie on the intersection of the locus corresponding to the lead state which passes through α^+, v_c^+, and the locus corresponding to the trailing state which passes through α^-, v_c^-.

22.6. Nonlinear Waves

The inlet step response solution in the previous subsection ignores the effect of the right-hand side of the system. If the initial conditions on the volume fraction and the velocity are sufficiently sharp that they may be treated as a discontinuity, then the solution will have large derivatives, at least for a short time after the initial instant. The shock and the rarefaction travel at different speeds, and the rarefaction weakens, so that a steep gradient shortly after the initial instant will become less and less steep. Eventually, the terms on the right-hand side will become important. In this subsection, we examine the situation that should be valid after a very long time. Specifically, we assume that the solution $\Psi = \Psi(\zeta)$, where $\zeta = x - ct$. We seek value(s) of the speed c to give nontrivial solutions satisfying

$$\frac{\partial \Psi}{\partial x} \to 0 \quad \text{as} \quad x \to \pm\infty. \tag{103}$$

Such solutions are variously called "solitons" or, more generally, traveling waves.

From the first of (70), we see that

$$(1-\alpha)(v_c - c) = k_0, \tag{104}$$

where k_0 is a constant. Thus,

$$v_c = v_c(\alpha) = c + \frac{k_0}{1-\alpha}. \tag{105}$$

From (66), we see that

$$v_d = v_d(\alpha) = v_c(\alpha) + \frac{j_0 - v_c(\alpha)}{\alpha}. \tag{106}$$

From the second of (70) we have

$$H(\alpha)\frac{d\alpha}{d\zeta} = C_2(\alpha), \tag{107}$$

where

$$H(\alpha) = K_1(\alpha)[v_d(\alpha) - v_c(\alpha)]^2 + K_2(\alpha)[v_d(\alpha) - v_c(\alpha)]\frac{dv_c}{d\alpha} - c, \tag{108}$$

and $C_2(\alpha)$ is the second component of the vector right-hand side of (70), with v_c and v_d replaced by the functions of α given by (105) and (106).

The solution of (107) is given by a quadrature, with

$$\int_{\alpha_0}^{\alpha} \frac{H(\alpha')}{C_2(\alpha')} d\alpha' = \zeta, \tag{109}$$

where the arbitrary location of the place where $\zeta = 0$ has value α_0. Note that (103) requires that, if $\lim_{\zeta \to \infty} \alpha = \alpha^+$ and $\lim_{\zeta \to -\infty} \alpha = \alpha^-$, we must have

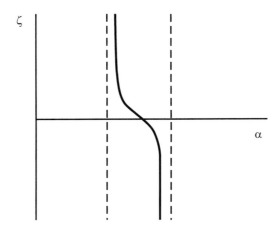

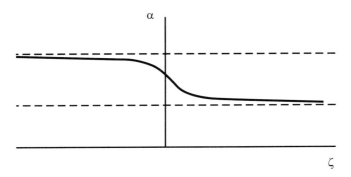

FIGURE 22.9. Traveling wave profile.

$C_2(\alpha^\pm) = 0$. If we examine C_2, we see that $C_2 = 0$ when

$$S(\alpha)[v_d(\alpha) - v_c(\alpha)] = (1 - \alpha)\rho_c g. \tag{110}$$

If we take $S = $ constant, and substitute (105) and (106) into (110), we see that the roots of C_2 are given by

$$\hat{S}[(j_0 - c)\alpha - k_0] = \alpha(1 - \alpha)^2, \tag{111}$$

where $\hat{S} = S/\rho_c g$. Thus, there are two, one, or zero roots between $\alpha = 0$ and $\alpha = 1$, depending on the values of j_0, k_0, and c. We also note that $d\alpha/d\zeta$ should not change sign in the interval in α. Thus, $H(\alpha)$ must have one sign. We further note that $H(\alpha) = 0$ when α and $v_c(\alpha)$ give a characteristic. We further note that whether α^+ is greater than, or less than, α^- must be consistent with the sign of the derivative $d\alpha/d\zeta = H(\alpha)/C_2(\alpha)$. See Figure 22.9.

The wave analyses done in the previous two subsections fit together in the following way. The solution to the conservation form of the equations (zero right-hand side) is correct for short times, starting from a sudden change in conditions.

The initial jump evolves into a shock and a rarefaction, which propagate at different speeds. The shock remains a shock "forever," while the rarefaction weakens until the terms on the right-hand side become important. The rarefaction evolves into a nonlinear traveling wave which propagates at the speed c, whose shape is described by (109) above.

References

[1] A. Alajbegović, Phase Distribution and Turbulence Structure for Solid/Fluid Upflow in a Pipe, Ph.D. Thesis, Rensselaer Polytechnic Institute, Troy, New York (1994).
[2] R. Aris, *Vectors, Tensors and the Basic Equations of Fluid Mechanics*, Prentice Hall, Englewood Cliffs, New Jersey (1962).
[3] R. Aris, *Elementary Chemical Reactor Analysis*, Prentice Hall, Englewood Cliffs, New Jersey (1969).
[4] G. S. Arnold, Entropy and Objectivity as Constraints upon Constitutive Equations for Two-Fluid Modeling of Multiphase Flow, Ph.D. Thesis, Rensselaer Polytechnic Institute, Troy, New York (1988).
[5] G. S. Arnold, D. A. Drew, and R. T. Lahey, Jr., Derivation of Constitutive Equations for Interfacial Force and Reynolds Stress for a Suspension of Spheres Using Ensemble Cell Averaging, *Chemical Engineering Communications*, **86**, 43–54 (1989).
[6] T. R. Auton, The Lift Force on a Spherical Body in a Rotational Flow, *Journal of Fluid Mechanics*, **183**, 199–213 (1987).
[7] A. B. Basset, *Treatise on Hydrodynamics*, Dieghton Bell, London (1888).
[8] G. K. Batchelor, The Stress System in a Suspension of Force-Free Particles, *Journal of Fluid Mechanics*, **41**, 545–570 (1970).
[9] R. M. Bowen, Part I, *Theory of Mixtures,* in *Continuum Physics*, A. C. Eringen, ed., Academic Press, New York, Volume III (1971).
[10] R. M. Bowen and C.-C. Wang, *Introduction to Vectors and Tensors*, Plenum Press, New York (1976).
[11] L. Brand, *Vector and Tensor Analysis*, Wiley, New York (1947).
[12] J.-P. Boehler, ed., *Applications of Tensor Functions in Solid Mechanics*, Springer-Verlag, Vienna (1987).
[13] H. Brenner, The Slow Motion of a Sphere Through a Viscous Fluid Toward a Plane Surface, *Chemical Engineering Science*, **16**, 242–251 (1961).

[14] H. Brenner, The Stokes Resistance of an Arbitrary Particle–IV. Arbitrary Fields of Flow, *Chemical Engineering Science*, **19**, 703–727 (1964).
[15] H. C. Brinkman, A Calculation of the Viscous Force Exerted by a Flowing Fluid on a Dense Swarm of Particles, *Applied Scientific Research*, **A1**, 27–34 (1947).
[16] Yu. A. Buyevich, Statistical Hydrodynamics of Disperse Systems, Physical Background and General Equations, *Journal of Fluid Mechanics*, **49**, 489–507 (1971).
[17] G. Capriz, *Continua with Microstructure*, Springer-Verlag, Berlin (1989).
[18] S. Carnot, *Rèflexions sur la puissance mortice du feu et sur les machines propres à développer cette puissance*, Bachelier, Paris (1824).
[19] P. Chadwick, *Continuum Mechanics: Concise Theory and Problems*, Wiley, New York (1976).
[20] S. Chapman and T. G. Cowling, *The Mathematical Theory of Non-Uniform Gases*, 3rd edn., Cambridge University Press, Cambridge (1970).
[21] L. Y. Cheng, D. A. Drew, and R. T. Lahey, Jr., An Analysis of Wave Propagation in Bubbly Two Component Two-Phase Flow, *Journal of Heat Transfer*, **107**, 402–408 (1985).
[22] R. Clausius, *Ueber die Energievorräthe der Natur u. ihre Verwerthung zum Nutzen der Menschheit*, Max Cohen, Bonn (1885).
[23] B. D. Coleman and W. Noll, The Thermodynamics of Elastic Materials with Heat Conduction and Viscosity, *Archive for Rational Mechanics and Analysis*, **13**, 167–178 (1963).
[24] B. D. Coleman, H. Markovitz, and W. Noll, *Viscometric Flows of Non-Newtonian Fluids*, Springer-Verlag, Berlin (1965).
[25] R. Courant and D. Hilbert, *Methods of Mathematical Physics*, Volume II, Wiley-Interscience, New York (1962).
[26] D. A. Drew, Averaged Field Equations for Two-Phase Media, *Studies in Applied Mathematics* **50**, 133–155 (1971).
[27] D. A. Drew and R. T. Lahey, Jr., Phase Distribution Mechanisms in Turbulent Two-Phase Flow in Channels of Arbitrary Cross Section, *Journal of Fluids Engineering*, **103**, 583–589 (1981).
[28] D. A. Drew and R. T. Lahey, Jr., Phase Distribution Mechanisms in Turbulent Two-Phase Flow in a Circular Pipe, *Journal of Fluid Mechanics*, **117**, 91–106 (1982).
[29] D. A. Drew and R. T. Lahey, Jr., The Virtual Mass and Lift Force on a Sphere in Rotating and Straining Inviscid Flow, *International Journal of Multiphase Flow*, **13**, 1, 113–121 (1987);
[30] D. A. Drew and R. T. Lahey, Jr., Some Supplemental Analysis on the Virtual Mass and Lift Force on a Sphere in Rotating and Straining Inviscid Flow, *International Journal of Multiphase Flow*, **16**, 6, 1127–1130 (1990).
[31] A. Einstein, Eine neue Bestimmung der Moleküldimensionen, *Annalen der Physik*, **19**, 289–306 (1906).
[32] J. L. Ericksen, Conservation Laws for Liquid Crystals, *Transactions of the Society of Rheology*, **5**, 23–34 (1961).
[33] H. Faxén, (Translated title) The Motion of a Rigid Sphere Along the Axis of a Tube Filled with a Viscous Fluid, *Arkiv för Matematik, Astronomi och Fysik*, **17**, 1–28 (1923).
[34] N. A. Frankel and A. Acrivos, On the Viscosity of a Concentrated Suspension of Solid Spheres, *Chemical Engineering Science*, **22**, 847–853 (1967).
[35] J. W. Gibbs and E. B. Wilson, *Vector Analysis*, Yale University Press, New Haven, Connecticut (1909).

[36] M. A. Goodman and S. C. Cowin, A Continuum Theory for Granular Materials, *Archive for Rational Mechanics and Analysis*, **44**, 249–266 (1972).
[37] A. L. Graham, On the Viscosity of Suspensions of Solid Spheres, *Applied Scientific Research*, **37**, 275–286 (1981).
[38] H. P. Greenspan, *Theory of Rotating Fluids*, Cambridge University Press, Cambridge (1980).
[39] M. E. Gurtin, *An Introduction to Continuum Mechanics*, Academic Press, New York (1981).
[40] M. E. Gurtin, On Thermomechanical Laws of a Phase Interface, Zeitschrift für Angewandte Mathematik und Mechanik, **42**, 370–388 (1991).
[41] J. Happel and H. Brenner, *Low Reynolds Number Hydrodynamics*, Noordhoff, Leyden (1973).
[42] E. J. Hinch, An Averaged Equation Approach to Particle Interactions in a Fluid Suspension, *Journal of Fluid Mechanics*, **83**, 695–720 (1977).
[43] G. Huisken, Flow by Mean Curvature of Convex Surfaces into Spheres, *Journal of Differential Geometry*, **20**, 237–266 (1984).
[44] M. F. Hurwitz, Hydrodynamic Interaction of Many Rigid Spheres, Ph.D. Thesis, Cornell University, Ithaca, New York (1996).
[45] M. Ishii, *Thermo-Fluid Dynamic Theory of Two-Phase Flow*, Eyrolles, Paris (1975).
[46] M. Ishii and N. Zuber, Relative Motion and Interfacial Drag Coefficient in Dispersed Two-Phase Flow of Bubbles, Drops and Particles, *AIChE Journal*, **25**, 843–855 (1979).
[47] M. Ishii and K. Mishima, Two-Fluid Model and Hydrodynamic Constitutive Equations, *Nuclear Engineering and Design*, **82**, 107–126 (1981).
[48] S. Itoh, The Permeability of a Random Array of Identical Rigid Spheres, *Journal of the Physical Society of Japan*, **52**, 2379–2388 (1983).
[49] D. J. Jeffrey and Y. Onishi, Calculation of the Resistance and Mobility Functions for Two Unequal Rigid Spheres in Low Reynolds Number Flow, *Journal of Fluid Mechanics*, **139**, 261–290 (1984).
[50] J. T. Jenkins and S. B. Savage, A Theory for the Rapid Flow of Identical, Smooth, Nearly Elastic, Spherical Particles, *Journal of Fluid Mechanics*, **130**, 187–202 (1983).
[51] D. D. Joseph and T. S. Lundgren, Ensemble Averaged and Mixture Theory Equations, *International Journal of Multiphase Flow*, **16**, 35–42 (1990).
[52] S. Kalkach-Navarro, The Mathematical Modeling of Flow Regime Transition in Bubbly Two-Phase Flows, Ph.D. Thesis, Rensselaer Polytechnic Institute, Troy, New York (1992).
[53] I. Kataoka, M. Ishii, and A. Serizawa, Local Formulation of Interfacial Area Concentration in Two-Phase Flow, *International Journal of Multiphase Flow*, **12**, 505–529 (1986).
[54] O. D. Kellogg, *Foundations of Potential Theory*, Julius Springer, Berlin (1929).
[55] Baron Kelvin (W. Thompson), On the Dynamical Theory of Heat, *Transactions of the Royal Society of Edinburgh*, Vol. XX, Part II (1851).
[56] H. Lamb, *Hydrodynamics*, Cambridge University Press, Cambridge (1932).
[57] P. D. Lax, Hyperbolic Systems of Conservation Laws and the Mathematical Theory of Shock Waves, Conference Board of the Mathematical Sciences, CBMS-NSF Regional Conference Series in Applied Mathematics, CB11, Society for Industrial and Applied Mathematics, Philadelphia, Pennsylvania (1960).

References

[58] M. J. Lighthill, The Image System of a Vortex Element in a Rigid Sphere, *Proceedings of the Cambridge Philosophical Society*, **52**, 317–321 (1956).

[59] A. E. H. Love, *A Treatise on the Mathematical Theory of Elasticity*, Dover, New York (1944).

[60] T. S. Lundgren, Slow Flow Through Stationary Random Beds and Suspensions of Spheres, *Journal of Fluid Mechanics*, **51**, 273–299, (1972).

[61] S. Marsh, M. Glicksman, and D. Zwillinger, Statistical Mechanics of Mushy Zones, *Modelling and Control of Casting and Welding Processes IV*, T. S. Piwonka and H. P. Wang, eds., The Metallurgical Society, Warrendale, Pennsylvania (1988).

[62] J. C. Maxwell, The Dynamical Theory of Gases, *Philosophical Transactions of the Royal Society, London*, **157**, 49–88 (1866). Also in *The Scientific Papers of James Clerk Maxwell*, Volume 2, Dover, New York (1965).

[63] I. Müller, A Thermodynamic Theory of Mixtures, *Archive for Rational Mechanics and Analysis*, **28**, 1–39, (1968).

[64] R. I. Nigmatulin, Spatial Averaging in the Mechanics of Heterogeneous and Dispersed Systems, *International Journal of Multiphase Flow*, **5**, 353–385 (1979).

[65] W. Noll, *Finite-Dimensional Spaces: Algebra, Geometry and Analysis*, Kluwer, Boston (1987).

[66] J.-W. Park, Void Wave Propagation in Two-Phase Flow, Ph.D. Thesis, Rensselaer Polytechnic Institute, Troy, New York (1992).

[67] J.-W. Park, D. A. Drew, R. T. Lahey, Jr., and A. Clausse, Void Wave Dispersion in Bubbly Flows, *Nuclear Engineering and Design*, **121**, 1–10 (1990).

[68] S. L. Passman, J. W. Nunziato, and E. K. Walsh, A Theory of Multiphase Mixtures, Appendix 5C, pp. 286–325 of C. Truesdell, *Rational Thermodynamics*, Springer-Verlag, New York (1984).

[69] S. L. Passman, Stress in Dilute Suspensions, *Two-Phase Waves in Fluidized Beds, Sedimentation, and Granular Flows*, D. D. Joseph, ed., Springer-Verlag, New York (1989).

[70] C. Pauchon and S. Banerjee, Interphase Momentum Interaction Effects in the Averaged Multifield Model, Part I: Void Propagation in Bubbly Flows, *International Journal of Multiphase Flow*, **12**, 555–573 (1986).

[71] C. Pauchon and S. Banerjee, Interphase Momentum Interaction Effects in the Averaged Multifield Model, Part II: Kinematic Waves and Interfacial Drag in Bubbly Flows, *International Journal of Multiphase Flow*, **14**, 253–263 (1988).

[72] J. Proudman, On the Motion of Solids in a Liquid Possessing Vorticity, *Proceedings of the Royal Society, London*, **A92**, 408–424 (1916).

[73] Lord Rayleigh, *The Theory of Sound*, Macmillan, London (1894).

[74] M. W. Reeks, On a Kinetic Equation for the Transport of Particles in Turbulent Flows, *Physics of Fluids, A*, **3**, 446–456 (1991).

[75] H. L. Royden, *Real Analysis*, Macmillan, London (1968).

[76] J. Rubenstein, Hydrodynamic Screening in Random Media, *Hydrodynamic Behavior and Interacting Particle Systems*, G. Papanicolaou, ed., IMA Volumes in Mathematics and its Applications, vol. 9, Springer-Verlag, New York (1987).

[77] I. Samohýl and M. Šilhavý, Mixture Invariance and its Applications, *Archive for Rational Mechanics and Analysis*, **109**, 299–321 (1990).

[78] J. A. Schouten, *Ricci-Calculus*, Springer-Verlag, Berlin (1954).

[79] L. E. Scriven, Dynamics of a Fluid Interface. Equation of Motion for Newtonian Surface Fluids, *Chemical Engineering Science*, **12**, 98–108 (1960).

References 301

[80] L. A. Segel, *Mathematics Applied to Continuum Mechanics*, Dover, New York (1987).
[81] J. C. Slattery, *Momentum, Energy, and Mass Transfer in Continua*, McGraw-Hill, New York (1972).
[82] G. F. Smith, On Isotropic Functions of Symmetric Tensors, Skew-Symmetric Tensors and Vectors, *International Journal of Engineering Science*, **9**, 10, 899–916 (1971).
[83] G. G. Stokes, On the Effect of the Internal Friction of Fluids on the Motion of Pendulums, *Transactions of the Cambridge Philosophical Society*, **9**, 8, (1851); reprinted in *Mathematical and Physical Papers*, **3**, 1–141, Johnson Reprint Corporation, New York and London (1966).
[84] G. I. Taylor, The Forces on a Body Placed in a Curved or Converging Stream of Fluid, *Proceedings of the Royal Society, London, A*, **120**, 260–283 (1928).
[85] K. Terzaghi and R. M. Peck, *Soil Mechanics in Engineering Practice*, Wiley, New York (1996).
[86] C. Truesdell, *The Kinematics of Vorticity*, Indiana University Press, Bloomington, Indiana (1954).
[87] C. Truesdell and R. Toupin, The Classical Field Theories. *Handbuch der Physik*, Volume III, Part 1, S. Flügge, ed., Springer-Verlag, Berlin (1960).
[88] C. Truesdell and W. Noll, The Non-Linear Field Theories of Mechanics. *Handbuch der Physik*, Volume III, Part 3, S. Flügge, ed., Springer-Verlag, Berlin (1966).
[89] C. Truesdell, *Rational Thermodynamics*, McGraw-Hill, New York (1969).
[90] C. Truesdell, *A First Course in Rational Continuum Mechanics*, Academic Press, New York (1977).
[91] C. Truesdell, Sulle Basi della Termomeccanica, *Accademia Nazionale dei Lincei, Rendiconti della Classe di Scienze Fisiche, Mathematiche e Naturali* (8) **22**, 33–88 and 158–166 (1957).
[92] C. Truesdell, I. The first three sections of Eulers [sic] treatise on fluid mechanics; II. The theory of aerial sound 1687–1783; III. Rational fluid mechanics 1765–1788: Editor's introduction to vol. II, p. 13, of *Euler's Works*. Orell Füssli, Zürich (1956).
[93] O. V. Voinov, Force Acting on a Sphere in a Inhomogeneous Flow of an Ideal Incompressible Fluid, *Journal of Applied Mechanics and Technical Physics*, **14**, 592–594 (1973).
[94] G. B. Wallis, *One Dimensional Two-Phase Flow*, McGraw-Hill, New York (1969).
[95] C.-C. Wang, A New Representation Theorem for Isotropic Functions, Part I, *Archive for Rational Mechanics and Analysis*, **36**, 3, 166–197 (1970).
[96] C.-C. Wang, A New Representation Theorem for Isotropic Functions, Part II, *Archive for Rational Mechanics and Analysis*, **36**, 3, 198–233 (1970).
[97] G. B. Whitham, *Linear and Nonlinear Waves*, Wiley, New York (1974).
[98] S. Whitaker, *Introduction to Fluid Mechanics*, Prentice Hall, Englewood Cliffs, New Jersey (1968).
[99] T. W. Lambe and R. V. Whitman, *Soil Mechanics*, Wiley, New York (1969).
[100] N. Zuber, On the Dispersed Two-Phase Flow in the Laminar Flow Regime, *Chemical Engineering Science*, **19**, 897–917 (1964).

Index

σ-algebra, 96

Acoustic model, 274
admissible, 37
algebra, 96
alternating symbol, 28
average, 17
    continuous component stress, viscous, 185
    density, 17
    dispersed component stress,viscous, 185
    Gauss curvature, evolution, 212
    interfacial force density, viscous, higher order , 187
    mean curvature, 212
    mean curvature, evolution, 212
    radius, 204
    volume, 202
average, conditional, 108
    ensemble, 93
    time, 114
    volume, 116
averaged dispersed component stress, viscous, higher order, 186
averages for inviscid fluids, 163

Balance equation, canonical, 89
    canonical, averaged, 121
    energy, averaged, 122, 126
    fluctuation kinetic energy, averaged, 130
    internal energy, averaged, 122, 127
    mass, averaged, 121, 126
    momentum, averaged, 122, 126
balance equations, 194
    kinetic theory, 55
    mixture, 83
    solutions, 60
balance of energy, 29, 30, 73, 89
    fluctuation kinetic energy, 130
    kinetic theory, 195
    mass, 22, 69
    material, 30
    mesoscale, 75
    microscale, 75
    momentum, 23, 28, 70, 72
    multicomponent, 74
    spatial, 30
balance of enthalpy, 30
balance of entropy, 36
balance of mass, 88
    kinetic theory, 194
    material, 23
    multicomponent, 69
balance of moment of momentum, 29

balance (*continued*)
  local, 29
  multicomponent, 73
balance of momentum, 24, 89
  kinetic theory, 195
  material, 27
  multicomponent, 71
  spatial, 28
Basset force, 229
Batchelor's approximation, 157
Bernoulli theorem, 162
blending function, 187
body, 14, 20
  force, 24
  force, averaged, 124
Boltzmann's equation, 48, 193
Borel sets, 96
boundary condition, inviscid, 162
boundary conditions, 140
  elastic, 44
  inviscid, 42
  Navier Stokes, 41
breakup, 201
  model, 202
bubble coalescence and breakup, 201
bubble resonance, 280
buoyancy, 177, 241

Calculated quantities, 197
Cauchy's second law, 27
Cauchy–Green tensors, 33
celerity, 276
cell model, 159
characteristic curve, 245
characteristic function X_k, 100
characteristics, 287
  inertial model, 252
  simple model, 248
  viscous model, 250
closure conditions *See* constitutive equations, 137
coalescence, 201
  model, 202
collision operator, 51
collisions, 172
comparison of averaging methods, 175
computing momentum exchange and stress, 182
computing the nearest-neighbor distribution functions, 109

conclusion, 152
conditional average, 156
conditionally averaged equations, 179
conservation of mass, 22
  spatial, 23
constitutive equations, 137, 140
continuous component Reynolds stress, 225
continuous component stress, 224
  Lundgren's, 182
continuous dependence, 243
convergence results, 113
coordinates, Eulerian, 20
  Lagrangian, 20
  material, 20
  spatial, 20
correlation assumptions, 195
cross product, 28
curvature, Gauss, 209
  mean, 209
cut, 26
  particles, 238

Deformation rate, 21, 182
dendrites, 199
density, 14, 22, 132
  averaged, 123
details, undesirable, 92
dimples, 199
Dirac delta function, interface, 101
disorder, 75
dispersed component elastic stress, computed, 167, 168
  fluctuation, 175
dispersed component heat flux, collision, 194
dispersed component pressure, 223
  fluctuation, 224
dispersed component stress, collision, 194
  kinetic theory, 195
dispersed component velocity, kinetic theory, 195
dispersed flow theory, 146
dispersion relation, 276
dissipation, 180
  multicomponent, 125
distribution function, multiparticle, 106
  nearest-neighbor, 107
drag, 189, 196, 226

drift-flux, 282
Dubois–Reymond Lemma, 16, 23, 69, 71

Effect of concentration gradients, 172
effect of particle velocity fluctuations, 172
effective fluid, 180
effective media, 159
effective viscosity, 180, 188
effects of rotation, 169
Einstein's viscosity, 180, 189
elastic collisions, 49
elliptic, 245
energy and entropy source assumptions, 79
energy flux, averaged, 123
energy jump condition, 31
energy source, averaged, 124
ensemble, 17, 87, 95
   average, 98
entropy, 75
   flux, averaged, 124
   inequality, 37, 89
   inequality, averaged, 122, 127
   inequality, mesoscale, 77, 149
   inequality, microscale, 77, 147
   source, averaged, 124
entropy, averaged, 123
   mesoscale, 132
   microscale, 132
equation of state, 273
equipresence, principle of, 141
ergodicity, 105
error, 113
evolution of curvature, 206
excluded volume, 56
existence, 243
expected value, 94

Favré average *See* mass weighted average, 132
Faxén force, 229
first grade, 145
"flake", 24, 131
flow regime, 200
fluctuation energy flux, kinetic theory, 198
fluctuation kinetic energy flux, 230
fluctuation kinetic energy, computed, 166
fluctuation, 175
   multicomponent, 125
   particle, 173
fractals, 199
frame, 31, 142
frame indifference, principle of, 141
frame of reference, 32
frame, change of, 32
frequency, 276

Gauss and Leibniz rules for time and volume averaging, 117
Gauss rule, 103
   for time averaging, 118
   for volume averaging, 119
generalized function, 100
Green's function, 43
Green's theorem, 171
guiding principles, 141

H-theorem, 53, 132
Hadamard problem, 246, 288
heat flux, 230
   conductive, 230
   kinetic theory, 51
   mesoscale, 75, 132
   microscale, 75, 132
   turbulent, 230
Heaviside function, 17
Helmholtz representation, 43, 170
hierarchical, 158
homogeneous flow model, 278
homogeneous wave speed, 279
hydrostatic, 242
hyperbolic, 247

Ill-posed, 243
impossible, 38
initial conditions, 140
interface, evolution, 199
interfacial area density, 122, 204, 211
   evolution, 205, 212
interfacial energy source, 231
   mesoscale, 75
   microscale, 75
   turbulent dispersion, 197
interfacial entropy source, averaged, 124, 126

interfacial force, 226
interfacial force and stresses, 179
interfacial force density, 171, 172
   computed, 165
   computed, fluctuation, 175
   turbulent dispersion, 197
interfacial heat source, averaged, 124
interfacial internal energy source, averaged, 126
interfacial kinetic energy source, averaged, 126
interfacial mass generation rate, 126
interfacial momentum source, averaged, 124, 126
interfacial pressure, 174, 223
   computed, 165
interfacial stress, computed, 169
   viscous, 185
interfacial velocity, computed, 164
interfacial work, 231
   averaged, 124
   computed, 166
internal energy density, 132
internal energy, averaged, 123
   mesoscale, 75, 132
   microscale, 75, 132
inviscid, 42
inviscid fluid, 42
irreversible, 38
irrotational flow, 162

Jacobian, 22
jump conditions, 127
   canonical, 89
   energy, 90
   energy, averaged, 127
   entropy, 90
   entropy, averaged, 128
   mass, 90
   mass, averaged, 127
   momentum, 90
   momentum, averaged, 127

Kinetic theory, 193
Kronecker delta, 32

Lag, 261
Langevin equation, 196
Laplace formula, 223
lead, 261

Leibnitz rule, 103
   for time averaging, 118
   for volume averaging, 120
lift, 169, 227
linear elasticity, 44
linear strain, 44
linearity, 99

Manipulations, 128
mapping, 14
mass, 14, 22
mass density, 22
mass jump condition, 31
material derivative, 21, 23
material frame indifferent, 35
Maxwellian distribution, 53
measurable, 95
mesoscale considerations, 149
microscale and mesoscale energy equations, 75
microscale and mesoscale entropy equations, 76
microscale considerations, 147
microscale solution, 162
mixture *See* solution, 59
model, continuum, 13
   corpuscular, 13
moment equations, 218
momentum jump condition, 31
momentum source, surface tension, 230
motion, 20
   multicomponent, 67
multicomponent mixtures, 65

Navier–Stokes equations, 41
Newtonian fluid, 41
nonpolar, 28, 73, 89
number density, 48, 193, 195, 201

Objective, 133
   scalar, 34, 142
   tensor, 34, 142
   vector, 34
observer, 32
other forces, 229

Parabolic, 247
particle diffusivity, 241
particle distributions, 155
particle paths, 132

particle-fluid mixture, dilute, 232
phase space, 48, 193
plane Couette flow, 259
plane Poiseuille flow, 260
potential flow, 43
pressure, computed, 165
probability, 97
   density function, 17, 172
process, multicomponent, 88

Quasi-equilibrium, elastic, 45

Radius of curvature, principal, 207
radius, equivalent, 201
Radon–Nikodym theorem, 22, 78
random motions, 48
rarefaction, 284, 290
   locus, 293
rate of deformation, 33
Rayleigh equation, 275, 278
realization, 17, 88, 93
rectilinear shear, 258
reference configuration, 20
relative velocity, 82
renormalization, 158, 186
repeatability, 88, 93
representation theorem, multicomponent, 144
reversible, 38
Reynolds energy flux, multicomponent, 125
Reynolds entropy flux, multicomponent, 125
Reynolds internal energy flux, multicomponent, 125
Reynolds kinetic energy, multicomponent, 125
Reynolds number, 42
Reynolds rules, 99
Reynolds shear working, multicomponent, 125
Reynolds stress, 225
   computed, 166
   computed, fluctuation, 175
   kinetic theory, 198
   multicomponent, 125
   particle, 173
rigid, 182
rigid solid, 45, 46
rigid spherical shell, 168

ripples, 199
rotation tensor, 33

Saturated, 82
second law of thermodynamics, 36
self-consistency, 159
separation of components, principle of, 141
shielding, 56
shock, 284, 290
   locus, 291
size distribution, 201
slip, 67, 82
solid component stress, 224
solids, 44
soliton, 294
solution, 59
speed of sound, 277
   isentropic, 277
state equations, 140
state variables, 35, 147
stationary, 105
statistical averaging, 108
Stokes drag, 185
Stokes flow, 44, 177
Stokes solution, shearing, 184
   translational, 184
strain, 33
stress tensor, 26, 132
stress, 222
   averaged, 123
   kinetic theory, 51
   mesoscale, 75, 132
   microscale, 75, 132
   mixture, 82
   multicomponent, 71
surface tension, 168
system, determined, 140

Temperatures, 77
   multicomponent, 78
test function, 100
tetrahedron, 25
thermodynamic process, 37
topological equation, 102
traction vector, 24
traveling wave, 294
treating generalized functions, 100
turbophoresis, 241
turbulent dispersion, 229

Uniqueness, 243
unsaturated, 82

Velocity, 21
  field, 21
  potential, 162
velocity, averaged, 123
  center-of-mass, 82
  center-of-volume, 82
  computed, 164
virtual mass, 227
viscosity, bulk, 41
  shear, 41

volume, 14, 22
volume fraction, 66, 82, 122, 202
  gradient, 172
volumetric velocity, 182
volumetric wave speed, 277
vorticity, 21, 42, 169

Wave equation, 273
wave hierarchy, 288
wave speed, 274
wave speeds, elastic, 45
wavenumber, 276
well-posedness, principle of, 141

Applied Mathematical Sciences

(continued from page ii)

61. *Sattinger/Weaver:* Lie Groups and Algebras with Applications to Physics, Geometry, and Mechanics.
62. *LaSalle:* The Stability and Control of Discrete Processes.
63. *Grasman:* Asymptotic Methods of Relaxation Oscillations and Applications.
64. *Hsu:* Cell-to-Cell Mapping: A Method of Global Analysis for Nonlinear Systems.
65. *Rand/Armbruster:* Perturbation Methods, Bifurcation Theory and Computer Algebra.
66. *Hlaváček/Haslinger/Necasl/Lovísek:* Solution of Variational Inequalities in Mechanics.
67. *Cercignani:* The Boltzmann Equation and Its Applications.
68. *Temam:* Infinite-Dimensional Dynamical Systems in Mechanics and Physics, 2nd ed.
69. *Golubitsky/Stewart/Schaeffer:* Singularities and Groups in Bifurcation Theory, Vol. II.
70. *Constantin/Foias/Nicolaenko/Temam:* Integral Manifolds and Inertial Manifolds for Dissipative Partial Differential Equations.
71. *Catlin:* Estimation, Control, and the Discrete Kalman Filter.
72. *Lochak/Meunier:* Multiphase Averaging for Classical Systems.
73. *Wiggins:* Global Bifurcations and Chaos.
74. *Mawhin/Willem:* Critical Point Theory and Hamiltonian Systems.
75. *Abraham/Marsden/Ratiu:* Manifolds, Tensor Analysis, and Applications, 2nd ed.
76. *Lagerstrom:* Matched Asymptotic Expansions: Ideas and Techniques.
77. *Aldous:* Probability Approximations via the Poisson Clumping Heuristic.
78. *Dacorogna:* Direct Methods in the Calculus of Variations.
79. *Hernández-Lerma:* Adaptive Markov Processes.
80. *Lawden:* Elliptic Functions and Applications.
81. *Bluman/Kumei:* Symmetries and Differential Equations.
82. *Kress:* Linear Integral Equations.
83. *Bebernes/Eberly:* Mathematical Problems from Combustion Theory.
84. *Joseph:* Fluid Dynamics of Viscoelastic Fluids.
85. *Yang:* Wave Packets and Their Bifurcations in Geophysical Fluid Dynamics.
86. *Dendrinos/Sonis:* Chaos and Socio-Spatial Dynamics.
87. *Weder:* Spectral and Scattering Theory for Wave Propagation in Perturbed Stratified Media.
88. *Bogaevski/Povzner:* Algebraic Methods in Nonlinear Perturbation Theory.
89. *O'Malley:* Singular Perturbation Methods for Ordinary Differential Equations.
90. *Meyer/Hall:* Introduction to Hamiltonian Dynamical Systems and the N-body Problem.
91. *Straughan:* The Energy Method, Stability, and Nonlinear Convection.
92. *Naber:* The Geometry of Minkowski Spacetime.
93. *Colton/Kress:* Inverse Acoustic and Electromagnetic Scattering Theory. 2nd ed.
94. *Hoppensteadt:* Analysis and Simulation of Chaotic Systems.
95. *Hackbusch:* Iterative Solution of Large Sparse Systems of Equations.
96. *Marchioro/Pulvirenti:* Mathematical Theory of Incompressible Nonviscous Fluids.
97. *Lasota/Mackey:* Chaos, Fractals, and Noise: Stochastic Aspects of Dynamics, 2nd ed.
98. *de Boor/Höllig/Riemenschneider:* Box Splines.
99. *Hale/Lunel:* Introduction to Functional Differential Equations.
100. *Sirovich (ed):* Trends and Perspectives in Applied Mathematics.
101. *Nusse/Yorke:* Dynamics: Numerical Explorations, 2nd ed.
102. *Chossat/Iooss:* The Couette-Taylor Problem.
103. *Chorin:* Vorticity and Turbulence.
104. *Farkas:* Periodic Motions.
105. *Wiggins:* Normally Hyperbolic Invariant Manifolds in Dynamical Systems.
106. *Cercignani/Illner/Pulvirenti:* The Mathematical Theory of Dilute Gases.
107. *Antman:* Nonlinear Problems of Elasticity.
108. *Zeidler:* Applied Functional Analysis: Applications to Mathematical Physics.
109. *Zeidler:* Applied Functional Analysis: Main Principles and Their Applications.
110. *Diekmann/van Gils/Verduyn Lunel/Walther:* Delay Equations: Functional-, Complex-, and Nonlinear Analysis.
111. *Visintin:* Differential Models of Hysteresis.
112. *Kuznetsov:* Elements of Applied Bifurcation Theory, 2nd ed.
113. *Hislop/Sigal:* Introduction to Spectral Theory: With Applications to Schrödinger Operators.
114. *Kevorkian/Cole:* Multiple Scale and Singular Perturbation Methods.
115. *Taylor:* Partial Differential Equations I, Basic Theory.
116. *Taylor:* Partial Differential Equations II, Qualitative Studies of Linear Equations.
117. *Taylor:* Partial Differential Equations III, Nonlinear Equations.

(continued on next page)

Applied Mathematical Sciences

(continued from previous page)

118. *Godlewski/Raviart:* Numerical Approximation of Hyperbolic Systems of Conservation Laws.
119. *Wu:* Theory and Applications of Partial Functional Differential Equations.
120. *Kirsch:* An Introduction to the Mathematical Theory of Inverse Problems.
121. *Brokate/Sprekels:* Hysteresis and Phase Transitions.
122. *Gliklikh:* Global Analysis in Mathematical Physics: Geometric and Stochastic Methods.
123. *Le/Schmitt:* Global Bifurcation in Variational Inequalities: Applications to Obstacle and Unilateral Problems.
124. *Polak:* Optimization: Algorithms and Consistent Approximations.
125. *Arnold/Khesin:* Topological Methods in Hydrodynamics.
126. *Hoppensteadt/Izhikevich:* Weakly Connected Neural Networks.
127. *Isakov:* Inverse Problems for Partial Differential Equations.
128. *Li/Wiggins:* Invariant Manifolds and Fibrations for Perturbed Nonlinear Schrödinger Equations.
129. *Müller:* Analysis of Spherical Symmetries in Euclidean Spaces.
130. *Feintuch:* Robust Control Theory in Hilbert Space.
131. *Ericksen:* Introduction to the Thermodynamics of Solids, Revised ed.
132. *Ihlenburg:* Finite Element Analysis of Acoustic Scattering.
133. *Vorovich:* Nonlinear Theory of Shallow Shells.
134. *Vein/Dale:* Determinants and Their Applications in Mathematical Physics.
135. *Drew/Passman:* Theory of Multicomponent Fluids.